Also by Allan Chapman:

Stargazers: Copernicus, Galileo, the Telescope and the Church (Lion Hudson, 2014).

Slaying the Dragons: Destroying Myths in the History of Science and Faith (Lion Hudson, 2013).

England's Leonardo: Robert Hooke and the Seventeenth-Century Scientific Revolution (Institute of Physics, 2005).

Mary Somerville and the World of Science (Canopus Press, Bath, 2004; reprint Springer, 2015).

Gods in the Sky. Astronomy, Religion, and Culture from Antiquity to the Renaissance (Channel 4 Books, Pan Macmillan, 2002).

The Medicine of the People. Popular Medicine in Britain before the NHS (Aeneas Press, Chichester, 2001).

The Victorian Amateur Astronomer. Independent Astronomical Research in Britain, 1820-1920 (Wiley-Praxis, 1998).

ALLAN CHAPMAN

Physicians, Plagues and Progress

THE HISTORY of WESTERN MEDICINE from ANTIQUITY to ANTIBIOTICS

LION

Published by Lion Books
an imprint of
Lion Hudson plc
Wilkinson House, Jordan Hill Road,
Oxford OX2 8DR, England
www.lionhudson.com/lion

ISBN 978 0 7459 6895 7
e-ISBN 978 0 7459 7040 0

First edition 2016

Acknowledgments
Extracts from The Authorized (King James) Version. Rights in the Authorized Version are vested in the Crown. Reproduced by permission of the Crown's patentee, Cambridge University Press.

A catalogue record for this book is available from the British Library

Printed and bound in Finland, September 2016, LH53

To Rachel: Wife, Scholar, and Best Friend
"… her price is far above rubies"
(Proverbs 31:10)

Contents

List of Illustrations

Chapter 11

Chapter 12

Chapter 13

Chapter 14

Chapter 15

Chapter 17

Chapter 19

Chapter 21

Chapter 22

Acknowledgments

I am indebted to many people, across several decades, for assistance in writing this book. What you will read in these pages goes back a long way, and the ideas they contain were researched and developed in preparing and delivering numerous lectures to academic, medical, and scientific institutions on both sides of the Atlantic. In particular, I thank friends in the University of Minnesota Medical School Department of Surgery, before whom I gave a series of annual lectures between 1986 and 2014, and especially John (Jack) D. Foker, MD, and John Najarian, MD, of that institution. As a historian not formally trained in medicine, I am indebted to these gentlemen and other medical professionals for advice and teaching on specific aspects of clinical technique, though I take full responsibility for any mistakes I may have made.

In June 2010, having first been alerted by my excellent dentist, Mrs Svetlana Dwyer of the Pendlebury Dental Practice, north Manchester, I was diagnosed as having cancer of the palate, a very rare condition for a lifelong non-smoker, non-drug-taker, and one who is very abstemious in alcohol consumption. But this disease opened up a whole new adventure for me, yes, mad as it might sound, an adventure. I became fascinated by the disease and wanted to fully understand it, scientifically and clinically. (A radiologist at Oxford University's Churchill Hospital said to me, when I asked to look at my scans and be given a dispassionate scientific analysis, "Dr Chapman, you are weird. *Seriously* weird." I got similar amusing responses when, going in for my 30 outpatient doses of radiotherapy some weeks after surgery, I joked about coming in for my "daily fry".)

Yet intellectual curiosity, in conjunction with a strong sense and acceptance of God's providence, trumped fear as I embarked upon a maxillofacial oncological learning adventure. Apart from reading the clinical literature, I am very grateful to the surgical team at Oxford's John Radcliffe Hospital "Max-Fax" Department, in particular Mr Steve Watt-Smith, Mr Steve Bond, and Miss Jennifer Wylie (Mrs Graystone), not only for the wonderful surgery they performed on me, but also for all the information they gave me about oncological diagnostics and step-by-step operating theatre technique. (Exactly *how* does a surgeon remove a patient's palate without splitting the head open?)

Following surgery, I am greatly indebted to Dr Sandip Popat and Dr Malcolm Berry, Robert Lawson, Michelle Nevitt, and colleagues, for superb reconstructive and prosthetic surgery to my mouth and ongoing dental care, and for teaching me so much, and even inviting me to visit the hospital laboratories. Similar thanks are owed to Dr Gywneth Hueter for her clinical expertise and teaching pertaining to the ear, nose, and throat system. I also give my thanks to Mr Peter Burge of Oxford's Nuffield Orthopaedic Hospital for his skilful operating upon my hand for carpal tunnel syndrome and, as the procedure was done under a local anaesthetic, for allowing me to watch the operation and even giving me something of a tutorial as he and his colleagues laid bare my median nerve and adjacent structures.

Also in Oxford, I must thank my anaesthetist friend Professor Andrew Farmery, of Wadham College and the Nuffield School of Clinical Medicine, for answering many clinical queries over the years, from how seventeenth-century surgeons identified *ranulae* ("little frogs" or stony lumps under the tongue) to how the Nuffield Vaporizer early anaesthetic machine worked –and his entertaining and instructive conversation. Likewise in Oxford, I thank Dr Stephen Goss, Dr Eric Sidebottom, and the late Sir Henry Harris of the Dunn School of Pathology for what they have taught me in conversation, especially about cellular pathology and oncology. In addition, I thank Dr Richard Whittington, a retired Coroner, for advice on understanding the causes of death of historical personages.

I am indebted, too, to Dr Christopher Jewell of Lancaster University for kindly supplying me with data concerning the blood pressure of sheep (Chapter 13), and my friend the epidemiologist

Professor Frank Cox, of The London School of Hygiene and Tropical Medicine, and Emeritus Gresham Professor of Physick.

It would be very difficult to list and thank personally all the general practitioners and dentists who have supplied me with information over the years, but I should mention Mrs Svetlana Dwyer, Dr Alan Brown, and Dr Chris Kenyon. Practical tips and guidance from working medical professionals have always been invaluable to a person as instinctively "hands-on" and technically minded as myself. But my first Oxford teacher of practical medicine was the late Dr Harold Harley, sometime of Wadham College, who qualified as a physician before 1930, and who only just missed the millennium. For Harold Harley, in the 1980s, taught me many traditional clinical techniques, such as how to diagnose pulmonary tuberculosis with a stethoscope, and how to size up a new patient at a glance.

And then, I have my research students over the years to thank, for while guiding their research theses in aspects of medical history, I learn so much from them. In particular, I thank Michael Clarke, Carole Brookes, Michael Osborne, Anna Simmons, Sarah Galloway, Amy Nicholson, Daniel Goward, Andrew Brocklehurst, Emma Andrews, and Joseph Hutchinson. I am especially indebted to Prianka Bose for her stunning account of assisting a local village shaman surgeon in north India in the performance of a cataract operation worthy of Sushruta Samhita in 600 BC (Chapter 4 and Appendix 1).

I owe much to museum and library staff members who have assisted me in so many ways. These include staff at the Royal College of Physicians, the Royal College of Surgeons Hunterian Museum, the Royal Society, and the Royal Society of Medicine, London; the Royal College of Surgeons, Edinburgh and Glasgow; and the Wellcome Medical Museum, London. In Oxford, my debts of gratitude go to the staff of the Bodleian Library, the History Faculty Library, the Radcliffe Science Library, The Queen's College Library, and the Museum of the History of Science, in particular to Tony Simcock. And my especial thanks for their ceaseless kindness go to Tim Kirtley and Francesca Heaney of Wadham College

Library, and to Dr Judith Curthoys, Dr Cristina Neagu, Alina Nachescu, David Stumpp, Angela Edward, and the sadly late Janet McMullin, of Christ Church Library.

I am also grateful to the Warden and Fellows of Wadham College, Oxford, and to the Dean, Chapter, and Governing Body of Christ Church, Oxford, for their encouragement, friendship, and provision of academic sanctuary over many years. I am especially indebted to the Governing Body of Christ Church for permission to reproduce plates from many of the rare books in the College Library, and to Alina Nachescu for photographing them for me. I also thank the Trustees and Governing Body of Gresham College, along with the Academic Registrar Mrs Barbara Anderson, in the City of London, for inviting me some years ago to take up a short-term *ad hominem* chair to deliver a series of lectures on medicine in London, 1600–1900. Thanks are due, too, to the Wellcome Unit for the History of Medicine, Oxford, where, many years ago, I undertook a research project examining the use of astrology in Tudor and Stuart medical practice. Likewise, I thank friends and clergy at my native parish churches of St Thomas and St Anne, Parish of Clifton, in Swinton and Pendlebury, now part of Salford, for their interest and support over most of my life.

I must offer my heartfelt thanks to the long-deceased John Pratt, a Swinton and Pendlebury, Lancashire, bus inspector, who in the mid and late 1950s ran the Boys' Section of the local St John Ambulance Brigade. It was Mr Pratt who first entranced a young rough-neck, recently thrown out of the local Wolf Cub Pack for being a mischievous little devil, with the wonder of bones, muscles, and the mechanics of the human skeleton. Given different educational opportunities in his youth, I am sure that this fine instinctive teacher could have gone on to great things. Thank you, Mr Pratt, for your inspiration.

My sincere thanks also go to Ali Hull of Lion Hudson, for all her encouragement, advice, and friendship, to Lawrence Osborn for his meticulous copy-editing of my text, and to Jenny Muscat, Jonathan Roberts, and Brett Rogers.

But my greatest debt of gratitude goes to my wife Rachel. She possesses all the organizational skills that I so conspicuously lack. For this excellent Durham and Oxford classicist has kindly typed my long-hand manuscript, and checked, corrected, and edited it, as she has many of my books and academic papers over our 25 years of marriage. Rachel is not only a mistress of linguistic scholarship and meticulous fine-tooth combing; she also knows where everything is, and can find in a trice the myriad things I lose or misplace on a daily basis. And that is to say nothing of her putting up with my swearing – "Oh, dear me!" – when I lose something. For without Rachel's skill, patience, and devotion, this book would have taken much longer to write. It is why I dedicate it to her.

Allan Chapman
May 2016

Preface

From earliest times, human beings have attempted to alleviate one another's sufferings: a circumstance demonstrated by surviving ancient skeletal evidence of badly set broken limbs, and even of the trephination of skulls, into which holes had been cut either to alleviate a head injury or to let out evil spirits.

In short, medicine is universal across all human cultures, and appears, from the earliest written records, to have been performed with relation to some sort of religious and wider beliefs. In this book, however, I confine myself to the Western medical tradition, for several reasons. Firstly, I have no first-hand knowledge of the Chinese or Indian languages in which the Chinese and Ayurvedic bodies of medical literature were written, and therefore cannot myself read their texts. Secondly, and quite simply, it is the medicine of classical Greece and Rome that lies at the heart of Western medicine as it now exists. For this tradition, coming into being somewhere around 500 BC, comprised a body of axioms, principles, and procedures which we might still recognize in a modern hospital.

I am not at all suggesting that Hippocrates, Erasistratus, and Galen "got it right", whereas writers in other cultural traditions "got it wrong"; for well back into the BC period, in China and India in particular, medicine had long ceased to be a mere body of empirical techniques and had already placed healing within wider and often sophisticated philosophical systems. These systems generally related to the context within which nature was believed to operate, as well as placing both patient and malady within a perceived physical, spiritual, and psychological framework. In short, medicine early on developed intellectual as well as practical dimensions.

So why was Western medicine significant? I believe it was a product of a wider set of circumstances taking place in the Greek world after *c.* 700 BC. For what we might style "scientific" medicine was but one aspect of a rich cultural flourishing which also produced political philosophy, mathematical astronomy and geometry, public theatre, sports contests, naturalistic art and statuary, city states and negotiated public space, civil law codes, and civic as opposed to tribal living. It also invented the "cult of the individual", as opposed to the idealized super-being. Just think of Socrates, Aristotle, Sappho, Aristophanes, Praxiteles, Galen, Alexander the Great, and others in philosophy, poetry, sculpture, science, medicine, and general-ship – real folk, warts and all, their ideas and deeds preserved in a well-attested written record.

I would suggest that this cultural package was conducive to natural enquiry, be that in astronomy, meteorology, or medicine. A realization that nature held together because of some unifying, creative *Logos* – word, being, or rational principle – which could suggest that the cause of a disease might lie in "the nature of things" rather than divine spite, a sin, or a ritualistic omission. And once the idea of *naturalism* came to develop, then could not one enquire into the perceived *circumstances* surrounding an illness, such as the patient's bodily condition, habits, diet, body fluids, and heredity when devising a cure? This suggested a fundamentally different approach to medicine from the one based upon the placation of spirits. As we shall see in Chapter 1, this was the approach that lay at the heart of the Hippocratic regime of medicine in *c.* 440 BC.

Another important, particularly Greek, contribution to Western medicine was the study of anatomy and physiology. Cures might depend not only on a careful external examination of the patient, but also upon a curiosity-driven urge to understand the internal structures and functions of organs, both human and animal. This led in turn to the beginnings of biology, most notably with Aristotle around 350 BC, as we shall see in Chapter 1, and with famous named dissecting anatomists such as Erasistratus, Herophilus, and Galen. For I would argue this curiosity-driven concern with dissection, vivisection, comparative anatomy, and experimentation was the hallmark of Greek medical thinking, along with a taxonomic approach to case histories, which it shared with Chinese and Indian medicine.

By the ninth century AD, when Greek texts were coming to be translated into Arabic, key elements of this tradition were absorbed into medieval Arabic medicine, as we shall see reflected in a galaxy of writers such as Rhazes, Avicenna, and Alhazen.

We must never forget, however, that philosophical foundations apart, medicine is at heart a deeply practical business, aimed at the curing of human ills. And the Euphrates river, flowing through modern Iraq into the Persian Gulf, along with the ancient land routes across the Asiatic Steppe linking the West with China, conveyed not only objects of trade in both directions, but also ideas: ideas about astronomy, geography, and practical medicine. Indeed, in both the classical Greek and Arabic traditions, we find evidences of useful items of medical knowledge being brought together from distant places, for doctors, as practical people, can be quite eclectic in their search for useful facts.

In addition to anatomy, this book contains much about surgery. Western evidence tells us that all cultures practised surgery in varying degrees, for wounds, broken bones, and nasty ulcers, and tumours are part of the human lot; and from the Egyptian Edwin Smith surgical papyrus of *c.* 1500 BC and the writings of Sushruta in sixth-century BC India, the history of surgery spans a good four millennia. It is in the Western tradition, however, with its passion for dissection, that surgery comes to be grounded upon a prior anatomical foundation, especially in the wake of Galen, as we shall see in Chapter 2.

This passion for the dissection of both human and animal cadavers really took off in the "new" universities of Christian Europe, especially in Paris, Montpellier, Bologna, and Padua after 1200. This flies in the face of the established myth that the medieval church banned all dissection and scientific experimentation, which is contradicted by plain, documented historical fact, as we shall see in Chapters 5 to 10.

If Greek independence and intellectual curiosity supplied one foundation for a specifically *Western* style of medicine, I would argue that the Judeo-Christian religion supplied another. Indeed, in many ways Christianity even ran contrary to the prevailing norms of the very Greco-Roman society in which the faith first flourished, for while the classical philosophers placed great stress upon justice, law, and friendship (especially between gentlemen), they were not over-concerned about the underdog, the beggar, the hungry, the homeless, women, and "charity" to the stranger. But Christianity was; and one of Christendom's enduring contributions to medicine lay in the provision of *hospitality* to the sick person: starting with the infirmaries of medieval monasteries and growing to a veritable flood of orphanages, refuges, old folks' homes, and hospitals. In Chapter 17 in particular, we shall see how this perceived duty of care gave Great Britain and Europe their basic infrastructure of hospitals, which survives to this day, along with their attendant medical schools, especially from the twelfth century onwards.

We shall also see, in Chapters 6 and 18, how Christian values underpinned a whole regime of mental health care, which, in spite of sometimes breaking down from sheer overload, attempted to provide *asylum* for the mentally distressed. And while not in any way suggesting that the Christian cultural tradition had an exclusive hold upon compassionate care for the sick and distressed, I would be so bold as to suggest that it was within Christendom that the most significant developments took place.

Yet what did all of this lead up to in practical terms? For even when Queen Victoria was born in 1819, life expectancy for a European, Chinese, Indian, or African person was not significantly better than it had been in the days of Julius Caesar. So where were the practical fruits of all the centuries of dissections, hospitals, and laboratories?

As we shall see, before medical progress could really produce large-scale, life-transforming treatments on a routine basis, it had a long way to go from the days when Hippocrates walked the shores of Hellas. Anatomy, physiology, disease taxonomy, and ingenious laboratory experiments were all essentially academic in character. Yet without the constantly expanding foundation of knowledge which they provided, those electrifying discoveries that came about in the nineteenth century – anaesthesia, antiseptics, microbiology, etc. – would never have been possible, and without them, the "wonderful century" described in Chapter 26 would not have transformed the human condition. One could say that medicine had to develop a critical mass of data and skills before large-scale healing became possible.

I wish to make it clear, however, that I am *not* a trained medical professional, and I beg forgiveness for any technical errors that I may have made in this book. I am a historian by formal training, yet with a passion for science, especially medicine and astronomy, that goes back to my junior school days. Technical detail, and how things work, have always captivated me, be the thing a steam engine, a human heart, a muscle, or a chronometer. In medical science, my especial love has always been anatomy, physiology, and surgery, and over the years I have lectured on surgical history to many professional bodies at home and abroad, most notably the University of Minnesota Medical School Department of Surgery, which, in 2004 and 2005, graciously presented me with formal University Certificates designating me a "Visiting Professor of Surgery". In some respects, I feel an affinity with Aulus Cornelius Celsus, author of the encyclopedic *De Medicina* (see Chapter 2). Celsus was a Roman gentleman who was probably not a trained doctor, but was fascinated by medical procedures, and knew a lot about anatomy and surgical operations, along with engineering and other aspects of classical technology.

The history of medicine is full of ingenuity, blind alleys, rapid spurts of progress, and human tragedies. It is populated by a rich gallery of individuals, amounting to what might be styled, after Geoffrey Chaucer, "God's plenty". These included inspired geniuses, brilliant teachers and researchers, flamboyant show-offs, silent, steady workers, engineers, public-spirited social reformers, compassionate carers, unprincipled thieves, rascals, and murderers, prostitutes, saints, clergymen, and even a twelfth-century court jester. They are all part of the story, and to meet them – read on!

CHAPTER 1

Physicians, Priests, and Folk Healers

*M*ódern medicine is rooted in science. It is based upon an understanding of the natural world in which precise observation, quantification, and internationally shared standards of experimental evidence are seen as the way forward. Over the past five centuries or so in particular, but with roots going back into the Middle Ages and beyond into classical antiquity, this approach not just to medicine, but to every realm of natural knowledge has transformed the perceptions of peoples exposed to it – from deep-space cosmology to microbiology to geology to the splitting of the atom. It bestowed a powerful cultural impulse upon Western Judeo-Christian civilization, whereby human beings have been able to transform both the environment and the human condition itself: via a progressive and science-based technology.

That science-based technology gave us everything, from medieval clockwork, via the great galleons of the Renaissance global explorers, to the steam engine, electric light, and on to mobile phones. Crucially, it gave us that capacity to alleviate not just our sufferings in the Western hemisphere, but also those of people in all places who could be reached by it: modern medicine.

That progressive medical understanding, and its ability to transform as many lives as it could reach, is the subject of this book. Yet unlike clocks or steam engines, living bodies are fiendishly complicated things, containing systems within systems, and the quest turned out to be a long one, as even modern-day medical researchers discover on a daily basis. For while we have come far, there is still a long, long way to go.

Medicine, therefore, has an ancient and a very diverse history, with many different strands coming into play over the millennia. But it was only over the last 150 years or so that medicine truly began to transform the human condition, first in the West, and then globally. So let us begin by looking at the roots of the healer's art.

ANCIENT DOCTORS

In ancient times, medical expectations were very different from those of today. It was acknowledged that the healer's art was much more limited than it is now, but beyond such stark practicalities, sickness itself was regarded very differently. Classical pagan, Jewish, and early Christian literature characterized disease in a wide variety of ways. There could be organic failure, such as blindness, caused by cataracts; accident, as when one fell off a speeding chariot and broke a leg; or death by poison, as in the case of snakebite. Yet even these visible maladies might be seen as the effects of deeper mysterious causes. Was one's accident or illness the result of some sin, or failure to perform an expected ritual correctly?

When it came to overtly mysterious diseases, such as deadly epidemics, madness, or even what we would call stroke – quite literally, "to be stricken" – or heart attack, it appeared perfectly rational, within the context of ancient culture, to attribute them to some sort of divine being or group of beings. Did not human bodily afflictions derive from the same source as crop failures, lightning bolts, devastating floods, and earthquakes? In these circumstances, was it not wise to find out who had done what wrong and then perform the correct rituals, sacrifices, and prayers to your gods, rather than wasting time taking pulses or asking philosophical questions about the workings of the natural world?

Ancient literature, especially that purporting to deal with religious, philosophical, or "meaning" matters, was very much concerned with that elusive yet supremely valued state of awareness: wisdom. Socrates, as he comes over in the *Dialogues* of his pupil Plato, was the supreme embodiment of wisdom in classical pagan antiquity, along with the later Stoic philosophers and Roman letter-writers

2

such as Cicero, Horace, and Seneca the Younger. And the Old Testament is full of wisdom literature: King Solomon's Proverbs, followed by Job, Daniel, Ecclesiastes, Wisdom in the Apocrypha, and other biblical books, where wisdom is often spoken of as a serene woman who must be "sought".

But what was "wisdom"? Generally, it was envisaged in Jewish, early Christian, pagan, and Far Eastern traditions as a "pearl beyond price" that enabled its possessor to see clearly, whereas his un-wise contemporaries saw only obfuscation. It could include the ability to set and solve riddles, to discern truth from falsehood, to interpret symbols, to know the natures of birds, animals, plants, and stones, to grasp the inner meaning of human motivations, to understand human illness, and to enjoy serenity. These were all properties conducive to a state of felicity and balance, as epitomized in the sayings and achievements of Solomon.

Wisdom, however, was essentially static, dealing with the perennial problems of the human lot, and largely inward-looking and contemplative by nature. It tended, in many ancient traditions, such as those mentioned not only in the Old Testament but also in Buddhist, Confucian, and other thought, to be of a higher value than what later ages would think of as "knowledge" or knowing the inner workings of nature from a more mechanistic viewpoint. Knowledge, or what the Greeks sometimes called *technë*, by contrast, was essentially outward-looking. It might be useful on an everyday basis – how to smelt iron, make wine from grapes, or splint a broken arm – but unlike wisdom, it was not especially *profound*. Beholding, as the Old Testament book Proverbs says, how a snake moved across a rock, or the way in which a man might try to seduce a woman, might be seen as a deep form of wisdom, yet such things were in no way related to the complex practical details of reptilian anatomy, human psychology, or neurology.[1]

This manner of thinking also applied to medicine, and, in some ways, has continued to do so right down to today. One still meets highly experienced academically trained physicians of the older generation who can put together a pretty good preliminary diagnosis for a patient on little more than first acquaintance,

3

although nowadays that diagnosis has to be backed up by subsequent scientific tests and analyses before treatment is attempted.

In the past, however, before modern tests and scans existed, all that a doctor possessed were experience, cultural sensitivity, a shrewd eye, religious and philosophical beliefs, and a high public regard based upon past successes. And the doctor, just like his patient, would approach each case with an established set of assumptions about the nature of health, disease, and well-being. Irrespective of whether that doctor was a Greek, a Jew, an Indian, an Egyptian, or even a Chinaman, he would most likely tacitly subscribe to a set of truisms about illness drawn from across the common lot of humanity. In this way of thinking, health was somehow about being in balance, and illness invariably came about when some natural bodily pathway became obstructed. That obstruction might be caused by the weather, the patient's individual temperament, the stars, heredity, or divine or demonic agency, and it was the job of the *wise* physician to identify the true cause and apply the correct remedy, be it a purge, a corrective diet or fast, a prayer, or a sacrifice.

The healer, therefore, was often a versatile figure, equally at home with bowel movements, spirit divination, and advising how best to keep on the right side of the local deities. All of these options, and a good few more, were on offer in the first century AD when St Paul spoke of Luke the "beloved physician" among those who accompanied him on his apostolic journeys.

While the New Testament says nothing whatsoever about Luke's professional activities beyond the fact that he was a doctor, we can, perhaps, risk a few guesses from writings attributed to him, namely his Gospel and the Acts of the Apostles. We know from Paul's remarks that Luke was not Jewish but an early Gentile convert to Christianity, that he wrote good Greek, assembled materials for and presented a series of early Christian narratives in a logical and well-thought-out manner, and corresponded with a man named Theophilus: almost certainly, judging from his Greek name "Lover of God", a fellow Greek, and probably of high social status.

One might guess from the above, therefore, that Luke was probably not a local folk-healer, but was familiar with the already widespread ideas of Hippocrates and other Greek and Latin medical writers, whose works we will examine below. These were all medical men in whose writings we see the beginnings of what might be called a scientific approach, founded upon careful observation, bodily examination, recorded case histories, and treatments based on physical criteria such as the study of lifestyle habits, food, and drug action. As a fluent Greek writer and reader, Luke would have had linguistic and cultural access to them.

None of this rational therapy would in itself exclude a belief in the miraculous, for in a world which saw the rational and the divine as intimately interpenetrating, there was no reason to think that just because a Hippocratic physician could not cure a fever, blindness, or insanity, God himself could not do so. And in his pre-Christian days, was Luke a devotee of the Greek god of healing, Asclepius? A benign deity, with his healing shrines across the pagan world, this god was instantly recognizable in art by his *caduceus*, or staff, around which wise serpents were entwined – the pagan patron god of pagan doctors, as Luke was destined to become a patron for Christian medical colleagues.

By the time of St Luke in the first century AD, there were already cultures familiar to any educated Greek or Roman, whose traditions already went back millennia. Some of them had been visited and written about by that redoubtable Greek tourist of the fifth century BC, Herodotus. And the country that most captivated Herodotus, about which he wrote the most in his *Histories* (or "Accounts"), was the already fabled and exotic land of Egypt.

MEDICINE IN EGYPT AND OTHER ANCIENT CULTURES

In the world of cultural truisms of the BC period, the mystical land of Egypt produced doctors possessing strange powers and insights. The quasi-legendary Egyptian vizier Imhotep, of *c.* 2700 BC, was said to have been a doctor, and wise Pharaohs and wise viziers (one of whom, in his own day a thousand years after Imhotep, would have been Joseph, of Genesis fame) gave order and health to the land of Egypt. This came about by facilitating a condition of *Ma'at*, meaning a state of order, peace, and well-being in the land. And the Egyptians had their own gods who could be called upon for particular medical conditions, such as the dwarf-god Bes, who assisted women in childbirth.

Part of the mystique of their medicine, however, derived from the Egyptians' skill in embalming their dead: actually a process of dehydrating the eviscerated corpse in powdered natron (sodium carbonate and other naturally occurring sodium compounds) for 40 days, prior to applying the mummy bandages: a process mentioned in Genesis 50, following the death of Jacob, and also in Herodotus. Yet while the 1611 Authorized Version of the Genesis account of Jacob's mummification specifies this task being performed or directed by the "physicians", this could hardly have been the case, for the ritual evisceration involved in mummification was a low-caste occupation and not necessarily a job performed by a medical man. This mummification procedure, in which the guts, heart, brain, and other organs were removed to accompany the mummy to the tomb in separate "Canopic" jars, each of which had its protective deity, has also led to the belief that Egyptians were skilled in anatomy. In reality, however, mummification required skills that were closer to those of the ritual or sacrificial butcher than those of the physician.

Around six medical treatises of varying length survived from ancient Egypt. Most significant, and both discovered and translated in the nineteenth century, are the Ebers and Edwin Smith papyri, named after the collectors who acquired them, and now preserved in Leipzig and New York respectively.

The Ebers Papyrus, of around 1500 BC, contains a mixture of things, including several hundred primarily magical medical recipes and rituals; yet it does discuss the apparently central role played by the heart, from which a series of pipes and tubes led to all parts of the body. This must in no way be thought of as implying knowledge of the circulation of the blood: a discovery that had to wait until AD 1628. Rather, it was an early expression of a subsequently long-standing classical and post-classical concept of the heart somehow being the centre of sentience.

6

The Edwin Smith treatise, also from *c.* 1500 BC but believed to be a copy of a treatise going back perhaps to the time of Imhotep of *c.* 2700 BC, is, in many ways, a much more interesting document. Essentially it is a surgical work, focusing upon 48 cases of injury and indicating a remarkable sophistication both of understanding and of practical technique. The treatment of broken bones and slash and puncture wounds is described, but the most interesting section deals with head injuries, even advising how an operator might tackle gaping head wounds in which the skull is split open, revealing the meninges and brain below. It is in the Edwin Smith papyrus that the brain is first mentioned in medical history. Whoever wrote the Edwin Smith papyrus was probably a well-seasoned military surgeon who, theoretical knowledge notwithstanding, had learned in the school of practical experience how best to deal with ghastly injuries.[2]

Fewer explicitly medical items have survived from Babylonian and Assyrian sources, written not on papyrus-paper (not a common plant in Mesopotamia, as it was in the Nile valley), but on clay tablets. Babylonian and Assyrian tablets, however, are extremely good sources for astronomical evidence, as astrological divination was an essential part of their political culture, and their precise recording of planetary positions has been of more use to historians of astronomy than of medicine. Yet in that world, the personal health of the ruler, the political health of the kingdom, and, by extension, the wider welfare of the people were – as with Egyptian *Ma'at* – seen as inextricably linked, and forewarnings about them could be read in the heavens.

In Babylonian and other ancient cultures, a very significant role was ascribed to the liver: a role that survived down through the millennia into medieval European and Arab culture, and even in folk and "alternative" medicine today. I have met out-of-sorts people who described themselves as feeling liverish, and as late as the 1960s the patent medicines manufacturers advertised the contents of their tins of Andrews Liver Salts as being "for inner cleanliness" on TV commercials. It is a tasty effervescent drink with mild purgative properties.

But why the liver? I suspect that it was due to that organ's invariable engorgement with blood when animals were butchered. Not only do Genesis and other biblical books tell us that "the life is in the blood", but so do the writings of the Jews' Egyptian and Mesopotamian neighbours. It is good common-sense pathology: without blood you die. Blood was seen as giving the body heat, nutrition, and substance, while it stood to reason that it made life itself. After all, when a pregnant woman failed to have her period, was it not because her blood and heat were now being used to nourish the growing foetus? After the birth, would not her "concocted" blood become breast milk?

As we shall see in later chapters, it was not until the seventeenth century that the true functioning of the heart, lungs, veins, arteries, liver, and blood would gradually come to be correctly understood. Why, after all, do writers, from antiquity to Shakespeare and beyond, always speak of the blood coursing through our *veins* (never *arteries*), and say that to be struck through the liver, with a spear, or even a curse, would bring instant death?

While it is true that the ancient pathologies, based as they were on the mysterious properties of blood and the liver, were simply incorrect, as were their often curse- and folly-related explanations for disease, their manipulative techniques, as used in simple surgery, could sometimes be effective. The successful binding up of broken limbs – as in the case of Pharaoh in Ezekiel, whose broken arm was splinted by a "roller" – the removal of external (hopefully non-malignant) tumours, and even cataract surgery, are all documented in ancient contemporary Middle Eastern literature.[3]

The sudden death of Pharaoh Tutankhamun around the age of eighteen in 1323 BC has been a subject of intense medical historical fascination, especially as constantly improving X-ray and CAT (Computer-Assisted Tomography) scans of his mummy over the years have revealed remarkable details of skeletal damage and broken bones. Some breaks even have traces of embalming substances still within them, suggesting that the young Pharaoh died soon after receiving compound fractures and other serious injuries to his head and limbs.

Did Tutankhamun, the dare-devil teenager, indulge a passion for dangerous chariot driving (as did the more fortunate King Jehu of Israel)?[4] Did he receive an assassin's blow to the head? Both now look unlikely, especially the chariot driving, as Tutankhamun appears to have had deformed feet, making balance difficult. But the modern coroner's jury sitting on the Pharaohic corpse is still out. Would we not love to know who his doctors were, what they did, and whether they were allowed to survive that god-king's death?

One thing for certain, however, is the personal danger to which a surgeon in Babylonia in *c.* 1700 BC could be exposed. The Hammurabic law code, engraved into a stone pillar and discovered in 1901, warns thus: if a doctor "treated a nobleman for a severe wound and has caused him to die... the physician's hands shall be cut off".[5] Modern medical litigation looks gentle by comparison.

But wounds and injuries apart, what did ancient medically related literature make of disease itself?

MOSES AND THE LEPERS: A SAGA FROM SINAI TO SCANDINAVIA

The Bible, especially the Old Testament, contains descriptions of a rich variety of diseases and afflictions. These include King Saul's and Nebuchadnezzar's episodes of mental delusion,[6] Hezekiah's "boil" cured by the application of a divinely suggested lump of figs (enzymes?),[7] and the bad King Jehoram's two-year stomach affliction, culminating in his bowels dropping out – a rapidly deteriorating hernia leading to the bursting of the scrotum, perhaps?[8]

Of all the medical conditions mentioned in the Bible, nothing is referred to more frequently than that affliction styled, in the Authorized Version of 1611, "leprosy": or as it is clinically defined nowadays, "Hansen's disease", after the Norwegian medical scientist Armauer Hansen, whose work will be discussed below.

Leprosy, as discoursed upon at length in Leviticus 13 in the context of Jewish spiritual purification rituals, is a difficult thing to pin down clinically. Levitical leprosy does not correspond to

9

the disease's modern pathology, which is why modern translations, such as the *New English Bible* (1963), tend to speak instead of a "malignant skin disease" or similar. Levitical leprosy is often sudden, short-lived, and naturally (or miraculously) healed. Sometimes it has white, red, or hairy spots, and the disease could even infect inanimate objects, such as buildings and clothes (moulds). Diseases as diverse as dermatitis, psoriasis, scurvy, and skin cancer, as well as true Hansen's disease, have been suggested to explain some of the Old Testament and New Testament references to the disease. "Leprosy" in Scripture is a portmanteau term used to describe an impurity, blemish, or affliction in a person who has in some way disobeyed Jehovah.

Yet why, within the vast sweep of surviving world literature, should "leprosy" be discussed at such length and in such detail by the post-Exilic Jews? Had the circumstance of spending 40 years wandering in the Sinai desert anything to do with it? Could a long-enforced diet of that strange food manna have been responsible? Without a doubt, the people following Moses *hated* the stuff: "our soul loatheth this light bread"[9] that tasted like coriander seed and honey. Manna was baked, fried, and eaten, day in day out, for nearly 40 years! There were the quails, but this vital protein supplement seems to have been short-lived and sent to punish the ungrateful children of Israel, who gorged upon the quail deluge until they made themselves sick. One presumes that there would have been meat available from the sheep, goats, and cattle which accompanied them out of Egypt.

Moses faced several food-related mutinies, as the liberated Jews were even willing to go back to their brick-making slavery in Egypt, where at least, in the fertile land of Goshen, east of the Nile delta, they had enjoyed melons, wine, oil, cereal bread, and other delights: a remarkably balanced and nourishing diet that would have been conducive to health. Did a sustained deficiency of vital vitamins and minerals, occasioned by an absence of fresh fruit, vegetable oils, and even wine, play a significant part in creating the skin and other conditions broadly described in the Pentateuch as "leprosy"? I have often thought that a diet of manna might have been similar to an unvarying diet of light, crispy, ice cream cone-like biscuits, day in day out; for 40 years!

Biblical "leprosy", however, gave us a priceless medical asset: the concept of isolating the infected from the well. Irrespective of the clinical nature of Levitical leprosy, that concept of isolation is still central to modern scientific epidemiology, and nowhere more, in recent times, than in combating the Ebola virus.

But what is Hansen's disease? It is what we now define clinically as true leprosy, as opposed to the medley of skin conditions described in ancient texts, being a bacterial infection caused by *Mycobacterium leprae* and *Mycobacterium lepromatosis*, and in the twentieth century it was even diagnosed in Egyptian skeletal remains. The disease is caused by both skin and water-droplet contact. It begins by attacking the peripheral, or non-brain, nervous system, generally through the face, feet, and fingers, and spreads from there. Different strains of the disease are known, producing the characteristic leprous lumps or "nodules" in some cases, along with an "anaesthetic" form, where peripheral nerve-endings die, robbing the patient of all sensation in fingers, toes, or feet. Anaesthetic leprosy can result in a dead-pan expressionless face, often accompanied by blindness, as the eyelids become incapable of irrigating the eyeball.

Contrary to popular legend, leprosy is not an easy disease to catch – many people having a natural resistance – and even when caught, it can take up to 20 years to develop and incapacitate its sufferers. It rarely kills – that is left to other diseases – but it causes hideous disfigurement and disability as it wears the patient away. The essential aetiology or clinical definition of the disease was established by three Norwegians in the second half of the nineteenth century. Daniel Cornelius Danielssen and Carl Wilhelm Boeck first identified and described the leprous conditions mentioned above in their *Om Spedalskhed* (Christiana – now Oslo – 1847), while it was Gerhard Armauer Hansen who, between 1873 and 1880, first pinned down its bacterial mode of transmission, and whose *Leprosy: In Its Clinical and Pathological Aspects*, which he co-authored with Carl Looft (English translation, 1895), began our modern understanding of the disease. (In 1980 I had the privilege of visiting Hansen's now

11

preserved laboratory at the St George's Hospital, Bergen, Norway, and even looking through his microscope.)

So one of the most notorious diseases in world history, which tormented Old Kingdom Egyptians, appears almost 70 times in the Bible, occasioned horror in medieval Europe, and still survived into twentieth-century Scandinavia and elsewhere in the world, was first scientifically defined in 1870s' Norway, and finally made amenable to cure in the post-antibiotic age, after 1950. A medical saga on a truly biblical scale!

HIPPOCRATES OF COS: RATIONAL MEDICINE, ETHICS, AND THE *OATH* OF *C.* 430 BC

The classical Greeks invented many of the things we now see as integral components of Western civilization. These include pure geometry and mathematics; philosophy and theology (or philosophical discussions about the nature of the divine); the theatre, tragedy and comedy; organized sports; linguistics; and mathematical astronomy – and the politics of negotiated urban living, as an alternative to the imposed absolute rule of a dictator or tribal mayhem. The Greeks also invented the culture of "personality", or fame: of the warts and all doings of real men and women – Pythagoras, Socrates, Aristophanes, Sappho – as opposed to behavioural exemplars.

In this post-sixth-century BC tradition of questioning and creative public thinking lay the origins of what we today reckon as rational medicine: medicine, that is, based not on magical, divinatory or simple empirical procedures, but upon careful physical enquiry. This is the study of illnesses as natural phenomena, complete with diagnostics, prognostics, case histories, and therapeutic decisions based upon observed facts in the natural world.

Whether Hippocrates, biographically a shadowy contemporary of Socrates in the late fifth century BC, actually invented this new approach to medicine, or whether – as is most likely – he became the first named embodiment of an emerging Greek medical movement, it is hard to be sure. But the books which he would write, or, perhaps in some cases, the books which his disciples wrote in his name, formed the foundations of modern medical practice. Hippocrates' *Prognostics, Epidemics, Aphorisms,* his *Airs, Waters, and Places,* and his other works have an astonishingly modern ring when it comes to assessing a patient. The Hippocratic physician would ask whether the disease ran in the family, what the patient's occupation was, and in what sort of environment he or she lived. What was the air and water quality like in the place where the patient was born or resided? Was he or she of a melancholic or easy-going temperament? What was the state of the patient's tongue, breath, skin, eyes, faeces, and urine, and were their bowels lax or costive? These factors, and information gained from other enquiries into the patient's physical condition, would be weighed in to the eventual diagnosis.

In his treatise *On the Sacred Disease,* Hippocrates further made ground-breaking innovations in what we might call psychology and neurology, as we shall see in Chapter 6. Hippocrates proposed that "sacred" epilepsy was not caused by some kind of divine possession, but by physical changes taking place within the brain: that epilepsy was a *neurological* condition. While a devout pre-Christian pagan, with the profoundest respect for Asclepius, the patron deity of medicine, as well as the eternal *Logos* which underpinned all rationality and of which the human intellect partook, he differentiated between mystical experiences and physical pathologies.

This distinction, I suspect, would also have been acknowledged by Luke and, after him, by a veritable galaxy of medical practitioners extending down to our own time.

Conjointly with Hippocrates' rational and inductive approach to illness ran an increasingly sophisticated theoretical interpretative structure. Central to the Greek natural philosophical view of things was a concern with definitive external fundamental truths. The pure and absolute truths of geometry lay at the heart of Greek astronomy and architecture; the actions of the rational, sentient soul supplied the defining excellences of poetry, number, and music; while pure justice, balance, and equipoise underpinned the whole of Greek political thinking. In many respects, all of them came together in Greek medical thinking.

13

In its basic rational assumptions, Hippocratic medicine saw the human body as resembling a self-contained "polis", or state. When all the parts worked together, and no attacks from outside impacted upon it, the person enjoyed health. Illness, however, was seen as a form of alien invasion or internal obstruction. It was the wise doctor's duty, therefore, to isolate the cause of the trouble, and restore the whole to good order.

By 400 BC, evolving Greek ideas about the nature of things were coming to see health as part of a wider whole. In pursuit of this attempt to understand the big picture, philosophers such as Heraclitus, Protagoras, and Empedocles were coming to think that all natural forces hinged upon the behaviour of one, two, and eventually four principles or Elements: namely, Earth (solidity), Water (moisture), Air (airiness), and Fire (heat). It was the constant intermixing of these Elements that produced everything that existed in the material world, from insects to trees to human bodies. These Elements were seen as possessing their own medical parallels in the writings of Hippocrates and his school.

These were the four bodily "Humours": Yellow Bile (hot and dry), Black Bile (cold and dry), Blood (warm and moist), and Phlegm (cold and moist), which Hippocrates outlined in his *On the Nature of Man*, Section 4.[10] Depending upon how they were "compounded" in each individual, these four Humours provided an explanatory rationale that could be used to account for every aspect of temperament, sickness, and health, from mental illness to cancer. They would survive through the millennia, not being seriously challenged until the late seventeenth century, and were still being echoed in attempts to explain mysterious illness when Queen Victoria ascended the throne in 1837.

Yet that section of the Hippocratic corpus which would make him a household name down 2,400 years, and which still springs to mind today in cases of medical misconduct, is his famous physician's Oath: the foundation stone of medical ethics. When one reads this ancient declaration of medical care, one is struck by its seeming modernity, for like so many other aspects of classical Greek culture, it has helped to frame many of the basic coordinates

14

of Western civilization and how we express our moral values. In the Oath, the Hippocratic physician swears that he will use his skills only to preserve life, and never to destroy it; he will never abort the unborn child; nor will he abuse, sexually or otherwise, his patients or their carers, even if they are slaves. The Oath also recognizes certain medical specialisms: the physician swears, for example, not to attempt the dangerous operation of lithotomy – or "cutting for the stone" – but to leave it to those who are accomplished in that procedure.

The true physician must also swear to be a good and noble teacher to his own pupils and students. And within the obviously pagan context of fifth-century BC Greece, in its deep concern with moral responsibility and almost love, the Oath enjoins the physician to live a "chaste and religious life", in a way that sometimes mirrors Christ's teachings in the Sermon on the Mount. The religion to which the Oath refers, however, is that of the followers of Apollo and Asclepius, the Greek gods of reason, order, goodness, and healing.

In addition to the ethics of the Oath, Hippocrates, on a more practical level, warns against showy quacks who are all fine talk, but who are likely to botch things when skilled treatment is needed; as in the case of resetting broken noses, as discussed in *On Wounds in the Head*, *On Fractures*, and *In the Surgery*.[11]

Most crucially, the Hippocratic Oath established that indissoluble bond between medicine and moral responsibility which still guides our ethical deliberations today.

ARISTOTLE (384–322 BC) AND THE NATURE OF LIVING THINGS

In Hippocrates and his school we find the beginnings of a rigorous, organized, naturalistic, investigative, and taxonomic approach to medicine, all directed towards the alleviation of human suffering. But Western medicine also developed a parallel strand of thinking: a sheer intellectual fascination with how living things work. How do bodies automatically sustain themselves, how do internal organs interact with each other to form living systems, and what *is* the life-process? We now call this *biology*: coined from the two Greek words *bios* ("life"), and *logos* ("word", "rational account"). Without that rational fascination with living things in themselves, creative medicine would have lacked its essential anatomical, physiological, and, by the nineteenth century, biochemical foundations.

Just as Hippocrates probably had something of an earlier, unwritten, "rational" tradition to draw upon – for it is unlikely that his maturity of thinking simply sprang from the void – so did Aristotle. Hippocrates in his treatises *On the Sacred Disease* and *On the Heart* had recorded some remarkably profound insights into the functions of the brain and heart, while even a thousand years earlier, the Ebers and Edwin Smith papyri had also contained some astute observations. But in Aristotle, one encounters a fundamentally different way of thinking.

Yet while he was descended from a medical family of Stagira, in northern Greece, his father Nicomachus being physician to the king of Macedon and a member of the Asclepian Medical Guild. Aristotle was, judging from his extensive writings, less interested in curing the sick than in the science of life. One might say that he was the father of laboratory doctors: people who work not with living patients at the sickbed, but who wrestle in the laboratory with the nature of organs, fluids, conception, decay, and inherited characteristics. (I once knew an eminent Oxford cellular pathologist – indeed, a modern-day son of Aristotle – who used to say "My patients come to me on microscope slides.")

Aristotle came to be referred to in Europe's medieval universities as "the Philosopher", for his intellectual range was breath-taking. In addition to straightforward biology and physiology, he wrote major treatises on politics, poetry, metaphysics, ethics, and the rational soul. Nowadays, we might say that these works pertained to psychology and sociology. They dealt with questions such as how our minds work, and how we best function as social beings. Aristotle's intellectual insights and techniques even inspired new ways of thinking in Christian theology; St Thomas Aquinas in the thirteenth century, in particular, was deeply influenced by "the Stagirite".

Medical insights are scattered throughout Aristotle's voluminous writings, while great treatises such as *Historia Animalium* ("An Account of Living Things") and *De Generatione Animalium* ("On the Generation of Living Things") helped to lay the foundations of functional biology, while his *Problemata* ("Problems") covered a miscellany of medically related topics, such as the nature of perception. In particular, he was concerned with the nature of conception, growth, and inheritance, together with how bodily organs functioned together. What, for example, was the relationship between food, digestion, blood, bile, the lungs, growth, reproduction, and disease?

Living as he did some 2,000 years before the discovery of the existence and biological function of cells, Aristotle thought in terms of fluids, solids, and airs. He also thought in terms of "innate" principles, both in body function and in the world in general – such as heaviness, lightness, and heat. Why, for instance, did spontaneously generated *internal* heat, related, it was believed, to the heart and the blood, sustain life, whereas when *external* heat (such as from a fire) was applied to a dead being, it merely hastened decomposition?

Mammalian life in particular, he suggested, began as a sort of curdling, when blood in the mother's womb spontaneously thickened upon contact with male sperm. In *De Generatione* Book II, Aristotle even saw an analogy between sperm and rennet, the fermenting agent that causes milk to curdle into cheese.[12] He was concerned, too, with the nature of biological inheritance, even touching upon what we might now ascribe to the action of recessive genes: why, for example, did skin, hair, or facial characteristics appear in a person when they had not been present in either parent?

In addition to being a brilliant and inspired observer of nature, Aristotle almost certainly dissected dead creatures. Whether he ever dissected humans is a moot point, but he was certainly living at a time when curiosity-driven men were opening up dead (and sometimes living) pigs and monkeys, the internal organ arrangements of which had parallels to that of humans.

17

Aristotle notwithstanding, the rising generation of Greek-Alexandrian medical researchers who came to prominence around 300 BC, and were probably influenced by the deceased Aristotle's teachings, certainly dissected in a big way. First Herophilus and then Erasistratus pioneered the use of both anatomy and vivisection experiments, performing public anatomies before students in Alexandria. This great city lay to the west of the Nile delta, founded by Aristotle's old pupil, Alexander the Great, after 331 BC – a city possessing not only a world-famous library but also the world's first great academic research centre, containing, along with books, research facilities for the study of medicine, astronomy, geography, and all the then known branches of learning.

There was both debate about and physical investigation into trying to establish which organ was most important in sustaining life. Was it the heart: Aristotle's seat of the sentient living soul and the focal point for a complex system of tubes and fluids running throughout the body? Or was it the brain, with its extensive system of nervous appendages? Either way, at least in the Greek world – and by extension, the world of expatriate, diaspora Judaism, for the open city of Alexandria would soon be the home of one of the biggest Jewish communities outside Palestine – Alexander's city would become a byword for educational excellence, science, and rational, experimental medicine, in addition to philosophy and all the arts.

And what kind of a doctor was St Luke? Had the author of the third Gospel and the Acts of the Apostles dissected cadavers, kept case records of patients, and asked preliminary diagnostic questions which would not have been out of place in a modern doctor's surgery? While we have no proof either way, I suspect that he had been, in the very best professional sense of the first century AD, *Dr* Luke: a man who wrote good Greek and could organize and present evidence in a precise and scholarly fashion, ask questions of his sources, and display all the signs of possessing a highly trained intellect. Or in Latin, he had been a *medicus*. Before, that is, he received, and responded to, an even higher calling.

CHAPTER 2

Galen: Surgeon to the Gladiators

$\mathcal{T}$wentieth-century cinema has given us a rather unbalanced view of the Roman "sports industry", or arena, with films such as *Ben Hur* (1959) and *Spartacus* (1960) emphasizing the ghastly barbarity of gladiatorial combats and chariot races in the Roman Colosseum and provincial amphitheatres. While the life of a gladiator was usually short, there were ameliorating circumstances. Surely young professional fighting men, whether slaves, victims of foreign wars, or state criminals condemned to death in the arena, were also an investment for their owners, and as such were surprisingly well-treated, given the circumstances of their condition.[1]

If they were stylish and popular in combat, and if they survived, gladiators could become sporting celebrities, win their freedom, and retire. But this was only for the lucky few, and life prospects for a gladiator tended towards the "nasty, brutish, and short". While they were being trained and prepared for combat, and – most crucially from our point of view, when they were wounded – however, things were not always terrible. As entertainers who needed to be in good physical shape, they were well fed, on a high carbohydrate diet, consisting of beans, barley, oatmeal, and fruit. They were also rested and massaged, and even banqueted by their owners and trainers, and would probably have looked rather chubby alongside modern-day athletes. The layers of fat covering muscles and other vital spots gave some protection, however, and a sword-swipe or a spear-thrust into adipose tissue might produce an impressive flow of blood to delight the crowd without causing major damage. And if he received skilful medical treatment, Felix or Marcellus might live to fight for a good few days, months, or even years to come.

Good doctors and surgeons could not only save gladiators' lives, but, more importantly by the standards of the time protect the long-term investment of their owners or fighting companies. And it was this circumstance that was to provide an opening for one of the most influential medical careers of all time: that of Galen.

AELIUS CLAUDIUS GALENUS OF PERGAMUM: SURGEON, SHOWMAN, AND PUBLIC ANATOMIST, AD 129–200/216

Galen was born into a wealthy patrician Greek family, his father Aelius Nicon being a successful architect, philosopher, and gentleman of wide culture of the southern Greek city of Pergamum (now Bergama, Turkey). From the autobiographical remarks contained in his writings, Galen had the greatest respect and admiration for his father, who spent liberally on his son's education, both in Pergamum and on his subsequent studies in Smyrna, Alexandria, and elsewhere.

Not only were father and son interested in philosophy and the world of ideas, but in one distinctive way Galen shared an aspect of his father's architectural interests, as the young man became increasingly fascinated by the architecture of the human body. How are we put together, how do our bodies work, and why do they not fall down? Why are there clear similarities of organic and skeletal design and function between creatures so seemingly different as humans, apes, and pigs? In that grand Greek anatomical and physiological tradition extending back to Aristotle, Herophilus, and Erasistratus, Galen was not only a human and comparative anatomist, but also a vivisectionist, in his attempts to fathom out how both human and animal bodies hung together mechanically and functioned internally.

By the time that he was 28, Galen was a wealthy independent gentleman of the world, his beloved father having died nine years previously and left him the family property. Scientific passion seems to have been his driving force, however, and in his twenty-eighth

20

year he applied for the vacant post of physician to the gladiators of his native city of Pergamum. While sports medicine may be a fast-growing branch of modern medicine, it seems odd that a gentleman possessing the wealth, status, and already significant reputation of Galen should apply for such a downmarket job, instead of enjoying the kudos of a comfortable private practice. It is my suspicion, however, that Galen was attracted by the intellectual and technical challenges which violent Greco-Roman sports medicine offered. One can understand this: not only would the young doctor face a gamut of fighting injuries, but there would have been dietary, psychological, and general health problems to deal with as well. It would have provided excellent all-round training for an ambitious young doctor: a bit like being an army doctor, without suffering the inconveniences of going on campaign.

How he got the job tells us much about Galen's wider style, temperament, confidence, and bravado. To show his superior skill over and against that of his rivals for the gladiatorial surgical job, he tells us that he eviscerated a living ape, and when his rivals were reluctant to try putting the poor creature back together again, he did so. If one could replace and stitch up the guts of an ape, one could perhaps do the same for a gladiator with a major abdominal trauma![2]

Galen was proud of the fact that during his time at the Pergamum arena, he lost only five gladiators, whereas his surgeon predecessor had lost an alarming 60. This may have been owing in part to his clearly superior anatomical knowledge, combined with the meticulous care he reputedly took in the management of wounds. For the scientific treatment of a wound involves far more than simple, empirical, binding or suturing. To control a severe haemorrhage, it is crucial to know exactly where the damaged blood vessels lie beneath the skin. Likewise, a thorough knowledge of the muscular structure of the wounded part of the body is essential if muscles, tendons, and ligaments are going to heal effectively. While Galen knew nothing about bacteria in AD 160, he would have known that a carefully handled wound, especially if well washed with wine, was more likely to heal, and to save the patient's life, than a wound dealt with in a careless fashion.

As a doctor who also considered himself to be a philosopher, therefore, Galen would have seen medicine and surgery within a context of responsible behaviour and a duty of care towards one's patients. This duty of care would not only be displayed in a technical operative capability, but it would also be rooted within a wider context of wise understanding with relation to the nature of things and to eternal truths. His time at the Pergamum arena would have given Galen an excellent grounding for what would become one of the most illustrious medical careers in antiquity, as well as for a writer whose works would constitute the canon of anatomy and physiology both in Europe and in the Arab world for the next 1,400 years. But what were Galen's enduring contributions to medical knowledge and understanding?

GALEN THE ANATOMIST AND PHYSIOLOGIST

As we have seen above, Galen had what in his time was already a 500-year-old tradition of rational Greek medicine to draw upon, and especially research methods pioneered by Herophilus and Erasistratus and their followers in his own *alma mater* of Alexandria. As his career progressed, from Pergamum on to fame in Rome, culminating in his appointment as physician to Emperors Marcus Aurelius, Commodus, and Septimius Severus after AD 168, Galen not only researched assiduously but wrote prolifically. He supplies us with by far the largest corpus of anatomical and to some degree physiological literature to survive from antiquity. For while Aristotle's writings were probably bigger in bulk, he wrote on a vast range of topics, and his medical works were but part of a larger output, Galen's surviving works – around 122 in all – are much more focused upon anatomy, physiology, surgery, and medical autobiography.

His anatomical writings would provide benchmarks of excellence for centuries to come. His writings were so precise and elegant that until the development of new experimental techniques in the sixteenth and seventeenth centuries, they were not easy to contradict. We must not berate the centuries falling between AD 200 and 1500 for being slavishly uncritical in their continuing use of Galen's works, for inevitable errors apart, they were so very good. His skill as a descriptive and taxonomic anatomist becomes even more breath-taking when one remembers that Galen possessed few research aids beyond a sharp scalpel, a sharp pair of eyes, and probably a photographic memory when it came to recalling and recording detail. So what did he say?

Fundamental to Galen's medical writings, as we have seen, was a fascination with the *architecture* of the skeleton – be it that of a human, a rhesus monkey, a Barbary ape, or a pig. We know that as a young man in Alexandria he had studied human skeletons, and as the years went by he did much of his research on the bodies of apes. Galen classified the bones into two broad categories: long bones, hollow and with marrow cavities; and flat bones, which were essentially solid. How did different bones join together to give different flexibilities, such as the hip bones, the finger bones, and the 24 pre-sacral spinal vertebrae? Galen leaves us one of the first really detailed descriptions of the bones of the skull and how they work together. Many of the osteological and joint names that Galen used are still current today, while his works established the anatomical rule of beginning a description of a living thing by delineating the skeleton: the coat-hanger, as it were, upon which all other parts are suspended.

From the skeleton, Galen progresses to a description of the muscle system, based largely on rhesus monkey and ape cadavers. It is the muscles that form attachments upon and between bones, to keep joints together, and, in boneless areas such as the abdomen, to keep the internal organs in place. He is less good on the brain and nerves, however, for while humans and all higher animals possess brains and a branching nervous system, there are major differences. Here Galen's sound, visually based anatomy blends into his often incorrect physiology, for one might characterize anatomy as a description of the parts, whereas physiology is more about trying to make sense of how, why, and to what purpose the parts move: a much more complex, and even philosophical, topic.

GALEN'S PHYSIOLOGY

One question becomes apparent when beholding the insides of a dead – or even a living – human or animal body: how on earth do all these organs, muscles, tubes, nerves, hollow, solid, springy, inert, and porous parts which are still in death, actually work in life? In this enterprise, Galen was initially guided by the writings of Erasistratus.

Central to Galen's physiology was the concept of nutrition. Food juices in the stomach become *chyle*, a whitish fluid which he believed was carried by the portal vein into the liver. There, by a semi-mysterious process and innate body heat, it became blood. The liver was also the seat of a basic life force, the "natural spirit", which suffused the new blood with a nourishing power of growth. The spirit-enriched blood then rose up the *vena cava* or "great vein", which had two major branches. One appeared to deliver blood into the entire venous system, to carry the food and spirit-enriched blood to all parts of the body to sustain body heat and growth. This seemed self-evident, for did not gluttons generate too much blood which, in turn, congealed into excess body mass – or fat? And the reverse happened to those who starved.

The other branch of the *vena cava* admitted a smaller volume of blood, through the tricuspid valve, into the right ventricle, or bottom chamber, of the heart (Galen was imprecise regarding the functions of the atria or auricles), from whence it passed down to the "artery-like vein" (pulmonary artery) to nourish the lungs. Correcting Erasistratus, however, Galen realized from animal vivisection experiments that a greater volume of blood came through the tricuspid valve into the right ventricle than the amount that passed down the pulmonary artery.

Galen therefore postulated that some of the blood in the right ventricle of the heart now passed through what were believed (incorrectly) to be tiny pores in the thick muscular dividing wall, or septum, of the heart to enter the left ventricle, and there to be infused with "vital spirit". This was in addition to its already liver-derived "natural spirit". These two spirits now combined to constitute a "life", "soul", or "breath" principle along with the inhaled *pneuma*.

24

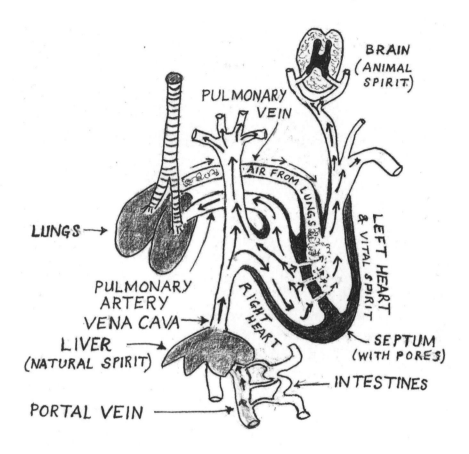

1. The Galenic heart. Blood from the liver enters the right heart from the vena cava vein, and goes on to nourish the lungs via the pulmonary artery. Additionally, a portion of this right heart-blood was thought to pass through the supposed "pores", or small holes, in the thick muscular septum, to enter the left heart. Here, it was thought to receive a vital life-force, breath, or *pneuma* from the lungs via the pulmonary vein. This newly-enriched blood then passed into the aorta, and on to nourish the brain and other parts. (Drawing by A. Chapman.)

To prevent the innately hot heart from overheating, however, it drew in cool air from the lungs via the pulmonary vein through the mitral valve. Because Galen believed that the two-leaved mitral valve was less tight-fitting than the three-leaved tricuspid, though, he suggested that it might allow a reflux of hot, smoky air from the heart back into the lungs to be exhaled as bad breath. (As we shall see in Chapter 11, however, the great sixteenth-century Flemish anatomist Andreas Vesalius demonstrated that the little pits in the septum muscle were no more than pits, and *not* pores, and that there was no way in which blood could pass through them.)

Galen argued that this volume of left-ventricle blood, enriched by "natural spirit", "vital spirit", and inhaled breath-of-life *pneuma*, now did two things. Rising up into the aorta, it filled the arteries with invigorated blood. Then a smaller volume of arterial blood rose up into the brain via the carotid arteries, each carotid branching into a separate *rete mirabile* or "wonderful network" of small vessels to nourish the left and right hemispheres. The brain blood was now mysteriously changed into a third, or "animal", spirit (from the Latin *anima*, "life"). "Animal spirit" (somewhat analogous to cerebrospinal fluid) was colourless, and a little salty to the taste, and seemed to be stored in the hollow lateral "ventricles" or chambers inside the brain. In a way, these were rather similar to a pair of cisterns in a plumbing system, for the left and right hemisphere lateral ventricles were thought to connect with the tubular nervous system which, even by Galen's time, was known to branch throughout all human and animal bodies. This made the brain a bit like a hydraulic control system receiving and sending signals through the body, although in reality Galen's physiology was *vitalistic* rather than mechanistic in its conception.

Yet just as Galen made major errors in his model of heart function and in his belief that the venous and arterial systems were distinct and unconnected, so he made another as far as the human brain was concerned. For while a variety of animals, such as sheep and pigs, do have *rete mirabile* structures at the base of their brains, humans and the higher primates do not. But it would take some 1,400 years for the error to come to light, when Thomas Willis of Oxford discovered in humans not a network of vessels, but a great, branching, circular artery. (But more on the "Circle of Willis" in Chapter 14.) And it would not be until the 1550s that Realdo Colombo (or Renaldus Columbus) of Padua unravelled the true heart–lung–blood relationship in his discovery of the lesser or thoracic (chest) blood circulation. William Harvey realized that this was but part of a general or whole-body circulation in 1628. Ironically, however, the first medical scientist to suggest a thoracic circulation was Ibn al-Nafis somewhere around 1270, as we shall see in the next chapter.

We will return to Galen and his impact upon medical thinking over the millennia shortly, but let us now look at the wider practice of surgery in the Roman Empire.

ROMAN SURGERY

Although we possess a number of written sources on Roman surgery, most notably Celsus and Galen, it has been archaeology that has done the most to widen our understanding beyond the written word. Surgical instruments, drug jars, and pieces of art and sculpture depicting medical scenes have been unearthed in archaeological excavations across the Roman world, but nothing compares with the riches that have come to light at Bingen, Germany, and especially Pompeii in Italy. There is nothing quite like being able to see, and even handle, 2,000-year-old instruments to get a feel for the art of their original, unknown, users.

One thing strikes you right away when examining these pieces: the seriousness and sophistication of the surgeon's calling. Be prepared to abandon all your cinema and historical novel assumptions about pre-modern surgery being performed by semi-uncouth brutes who hacked and chopped in ignorance and invariably subjected their patients to hideous deaths. When an instrument is exquisitely made, fits beautifully into the operator's hand, and, in many ways, looks almost as though, with a bit of polishing, a modern-day surgeon could use it in theatre, it tells you a good deal.

There are similarities between the artefacts found in the two hoards (as there would be in operating theatres in London and Beijing today), which include not only beautifully made lancets, probes, and forceps, but also more complex instruments, some with moving parts. The Bingen excavations also brought to light a beautiful set of bronze cupping vessels. These were used to draw out noxious humours from the patient's body to correct imbalances between the Blood, Phlegm, Yellow Bile, and Black Bile which were seen as lying at the heart of health and disease.

In Pompeii, however, which provides us with a complete time capsule of a prosperous community that came to a sudden end in August AD 79, there are two medical collections in particular. One is that of the house of A. Pumponius Magonianus, whose name still remains painted on the wall outside.[3] Inside his *taberna medica* or private hospital on the Via Dell'Abbondanza were found many instruments, drug jars (some containing the dried-up remains of their ancient contents), and other medical artefacts. The other collection is the instruments found in that house on the Pompeian Via Consolare now referred to as the "House of the Surgeon" owing to the rich variety of medical objects within.

In addition to the beautifully made probes, lancets, and forceps, the Bingen and Pompeian collections contain physical examples of sophisticated instruments mentioned in Roman medical literature. These include trephine "circular saw" drills, or *modioli*, some with screw-in parts to be attached or removed at different stages of an operation. Such trephines would generally have been used to treat serious head injuries, such as depressed fractures of the skull, where a blow to the head had driven in part of the cranium. Even more impressive was the bronze quadrivalve, or four-point, dilator, used to enable a doctor to examine a woman's uterus. The four smooth and rounded "points" would have been inserted closed, when their combined thickness was scarcely more than a pencil. Then, using a handle, the surgeon would have employed a precision screw mechanism to gently draw them apart, so that an inspection could be made.[4]

As one might expect from an imperial power like Rome, there are many examples of the skill of military surgeons. Classical literature and Pompeian art works swell our knowledge in addition to what Galen and Celsus tell us in more explicitly medical texts. The Greek historian Plutarch, for instance, praised the skill of the army surgeon Cleanthes, who successfully replaced the intestines of a badly wounded soldier: a procedure reminiscent of Galen's

28

poor monkey at Pergamum;[5] while in his *De Medicina*, Celsus describes how to successfully perform such an operation. It must have been a common procedure, in a military world dominated by sword slashes and javelin attacks. A beautiful wall-painting in Pompeii depicts the legendary Aeneas having what was probably an arrowhead removed from his thigh by a dexterous surgeon using an elegantly fashioned pair of forceps.

2.   A Greek surgeon uses a pair of long-handled forceps to remove an arrow head from Aeneas's thigh. (Mural painting, Pompeii, first century AD, in Ralph Jackson, *Doctors and Diseases in the Roman Empire* (British Museum, 1988), plate 33. Copied by A. Chapman.)

3. Bronze long-handled forceps similar to those depicted in Figure 2.(First century AD, Pompeii. Drawn by A. Chapman.)

In fact, Roman art, including sculptured stone panels and Pompeian wall-paintings, depicts a variety of medical scenes in addition to military injuries, such as doctors in their consulting rooms with various instruments on display; women in childbirth; cataract operations; and human limb deformities. For Rome had both a great literary culture and a deeply visual one; and both of these cultures were as much civil as military, showing medicine – among many other Roman activities – as part of life. Soranus, for instance, wrote on obstetrics and gynaecology, while there exist excellent accounts of Roman dental techniques, along with surviving examples of gold bridge-work. The writings of Hippocrates were revered, commented upon – by Galen among others – and expanded by Roman doctors, as were his medical ethics and comments upon the healer's duty of care to his patient, which I will discuss below.

Pharmacy, in an organized and coherent form that would survive for the next two millennia, was also a product of the Roman world: Pedanius Dioscorides' *De Materia Medica* ("On Medicine and Drugs") came out in five volumes in the AD 60s. It was destined to dominate Latin, Greek, and Arabic medical thinking about pharmacy (the collection and preparation of drugs and medicinal substances) and be translated into English and a variety of European vernacular languages from the sixteenth century onwards.

But one of the most fascinating medical documents to survive from antiquity is *De Medicina* ("On Medicine"), the great multi-volume encyclopedia of Aulus Cornelius Celsus.

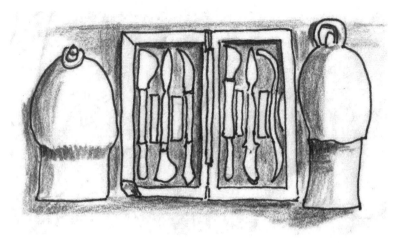

4. Surgical instruments, probably Roman, depicted in a stone carving at the Asklepieion, Athens. The elegant hinged wooden case contains a set of scalpels and, perhaps, a bone lever. Two "cupping vessels" are also shown. These would be used to raise "blisters" and supposedly extract poisons. (Ralph Jackson, *Doctors and Diseases in the Roman Empire* (British Museum, 1988), plate 28. Copied by A. Chapman.)

CELSUS AND HIS *ENCYCLOPEDIA* OF C. AD 30

As far as we can tell, Celsus was not a doctor by training or profession. He appears to have been a Roman gentleman of culture, perhaps a native of southern Gaul, in love with learning, and very obviously fascinated by technical subjects and ingenious devices. Scholarly evidence suggests that his wider, and sadly lost, writings embraced a whole variety of subjects, including agriculture, rhetoric, military technology, and philosophy, though all that survives is his great medical treatise *De Medicina*. Yet that work in itself was a milestone, and the first great Latin (as opposed to Greek) treatise on medicine and surgery. It provides us with the golden key to first-century AD Roman medicine and surgery from over 100 years before Galen.

Celsus, the gifted and informed amateur, divides his medical encyclopedia into eight books, or divisions, and covers a therapeutic range extending from how to use a catheter tube to relieve a urinary blockage to how one might deal with mental illness, depression, delusions, and phobias. In Book III, in the section dealing with

31

disorders of the mind, Celsus is the first writer to use the word *insania* – insanity – in a modern clinical context. Like Galen, Celsus had a profound respect for Hippocrates, the Greek Asclepian medical tradition, and the doctor's duty of care. (Note that Aulus Cornelius was *not* the same Celsus as the anti-Christian philosopher of the same name who was countered by Origen: he lived around AD 180, long after Aulus Cornelius Celsus was dead.)

De Medicina Book VII is especially useful, as it begins with a history of surgery as it was understood in *c.* AD 30. The surgical sections of the Hippocratic corpus are praised, and Hippocrates is seen as the founding father of rational medicine. Celsus then goes on, in Book VII, to look at a variety of "elective" surgeries (i.e. non-emergency, unlike military surgery), such as the treatment of hernias, bladder problems, and fistulae, which are wounds or holes in the body that will not heal and tend to discharge. (The woman who had suffered "an issue of blood" for 12 years, which Christ healed miraculously, probably had such an unhealed fistula, going back, possibly, to a protracted and complicated childbirth.[6])

Fistulae and sores and lacerations that would not heal were an ever-present surgical danger right down the millennia, until Lister developed his techniques of antiseptic surgery after 1865. In the meanwhile, however, Celsus records several classical "truisms" about wound-healing. If a patient can sleep well after surgery, has a good appetite, breathes easily, and does not develop a raging thirst, then things are going well. And much hinges on the pus discharge from the wound, for in an age ignorant of bacteria most surgical wounds ended up infected, irrespective of the care that the surgeon had taken to clean his instruments beforehand: such as by breathing and spitting on them, then using his handkerchief to bring them up to a good polish. Such polishing techniques would still have been used by a surgeon's assistant in a London or a Parisian hospital in 1840.

Good, clean, relatively odourless whitish pus was fine, for it was a "natural" corollary of healing. But if the pus stank and was bloody, things were going wrong. The "naturalness" or inevitability of post-surgical pus would also be discussed at length by Galen, who would style it "laudable", or praiseworthy, and a sign of hope in healing. Yet as was to be discovered by the late nineteenth century, *all* pus is actually a sign of infection, and only once we had learned to kill germs or exclude them from the surgical site could wounds heal quickly and cleanly.

In Book III, where he deals with specific diseases and the humane management of inflammation, such as that following surgery, Celsus coined a maxim that was to ring through the centuries down to Queen Victoria's time: *Notae vero inflammationis sunt quattuor: rubor et tumor cum calore et dolore*, or, in English, "Now the signs of an inflammation are four: redness and swelling with heat and pain."[7] Also, sometimes, there is fever. Wounds tend to heal without infection, however, if they have been cauterized with a hot iron, for as we now know, heat kills germs. And if we take away our modern-day antiseptics and antibiotics, they still do!

What is so fascinating about Celsus, the gentleman amateur, is his love of technical detail: he is one of antiquity's greatest technical writers. *De Medicina* describes a variety of surgical instruments, including the *modiolus* or circular drill mentioned above, with its detachable and interworking parts, used for boring holes in injured skulls (examples from the time of Celsus survive from the Bingen and Pompeian excavations). Celsus also describes a variety of surgical procedures, including abdominal wound surgery, advising that, after being put back together and bandaged, the patient should be made to lie on his back, thereby reducing pressure on the damaged muscles.

One can only assume that Celsus must have been deeply learned in the Greek medical literature of his day and have had plenty of doctor friends, for his grasp of technical detail strongly suggests that he had seen, and may even have assisted in, many of the procedures described. There were no legal regulations on the practice of medicine in ancient Rome, and anyone whom the patient might trust was welcome to have a go. The same applied to some degree in England until the Medical Act of 1858 began to separate the sheep from the goats, clinically speaking; and as we shall see in later chapters, numerous well-educated gentlemen –

usually Anglican clergymen – practised charitable medicine in their parishes between the sixteenth and nineteenth centuries.

As a medical layman myself, with a passion for technical details of all kinds, who is untroubled by the sight of blood, and who has many doctor, surgeon, and other scientist friends, I have always felt an affinity with Aulus Cornelius Celsus. And would we not love to have known what he wrote about in the sadly lost, non-medical sections of his encyclopedia, dealing as they reputedly did, with agriculture and various military and civil technologies.

Celsus was by no means the only Roman layman to display a knowledge of contemporary medicine. Later in the first century, there was the scholar and encyclopedist Gaius Plinius Secundus, or Pliny the Elder, whose *Historia Naturalis* (*c.* AD 78) contains a staggering range of observations and opinions relating to natural phenomena.[8] The work also devotes some nine chapters, numbers 20–29, to medical topics, including a discussion of drugs and medicinal substances. Book 29, on the medical profession itself, is highly informative. Very few Roman citizens have practised medicine, he tells us (Section 17), in spite of medicine's profitability, for Greek doctors and Greek techniques seem to predominate. (Galen, Soranus, Dioscorides, and St Luke, to name but a few Roman Empire doctors, were all either Greeks or of Greek ancestry.)

Rather cynically, Pliny tells us (Section 18) that only doctors can kill with impunity, and try out their theories and their cures upon their hapless patients. In Book 29 Section 14, Pliny even cites the formidable Greco-phobic lawyer Cato, in a portmanteau condemnation of doctors as a body of men who have sworn to kill all foreigners (i.e. Romans) with their medicine, but do so for a fee. How much of this assessment stems from Pliny's xenophobic patrician prejudices?

Between them, Galen and Celsus opened up a medical, and especially an anatomical, physiological, and surgical, world which is unique in the surviving literature of ancient science. When one adds to this the learned and influential gynaecological writings of Soranus and the pharmaceutical works of Dioscorides, one gets a strong sense of the sophistication of medical understanding by the first century AD.

In this unlicensed open-market medical world, one finds not only male practitioners, but female healers as well: styled in books and on tombstones *medicae*, or women doctors. Pliny, in his *Historia Naturalis* Book 25, made disparaging remarks about female folk healers and empirics, yet firm documentary and even archaeological evidence survives of women who were serious medical practitioners. There was, for example, Antiochis of Lycia – another "Grecized" foreigner – of whom Galen spoke highly, while both Juvenal in his *Satires* and Martial in his *Epigrams* mention high-society female practitioners. Many of these women would have treated female patients for a variety of gynaecological ailments and children's diseases. When one adds women and small children together, one has a total of more than half the Roman population: a major patient catchment. While some women may have been folk healers and empirics – as many male quacks were – there is sound surviving evidence of Greek and Roman women who practised rational or at least common-sense medicine and used the sound, scientific techniques of the age. Their professional daughters and granddaughters would continue doing so with remarkable success down the millennia, eventually phasing, by the late nineteenth century, into those European and American women who finally gained entrance to academic medical schools. We will meet them in later chapters.

GALEN'S INFLUENCE: MEDICINE, ETHICS, RELIGION, AND TEACHING ACROSS FIFTEEN CENTURIES

Why were the teachings and practices of Galen so pervasive and long-lasting? After his settlement in Rome after AD 162, leaving the gladiators of Pergamum to the care of his successors, he rapidly won renown, as a medical practitioner, a brilliant teacher, and an anatomical lecturer. Galen became physician to the army, then to the imperial court, as well as doctor to the Emperor Marcus Aurelius and then his heir Commodus. His professional career, which began beside the bloody operating tables beneath the Pergamum arena, ended among the very highest echelons of Roman society.

But what was it about both his writings and his personal example that made this rather flamboyant Greek so persistently appealing to later Roman doctors, to medieval Arabs during the "golden age" of their civilization from Spain to Persia, and to the professors of Bologna, Montpellier, and Paris in Europe's new post-1150 universities, and would inspire a diversity of Renaissance Europeans and beyond – such as Vesalius and Harvey – to challenge his theories? While they may have found errors in Galen's physiology, and corrected them, to lay the foundations of post-Renaissance Western medicine, no one could fail to stand in awe of Galen's passion for exact taxonomy, lucid description, and the concept of *method* and *technique* he bequeathed to later ages.

As did his master Hippocrates, Galen revered the Oath and respected the wider moral and ethical duty of the healer to his (or her) patient. Likewise, he revered Plato and Aristotle, for Galen always saw himself as a *philosopher*-doctor, placing great stress upon reason and logic as a means of understanding. His thoroughgoing empiricism, however, led him into disputes with the Stoics and other philosophical schools, particularly regarding the nature of the soul. Unlike the Stoics, Galen saw the soul not as a separate entity, but as a functioning, integral aspect of the material body, the nerve-centre of which was the brain. He did, however, regard the soul as possessing three aspects: vegetable, animal, and rational.

The "vegetable" soul corresponded to functions such as digestion, the "animal" soul to the capacity of movement, and the "rational" soul to the intellect. One sees here to some degree a parallel with his three "spirits", *natural, vital,* and *animal,* in liver, heart, and brain respectively, which were functional cornerstones of his physiology. Of the highest importance was *pneuma*, or that air life-force which the trachea sucked down to form the breath of life and energize the lungs, heart, liver, brain, and other organs: a sort of special fuel that transcended common food.

One might have expected the very theistic post-classical medical culture of the Christians, Jews, and Muslims to have been put off by

Galen's apparent materialism. Yet this was far from the case. Quite apart from the sheer volume of invaluable medical knowledge that was contained in his voluminous writings, key aspects of his system enabled later ages to adapt him to more obviously theistic, mind–body–God interpretations.

We can see a quintessential *purposiveness* in Galen, as bones, nerves, organs, fluids, heart, and brain all work beautifully together to form a functioning, self-sustaining system: a healthy being. This structure and function relationship in Galen set him at odds with those pagan philosophers, such as the followers of Epicurus and Lucretius, who argued that all order was but a brief illusion of atoms chancing to coalesce in a greater meaningless void. While the Stoics believed in the logical reality of the world, their wider sense and acceptance of human impotence in the teeth of stern inevitability did not really chime in with the very purpose of medicine: namely, bringing about deliberate change to a natural process, and using philosophy and ingenuity to snatch a sick person back from the gates of death. Furthermore, Galen's admiration of Hippocrates, and his endorsement of the doctor's noble calling enshrined in the Oath, would also endear him to later theistic thinking.

So, in the same way that St Thomas Aquinas and other medieval academics transformed pagan Aristotle into "a good Christian" from the ideas implicit in his wider philosophy, so medical thinkers would do the same for Galen. And in the wake of the new religious force that exploded across the Middle East and into Spain after AD 622, Galen's thinking, along with that of Aristotle, would be adapted to suit the requirements of Islam as well.

CHAPTER 3

Arabia: The First Fruits of Medieval Medicine

*T*he causes behind the dying away of the once great Roman Empire have been discussed at length by scholars across the centuries. Among the various economic, military, and other causes, a series of diseases and disasters hit Europe by the fifth century AD onwards. These included cycles of epidemics, culminating in the plagues of Justinian after AD 541, and famines, especially after AD 400, not to mention the more familiar barbarian invasions. Irrespective of what triggered what, however, it was clear that the Roman imperial system was falling apart by AD 500.

Yet the empire did survive, in a new form, and the post-AD 330 *Christianized* Roman Empire provided the structure for the Roman Catholic Church, canon law, and the still classically based culture of Western Europe. Following Constantine's removal of the heart of the empire from Rome to his new city of Constantinople in AD 330 – refounding it from the earlier Greek city of Byzantium – the "Eastern" empire of the New Rome was born. As the centuries rolled by, these two increasingly separate Western (Rome) and Eastern (Constantinople) empires came to constitute *medieval* Europe: or as it saw itself, differences of politics and theology notwithstanding, Christendom. This domain extended from the Outer Hebrides to Mesopotamia and, before AD 622, down into the Holy Land, Egypt, Sinai, North Africa, and eastwards towards the Asiatic Steppe.

BAGHDAD AND THE HOUSE OF WISDOM

Then, after AD 622, a new force hit the Middle East, in the wake of the Prophet Mohammed's visions. After his initial military conquests in Medina and of pagan, polytheistic Mecca, down in the Arabian Peninsula, or Yemen, Mohammed brought his armies north into Egypt. Mohammed died in Medina in AD 632, but his followers went on to burn what was left of the great Alexandrian Library in AD 641, force Byzantine Jerusalem into submission, carry Islam ("Surrender" or "Submission") into Mesopotamia, and in AD 711, into Spain and southern France. Mohammed's *caliphs*, or chieftains, would go on to fall out among themselves and eventually kill each other, thus splitting the faith into its Sunni and Shi'ite sects.

It was from this beginning, however, that the first great scientific and medical flowering in the post-classical world arose. In AD 762, the newly Islamicized region of the upper Euphrates valley became the birthplace of one of civilization's great centres, for in that year the Abbasid Caliph Al-Mansur established the new city of Baghdad. Its site already had its thriving local settlements, as it occupied an important geographical location, both as an entrepôt on the east–west caravan route, and as a commercial centre for the mighty Euphrates–Tigris river system. The new city of Baghdad would rapidly become not only a commercial but also a vibrant cultural centre, where east met west.

It also appears to have been, considering the conditions of the time, a tolerant place and in that cultural centre which grew up there, and came to be styled "the House of Wisdom", one found scholars from a variety of faith traditions working together. In this heartland of the ancient Babylonian–Assyrian–Persian empires, many languages and traditions survived the coming of Islam, and the less fanatical religious and military leaders recognized their value. Here were diaspora Jews, Persians, Sabbaeans, Zoroastrians, Syrians, and also Nestorian Christians.

Translation became the driving force behind the House of Wisdom, as the new, more cultured Islamic rulers were interested in the ancient civilizations that had occupied or passed through these regions in past millennia. As scholars have a habit of picking up languages, they began to translate the wisdom of antiquity into the new Islamic (yet already ancient before Mohammed) lingua franca of Arabic. Nor must it be assumed that Baghdad was the only such city in the new Islamic domains to recognize the treasures of antiquity and their value not only culturally, but also practically: for Damascus, Cairo, Fez, Jundishapur in India, then later Seville and Granada in the far west, would also become major cultural centres. Sciences such as astronomy (and astrology), surveying, chemistry, and gold and other metal working were downright *useful*. And medicine in particular promised to be very useful indeed.

Arabic medical and scientific writers would add their own original discoveries to those found in classical and earlier sources. These men were much more than mere translators and purveyors of what had gone before; nowhere more so than in chemistry and its links to medicine.

FIRE AND WATER: TRANSFORMATIVE FORCES

It is hard to say when humanity's fascination with what would later be styled "chemistry" first began. And the same can be said about medicine. For both of these subjects are extremely useful in so many ways. Medicine was pursued in the hope of alleviating human suffering, and when it had advanced to such a state where hopefully useful medicinal substances – such as herbs or animal products – were not simply administered raw, but were pounded, dissolved, and cooked beforehand, then medicinal chemistry was born.

Yet even from earliest times, chemistry was more than the simple handmaiden of medicine, and was quickly recognized as being both an intellectually interesting and a practically useful discipline in its own right. Chemistry not only offered a possible key to health, but also promised to unlock the hidden nature of substances. Metal-smelting, paint-making, adhesives, glass-making, and cooking all go back into a pre-legendary time, as modern archaeologists have discovered in cave and other remains. All of these skills involve primary chemical manipulations, such as modifying natural products by those two agents which lie at the very genesis of *Homo faber*'s ("man the maker's") achievement: making solutions and heating things.

Liquefaction, using water, with or without other additives, is a mighty technology, allowing natural products to be separated out into different parts, which might next, by means of drying, be turned into useful powders or crystals, such as common salt. Fire, the making of which goes back to our ancestors on the African savannah, gave its possessors unprecedented powers: to cook, keep warm, frighten away predators, extract hard metal from stone, or, if you lived in a place blessed with the right naturally occurring minerals, make glass. All of these manipulations, along with plenty more, represent the first infant steps of chemistry, and by extension, pharmacy.

One of the compelling features of chemical action is its ability to *transform* things. If certain stones could be smelted into metal, then could not one metal, perhaps, be transmuted into another? Could not chemical action even bring the sick back to health? It is hardly surprising that this "art" seemed full of ancient mystery and deep wisdom. It also helps explain why its origins are lost in legend.

Some traditions said that Miriam, the sister of Moses and Aaron, was the first alchemist, having learned the secrets of gold-working from the ancient Egyptians during the Jews' slavery in that land. Other traditions spoke of Hermes, Apollo, and other legendary beings. The Greek Byzantine writer Zosimus of *c.* AD 300 also spoke of the "art", as chemistry was often styled, but it was, without doubt, the Arabs after *c.* AD 750 who first seriously practised and wrote upon it.

JABIR (GEBER) AND RHAZES: CHEMISTRY AND
MEDICINE

Two Arabic physician–chemists were among the first sustained organized writers: Jabir ibn Hayyan (*c.* 721–815) and Muhammad ibn Zakariya al-Razi (*c.* 854–925), who came to be known in the West under their Latinized names, Geber and Rhazes. It was said that Jabir – a Persian from Tus, Iran, but working at Kufa near

41

Baghdad – was so fascinated by the power of chemical action and the way that fire had the ability to transmute and enliven things that he even attempted to make living beings! He seems to have tried making snakes and scorpions in his laboratory, and wondered whether one might even be able to make a human by artificial means. One might think of this as making him an antecedent of the fictional Dr Frankenstein; yet medical alchemy would have a long-term love affair with attempting to transmute dead or inert things into living beings, and the early sixteenth-century Swiss alchemist Paracelsus and his followers also spoke of creating the *homunculus* – "little man" – or proto-human-being in a chemical flask in the laboratory. In many respects, however, it made logical sense, for if the physician–chemist could *save* life by substances prepared in the laboratory, was it not only a few more steps down the same road to creating life *ex chemico*? And Paracelsus's followers were familiar with both the authentic and the apocryphal Jabir in Latin translation.

Such lines of thinking posed all sorts of problems, for surely only God had the power to create new life? This was an area where science could truly impact upon theology, irrespective of whether the operator were a Jew, Christian, or Muslim. The ethical problem is still very much with us today, and nowhere more so, especially in the public imagination, than in genetic engineering.

Jabir's ideas about the possibilities of making life, however, were of a piece with his wider concern with the chemical transmutation of metals, which he developed out of the Arabic translations of Aristotle and his theory of the four Elements, Earth, Water, Air, and Fire. Jabir went beyond Aristotle to develop a chemical model, which argued that all the then known metals – gold, silver, mercury, copper, tin, iron, and lead – were not fundamental, natural agents, but compounds or mixtures. The fundamental creative power lay in mercury and sulphur: but not "common" mercury and sulphur, so much as their "philosophical", mystical, life-energizing counterparts – *alchemical*, in so far as mercury represented the principles of volatility and liquidity, while sulphur represented combustibility. Later alchemists would hold incombustible, non-volatile salt to characterize stability, to constitute the *tria prima*, or "first three": mercury, sulphur, and salt.

This theory would then provide the rationale behind alchemy, or the attempt to transmute base metals into gold. In a mode of thinking that would last a millennium, true gold was obtained only when the correct mercury–sulphur balance was achieved. And as mercury is a very useful metallic flux and, it was believed, could be made to blend with base metals such as copper or tin, it seemed to be the primal "seed" of all metals.

Stemming, quite possibly, from Jabir, and profoundly influential first in Arabic then in Western alchemy, was the search for the secret of metallurgical transmutation. Far from being the domain of crazed fortune-hunters – as the so-called Enlightenment portrayed it – the quest for the transmutative agent could only be successfully accomplished by a true philosopher, of pure heart, mind, and soul, who pursued wisdom and held mere wealth in contempt. It needed a man whose driving force was to understand profound secrets and who, in the deepest sense, was truly *wise*.

The key ingredient necessary to accomplish the transmutation was variously referred to as the Philosopher's Stone, or the Elixir. The word Elixir carried powerful medical overtones, for it was recognized as the universal, quasi-magical preparation that would cure all ills for ever and even grant immortality. In this medieval culture, which placed great emphasis upon purity and transformation, the Elixir or the universal medicine that could cure all diseases through its perfect humour-rectifying powers was next of kin to that stone which turned all base metals to perfect gold.

In this respect, the intimate relationship between chemistry, medicine, metallurgy, and religion was central to subsequent Arabic and later Western alchemical thinking, and in their true medieval context, be it in Arabia or Europe, these four were inseparable. It was the Arabs of the ninth to eleventh centuries who first grounded the quest for medicinal agents in the chemical laboratory. Very conveniently from the historian's point of view, Jabir left a list of the books he had written, some of which began to be translated into Latin in the thirteenth century by Robert of Chester and Gerard of

Cremona. Jabir became the victim of his own posthumous fame, and his name would become such a byword in chemistry across the centuries that we are sometimes sadly uncertain which sections of his works were actually by him, and which were attributed to him by later chemical writers.

Was Jabir the first chemist to prepare nitric acid by distilling a mixture of saltpetre (potassium nitrate) and copperas (iron sulphate), or was this ascribed to him by his thirteenth-century Western followers? Did he know how to prepare pure spirits of wine – al-kohl – by distillation from wine? It is ironic that the name of that substance forbidden to Muslims should have an Arabic derivation and may first have been isolated as a pure chemical substance in a Middle Eastern laboratory! Either way, in addition to writing on elixirs and transmutations, Jabir was a highly skilled chemical laboratory operator and experimentalist.

No matter who first isolated them chemically, it is hard to imagine the future development of chemistry without powerful mineral acids such as nitric, or of medicine without the wonderful organic solvent properties of pure alcohol. Both were familiar to thirteenth-century European alchemists, who tended to attribute them to the true or to the apocryphal Jabir, or perhaps to Rhazes, or another Arab experimentalist.

Rhazes, who was 30-odd years younger than Jabir, was also Persian, coming from Rey, though working for many years as a doctor in Baghdad. Like many scholars of the age, both Muslim and Christian, he was an encyclopedist, fascinated by the collection and organization of knowledge, which helps to explain why Arabic medical writers made such an outstanding contribution to the taxonomy of disease (as we shall see later with Avicenna).

Rhazes showed himself to be an accomplished laboratory alchemist in works such as his *Secretum Secretorum* ("Secret of Secrets") and his medical treatise, the *Liber Continens* ("Comprehensive Book on Medicine"), both translated into Latin in *c.* 1280, among many others. Familiar with Arabic translations of the works of Plato, Aristotle, and Galen, Rhazes took issue with the already well-established Greek doctrine of the four Elements and four humours lying at the heart of both chemistry and medicine. Besides the four Elements, Rhazes suggested – not unlike Jabir before him – that two active principles were at work in all chemical action: "oiliness" and "sulphurousness".

Oiliness is a characteristic of many organic substances, such as fats and greases. One even found it in minerals, such as the black crude petroleum oil which, in certain parts of the Middle East, once bubbled up to the surface. It has been suggested that Rhazes distilled crude oil to produce an impure form of paraffin, but this is unproven. Sulphurousness was seen as an innate state of potential inflammability or the emission of an intangible spirit, what a modern chemist might call a phase-change in a substance. Just try heating some sulphur, and watch the yellow mineral powder begin to *burn*, produce a thick, brown oily sludge, and then emit clouds of white, choking, stinking *fumes*! This could be viewed as posing an experimental challenge to a straightforward four-Element concept of matter.

Rhazes divided mineral substances alone into six separate classes: metals, volatile spirits, stones, vitriols (sulphates and acids), sodas (alkalis), and salts. We still use the Arabic-derived word alkali to designate bases, or non-acidic substances. Just like Jabir's, Rhazes' medical and chemical concepts interpenetrated each other, as he further posed an experimental challenge to the Hippocratic–Galenic theory of the four bodily humours. Just think of this: if a normal, healthy person has their four humours in perfect balance, then why, if you give them a drink which is much hotter than body temperature, do you not suddenly make them ill by upsetting their humoral balance? You don't even need thermometers to test that idea: just give a healthy person a drink of hot liquid, which touch alone tells you is much hotter than their forehead, and observe the body's failure to go into convulsions or suffer a violent reaction!

Rhazes, after Jabir, continued to perfect that very largely Arabic chemical technique of distillation. Evidence from surviving sources suggests that it had been Rhazes who had perfected the distillation process, by devising a flask with a long swan's neck type of spout, known as an alembic or retort. Composite liquids, such as wine, crude oil, or human body products could be put into the alembic, which was then heated over a laboratory furnace. As the heat increased, different volatile parts of the original mixture in the retort would be given off. But as they entered the long, angled, tapering neck, they would cool, until liquefaction was reached – at which point the condensed substance would drip out into a receiver vessel as a fluid.

In this way, al-kohl could be separated out from wine, or nitric acid distilled and condensed from the bubbling mass of saltpetre (nitre) and iron sulphate, to produce "spirits of wine" and "spirits of nitre" respectively. The term "spirit" came to be used to signify what was believed to be the light and volatile parts of a reacting substance, which flew away when its parent mass was breaking down in the alembic, in the same way that the spirit flew from a dead person. In popular parlance, we still use the term spirit to signify an odorous, volatile substance. Think of methylated spirits, or even the distilled gin (once called genever, or juniper spirit) that one might buy in a pub.

Rhazes was also an extremely astute clinician. It was he who drew the first clear diagnostic differentiation between smallpox and measles: in measles, all the rashes tend to appear at once, whereas in smallpox each spot appears separately. He also drew attention to the occurrence of back pain and nausea in diagnosing the two diseases. Mental illness, however, Rhazes tended to ascribe to demonic influence.

Alchemy, commentaries upon and expansions of the Greek writers, astute clinical observation, the taxonomic arrangement of both diseases and medicines, and a willingness to absorb useful ideas from Indian and even Far Eastern sources all became part of Arabic medicine. Perhaps the most influential of them all was the precocious son of a Persian tax collector in *c.* 980, whose name would become synonymous with medical wisdom across medieval Europe. This was Ibn Sina, or as he was known in the West, Avicenna.

"I SUPPOSE THAT AVYCEN / WROOT NEVERE IN NO CANON…" (CHAUCER)[1]

Such was Ibn Sina's, or Avicenna's, medical fame by *c.* 1381 that the English poet Geoffrey Chaucer could confidently throw his name as an acknowledged expert on pharmacy into a squalid tale about poisoners and thieves – "The Pardoner's Tale". And the "canon" to which Chaucer referred was Avicenna's *Kitab al-Qanun*, or in its already familiar Latin translation, his *Canon*, or rule of medicine. It is a clear sign of how many Arabic medical, alchemical, astronomical, and other scientific works were circulating around Europe by 1380, that Chaucer, the London merchant's son and literary civil servant and courtier, mentions several in his entertaining *Tales*: "Razis" (Rhazes) is there too, along with "Averrois" (Averroes). While many literary scholars ignore or dismiss the science in Chaucer, there is a great deal of it to be found: medicine, alchemy, astronomy, astrology, scientific instrumentation, and even the connection between the moon and the tides. Chaucer also possessed a fascination with, and a skill in, technical subjects.

But who was "Avycen"? We saw above that he was the son of a Persian tax collector, whose youthful intellectual precocity led to his beginning his medical career at the age of 16. He also appears to have been something of a showman, flamboyant in manner, who had a way of impressing people. These are useful traits in any profession, but especially for a doctor, particularly at a time when most serious diseases were incurable, and winning the patient's confidence played a vital part in the therapeutic battle.

Avicenna was both a fine clinician and a medical taxonomist, in the already well-established Arabic tradition. His magisterial *Canon* of medicine would become a medical classic, with his case histories, observations, and techniques surviving him down through many centuries. In his *Canon*, Avicenna divided his material up into four sections or Books. Very conspicuously, his whole way of thinking is rooted in the tradition of the great Greek clinicians and systematists: Hippocrates, Dioscorides, Aristotle, and especially Galen underpin his approach to medicine. One reason why Avicenna won such enduring renown was that he collated, systematized, and interconnected the Greek writers, and combined their works into one coherent, rational whole.

His *Canon* advances across medical science, starting from the basic principles of physiology, hygiene, and humoral pathology, and moving on to the art, or emerging science, of pharmacy. On the way, he deals in detail with *materia medica*, listing 760 useful medicinal and toxic substances, puts together a catalogue of all manner of diseases afflicting the human frame, and makes a study of specific maladies, such as different types of fever, ulcers, abscesses, and even what we today might class as psychological and metabolic disorders. Thinking back to "The Pardoner's Tale", it was almost certainly to Book 2, with its list of herbal and mineral substances and their effect upon the human body, that Chaucer was referring.

Avicenna wrote on that universal therapy of the pre-modern medical era: phlebotomy, or bloodletting. It may seem irrational to us today, even *dangerous*, to drain the sick body of a pint or more of oxygen-bearing haemoglobin and other vital substances; but in the Galenic humoral tradition, it made eminently good sense. If a patient were in a state of burning fever, and blood was seen as a "hot" humour, then could it not be beneficial to attempt to bring down the body temperature by removing hot blood? The exhaustion induced by heavy bloodletting (or by purging the bowels) might similarly appear beneficial in the treatment of the severely mentally ill, because it quietened them down. Even wounds would be bled in accordance with the (as we now know) misguided logic that the blueness of an inflamed wound was caused by poisons, which needed to be drained away. Besides, as every physician knew, patients often felt more relaxed after being bled: caused, as we now know, by the release of endorphins in the brain.

Avicenna seems also to have been aware of diabetes mellitus, from the sweet taste of the patient's urine: a discovery that was re-made – and named – in the West in the seventeenth century. Revolting as it may strike us today, physicians not infrequently tasted body products as a routine part of diagnosis, at a time when very few decisive chemical tests were available.

While it was by far the most influential of his works, the *Canon* of medicine was only one of Avicenna's treatises. As a Persian who spent most of his career near modern Tehran, he was well placed to encounter and learn from Arabic translations of Persian, Indian, and other medical traditions, but it was undoubtedly to classical Greece that he looked for his guiding stars. Through Avicenna, Rhazes, and many more of their like-minded colleagues, the treasures of an *amplified* and *systematized* Greek medical tradition passed into Latin Christendom in a flood tide, especially in the twelfth and thirteenth centuries. What came with these Arabic and Latin translations were not only medical facts and theories, but also the wider concept of the noble physician: the man who, walking in the footsteps of Hippocrates, Aristotle, Dioscorides, Celsus, and Galen, was also a learned *gentleman*, in whom medicine blended effortlessly with philosophy, poetry, astronomy, alchemy, architecture, ethics, and religion. The Arabs, in so many ways, created that enduring tradition of the "learned physician" which would ring down the ensuing centuries.

But what of surgery during the Arabian "golden age", between *c.* 800 and 1400?

ALBUCASIS AND ARABIC SURGERY

Until the development of anaesthetics and antiseptics in the mid nineteenth century, surgery was always the forlorn hope of the medical arts: resorted to when all else had failed in elective medicine, or in desperate necessity in the case of battle or other injuries. Yet because wounds of all kinds, along with tumours, ulcers, and cataracts, are part of the human condition, procedures to deal with them go back into the mists of time. And given a steady hand, a strong patient, and a large dose of good luck, these procedures could give the patient a life-opportunity. Ancient Egypt and Babylon already had their surgical texts, while archaeologists have found clear cases of flint-tool surgery in prehistoric human skeletons.

Ancient India had its own medical and surgical traditions, incorporated into the texts of the Brahmin Ayurvedic or Vedic

Sanskrit culture, which, like later parallels in Hippocrates, placed great stress upon the relationship between health, lifestyle, and religiosity. The pre-eminent Indian practitioner and writer in this tradition was Sushruta of *c.* 750 BC. The *Sushruta Samhita*, or "Sushruta's Compendium" or "Collection", emphasizes the medical practitioner's noble calling and the high standards expected of him no less than does the Hippocratic Oath. But where Sushruta is significant in the present context is in the attention that his *Samhita* gives to surgery, for all the key pre-anaesthetic procedures are there, such as "cutting for the stone", cataract operations, and draining ulcers.

While the Arab medical writers whom we met above saw themselves in the Hippocratic and Aristotelian Greek traditions, one might speculate how much Indian medicine was incorporated into their therapeutic regimes. India was not far away from Persia and Mesopotamia: just sail for 1,000 miles or so straight down the Euphrates, go through the Persian Gulf into the Indian Ocean, and keep left until you arrive at the estuary of the Indus. Archaeologists have found that trade goods had moved this way over the millennia, and ideas and skills often accompanied the articles of commerce, for people instinctively talk to each other. This brings us to the work of Abu al-Qasim al-Zahrawi (AD 936–1013), known in the West as Albucasis: the most prolific and profound Arabic surgical writer.

Unlike so many of the great Arabic medical writers, Albucasis came from the opposite end of that world from Mesopotamia. He was of North African descent and lived in Arabic southern Spain, at Cordova. As a physician in the noble philosophical tradition, Albucasis embedded his surgery within a wider corpus of noble, learned, and humane medicine, although the pervasive traits of the surgeon shine through him. These are traits of self-confidence, ingenuity, virtuosic manual dexterity, and showmanship, which one also finds in Erasistratus, Celsus, and the Roman surgeons, in Galen, and in later practitioners through the Middle Ages – and down to today. Surgeons tend, by temperament, to be men (and nowadays women) of action rather than given to pondering; if a patient is bleeding to death, a polite discussion about diets and lifestyles is not likely to do much good! Surgeons also tend to be interested in how things are put together and how they work; a surgeon who is not fascinated by the functional machinery of the living body is not likely to get very far. And one finds evidence of all these traits in Albucasis's *Kitab Al-Tasrif*, a vast encyclopedic work in the grand Arabic tradition, covering most aspects of medicine. It is, however, the thirtieth treatise of the work that concerns us here: *On Surgery and Instruments*.

Albucasis's cutting and probing instruments would not have been that different from those of the surgeon of Pompeii – for they were devised to serve similar functions – but it is when Albucasis displays his astute clinical nous that we feel compelled to take notice.

When dealing with small wounds in the intestines, for instance (not uncommon, as we saw with Celsus and Galen, in a sword-fighting culture), Albucasis suggests the introduction of live *ants* into the cut, for their sharp jaws can help to suture lacerations. He also records a careful study of gangrene: a 'creeping corruption of the limb, consuming it as a fire consumes dry wood'.[2] Like any classically trained physician, he sees poor-quality blood as lying at the root of many diseases, especially those giving rise to ulcers, fistulae (non-healing, suppurating sores), nasty tumours, and gangrene. Part of the cure for such patients lay in trying to "strengthen" their blood, yet good timely surgery could often put them on the way to recovery.

Albucasis's self-confidence and professional excellence shines through one of his cases in particular. This involved a young man with a deteriorating leg ulcer. Several physicians had failed to arrest his decline, until "eventually he sought me", he tells us, with knowing self-certainty. Albucasis then operated twice on the young man, opening up the ulcer and removing necrosed bone, before managing the post-operative patient back to health.[3] And just like his Greek role model Hippocrates, Albucasis was of the opinion that place and location were important, for there are some places "where diseases tend to linger and tumours are naturally slow to heal".[4]

Like Hippocrates, who in *On Wounds, Surgery, and Fractures* had condemned showy quacks who mis-set broken jaws and noses,[5] so Albucasis is quick to condemn ignorant operators who botched up the treatment of fractures so that the bones did not join correctly.[6]

One of Albucasis's favourite techniques was the application of the hot cautery iron: a technique that would remain in use down the centuries, electric cauteries even being used to join small wounds in modern-day surgery. Albucasis especially recommended cauterization in cases of gangrene for, unbeknown to him in that pre-bacterial age, a red-hot iron could destroy bacteria and cleanse a wound. Three days after cauterization, the still sore wound would be treated with sulphur and oil-healing salves. The cautery may have been momentarily very unpleasant for the patient, yet on purely empirical grounds it gave the wound the best chance of healing.

ARABIC MEDICINE IN RETROSPECT

The contribution of the Arabic people to science and medicine, from Spain to Uzbekistan and down into modern northern India, during their cultural golden age, was truly impressive, as was their work on astronomy, geometry, and other physical sciences. Arabic contributions to the medical, chemical, and pharmacological sciences constituted much more than a bridge between classical and later Western science, as some have suggested. They furnished a rich and a dynamic influence all of their own; while the Arabic passion for taxonomy and classification led them to devise names for numerous chemical processes and substances which have been, often unknowingly, absorbed into the everyday usage of the modern world: alkali, alembic, camphor, sherbet, borax, elixir, talc, coffee, and alcohol, to name but a few.

I do not remotely claim to do anything more than review Arabic medical achievements in medicine and chemistry: there is so much more. Take, for example, the Arabic fascination with light, optics, and perception. In the next chapter, we will examine the enormously formative part that the Cairo physician Hasan Ibn al-Haytham (Alhazen) played in the birth of scientific optics. Then just think of Ibn al-Nafis (1200–88) of Damascus and Cairo, whose physiological researches led him to reconsider Galen's doctrine of the blood passing through (non-existent) pores in the heart's septum wall. Instead, he argued, around 1270, the blood passed down what we now call the pulmonary artery into the lungs, to mix with air. What he had discovered was what is known as the "lesser" or "thoracic" circulation, the movement of the blood between the heart and the lungs. Its quite independent discovery in mid sixteenth-century Europe, by Michael Servetus and Realdo Colombo, would eventually supply a vital physiological component to William Harvey's discovery of the whole-body blood circulation in 1628 (Chapter 12). Yet Ibn al-Nafis's discovery would remain unknown in the West until modern scholars came upon it in his Arabic writings in 1924.

With all this scientific and medical achievement behind them – and equally major contributions in poetry, philosophy, and architecture – why had the Arabic golden age come to an end by 1400? Many scholars have pondered that question. It is true that the Mongols sacked Baghdad in 1258 and 1401, and that the Black Death of the mid fourteenth century wreaked havoc across the Arab world; yet it wreaked no less havoc in Europe. But Europe's civilization continued to flourish and ascend, whereas that of Arabia stagnated; and while numerous Arabic original works and Arabic translations of Greek and Syriac classical writers were translated into Latin for European minds hungry for ideas, I am not aware of any of the great post-fourteenth-century European scientific works being translated into Arabic.

So why was this the case? One possible reason is the increasing formalization of Islamic theology, as the *Koran*, the sayings and teachings of Mohammed, and an emerging Sharia legal system developed integrated structures. And as this Islamic canon came to be increasingly self-sufficient and self-referential, the pursuit of external knowledge could be seen as an irrelevance, or even a threat. Either way, one cannot help but be struck by how the development of this Islamic canon coincided with the Arab world's decline in intellectual distinction and curiosity by *c.* 1400.

Yet at the same time, Christian Europe was avid for fresh ideas and for new useful knowledge, be they from Judaism, pagan classical antiquity, or Muslim Arabia. And one early manifestation of this curiosity lay in a fascination with light, optics, and the physiology and psychology of perception.

CHAPTER 4

Divine Light: Seeing and Perceiving
in the Middle Ages

*O*f all natural phenomena, light has always been the one that has possessed the most powerful spiritual resonances. At the creation itself, so Genesis tells us, God separated the light from the darkness. The Buddha, around 500 BC, spoke of paths to "enlightenment"; while Pythagoras and his followers, also *c.* 500 BC, were captivated by the mystical relationship that seemed to exist between light, fire, and motion.

Light itself appeared to originate in the heavens and have an inextricable connection with astronomical bodies. The sun was clearly the primary source of light, and even very ancient cultures were aware that the moon's phases indicated that it only shone by reflecting the light of the sun. Regarding the stars, however, there had been uncertainty in remote antiquity whether they shone by reflecting sunlight, or whether they were self-radiant, possessing individual lights of their own. Yet either way, the fact remained that when the sun was above the horizon, one's world was filled with light, and after it set, the light vanished. What was beyond dispute was light's mighty power, practically, metaphysically, and spiritually. Medieval Arabic scholars would draw upon ancient Jewish and Christian sources to develop their own metaphysic of light, while St John tells us at the opening of his Gospel that Christ, the "Logos", the Son of God, came into the dark world, to be the "light of men".

In addition to the theology and metaphysics of light, however, there was the emerging science of optics, which further led to a fascination with the physiology of the eye and the nature of perception.

THE ANATOMY OF PERCEPTION: WHAT WAS "SEEING" BELIEVED TO BE?

As we have seen with other organs of the body, such as the heart, liver, brain, and blood, anatomy, philosophy, and even religious thinking were never far apart. How did structure, function, and purpose interact? In many ways, the eye presents even deeper mysteries, for what exactly are *seeing* and *perceiving*, and how do light, flesh, and states of mind interrelate? Questions about how those "lights of the body" form our inner and outer worlds had emerged well before any coherent anatomical studies of the eye were made. For several centuries during and after classical antiquity, two theories of perception were on offer. Did the "sensitive" eye radiate feelers of some kind that sensed the world beyond the body and helped us form both concrete and metaphysical conceptions of it? Like a form of radar, in fact?

This "extramission", or outward-transmitting, theory of eye-perception was favoured by Plato and other early Greek philosophers, although it was concerned less with the eye as an *anatomical* structure than as a sensitive perceiver. From a scientific point of view, it had a number of obvious shortcomings: for instance, it did not explain why we are unable to see in the dark. If perception is dependent upon the eye somehow "feeling" our environment, then we should be able to see at night. Could it be that these sentient "feelers" somehow "sense" the divine light flooding in from the heavens, rather than generating light in their own right? One appealing feature of the extramission theory was the shining, crystalline glow of the eye itself. How do we see the whites of each other's eyes – and most notably, those of cats – in the dark, if they do not emit a sort of glow that an external observer can see? The lens, cornea, and aqueous humour of the eye were seen by the early Greeks as being somehow like ice, gemstone, or *krustallos*: crystal.

The alternative theory of the eye was warmly espoused by Aristotle. This was the *intromission* theory, in which, instead of the eye somehow generating its own light, it acted as a receiver for external light coming into the body. We now accept this as the correct interpretation of eye function, although it would take many centuries to unravel how the eye actually worked anatomically and physiologically.

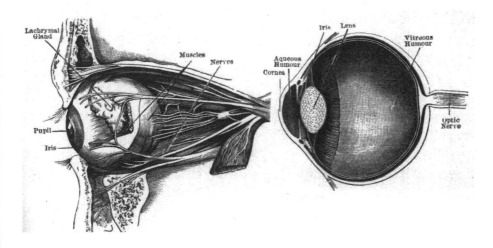

5. The human eye. The right-hand image depicts the cornea, aqueous humour, iris, lens, vitreous humour (plus retina), and optic nerve. The left-hand image illustrates the muscular and nervous structures that control eye movement within the socket. (*The Family Physician* (undated, *c.* 1905), Vol. III, frontispiece plate. A. Chapman collection.)

On one level, understanding the anatomy of the eye may seem straightforward: just go around to the slaughterhouse, obtain a few eyeballs from some recently killed large animals, such as cattle, and then dissect them. But as with most things in anatomy, when we try to forget what we now know and think ourselves back into the world of 350 BC, things begin to get more puzzling. Eyes, like most bodily organs, have a tendency to resemble amorphous lumps of wet stuff when seen close up, isolated, and *dead*. And soon after the death of their four- or two-legged owner, the crystalline parts of the eye begin to cloud over.

In its basic parts, the eye – of a human, sheep, pig, or cow – comprises three functional zones contained within the spherical structure of the eyeball. At the front, directly behind the transparent cornea, is the aqueous (from Latin, "watery") humour, occupying a relatively small part of the whole eyeball. The greater part of the eyeball volume, behind the aqueous humour, is taken up by the vitreous (Latin, "glassy") humour. Separating these two unequal spaces are the lens and iris. In a healthy living creature, all these regions, aqueous humour, lens, and vitreous humour, should be perfectly transparent, while the iris, by expanding and contracting, regulates the intensity of light entering the lens and passing through into the vitreous humour. Except for the transparent frontal cornea and transparent inner regions, the rest of the eyeball on simple inspection is wrapped within an opaque spherical membrane with what we today know to be the optic nerve exiting at the rear.

The early Greeks believed the eye "saw" because the light coming through the transparent cornea somehow acted upon the transparent fluid parts of the eye and then passed up hollow tube-like vessels at the rear to act upon the brain, thereby producing those impressions which we call sight. Celsus, however, in *De Medicina* is really the first medical encyclopedist to supply us with an approximately realistic account of the anatomy of the eye, although his account is often confused, and he misunderstands or misses altogether certain key structures. Celsus was probably drawing upon the writings of Herophilus and the Alexandrian School of some three centuries earlier, none of whose original anatomical manuscripts survive, although they were in circulation during the Roman period. Where Celsus (or Herophilus) is significant, however, is in his recognition of that "crystalline" drop-like structure we call the lens. Whether he had any idea of its light-projecting function is a moot point, for lenses as optical devices, as opposed to curiously polished pieces of transparent gemstone intended for jewellery, did not exist in Roman times.

Rufus of Ephesus (*c.* AD 70–120), who appears to have learned his anatomy in Alexandria, is credited with making the next serious contribution to our understanding of the anatomy of the eye. He described several ophthalmic structures, such as the retina and conjunctiva, yet without having any clear understanding of their function.

It was Galen in the second century AD, who had clearly dissected animal (and perhaps a few human) eyes, who was to provide both the physical description and the theory of function of the eye that would survive down to the sixteenth century. Galen described the basic anatomy of the eye in considerable detail, recognizing all the principal structures, such as the cornea, aqueous and vitreous humours, iris, and lens. Galen regarded the lens or "crystal" as the principal structure of the eye, its shape, properties, and focusing power being brought about by the action of *pneuma*, or a life-agency, acting upon it in the aqueous humour, and hence altering its focus. He also described the lens's attachments to the surrounding eye-structure, or choroid, and the principal surrounding muscles which secure the eye in its socket and enable it to move. Galen even recognized the significance of the tear ducts and conjunctiva. Additionally, he described the solid structure of the optic nerve, exiting from the rear of the eyeball en route for the brain, although he still regarded that nerve as a conduit for "humours" which somehow conveyed the eye's sensitive impressions to the brain to enable us to "see".

Galen would provide the most detailed account of ophthalmic anatomy to come out of the entire ancient world. How much was entirely his own, and how much he had learned from then extant but since lost prior texts, it is hard to be certain. Yet with ophthalmic as with other branches of anatomy, Galen certainly made references to earlier researches, such as those of the Alexandrian School, and we should be thankful to him, for in many instances he is our only detailed source for some of his Greek predecessors.

Priority for specific discoveries apart, Galen would, more than any other writer, supply the definitive framework for ophthalmic understanding for the next 14 centuries. In particular, he inspired a rich tradition of Arabic ophthalmological writing, as did Aristotle. These included figures like Hunayn ibn Ishaq, Al-Kindi, Rhazes, Alhazen, and Avicenna.

Arabic writings on the eye tended to be of two broad types: they dealt either with the practical treatment of ophthalmic ailments, such as cataract (which we will discuss below), or with the metaphysics of perception, or how we see. Central to much Arabic writing was the classical concept of the eye's quasi-metaphysical "sensitivity", as the seat of vision in a spiritual rather than a merely physical sense. In particular, this sensitivity was centred on the transparent lens and humours, with less attention being paid to the significance of the retina and optic nerve. While it was obvious that the eye dealt in some way with light, exactly *what* it did remained an intractable mystery, as physiology, optical physics, spirituality, and religion all coalesced together in one great tangle.

On the whole, the Arabic writers tended to favour the extramission theory – the eye somehow projecting sensitive "feelers" into the world – though some clinicians disagreed on observed practical grounds. Rhazes, for example, noted that the eye responded to and could even be damaged by incoming bright light, such as looking at the sun, as did his colleague of a century later, Alhazen. And Alhazen of Cairo, the greatest optical writer of the earlier Middle Ages (*c.* 965–1039), was not only a supporter of the incoming-light, intromission theory, but would profoundly influence the future course of optical writing, especially after his *Optical Thesaurus* became available to European readers in Latin translation around 1170.

It would be Alhazen's writings that would inspire optical studies in Western Europe after *c.* 1200, and guide it towards a predominantly, but no means exclusively, "intromissionist" way of thinking. And medieval writers would leave some elegant and remarkably exact descriptions of the structure of the eye, based largely on translations of Galen and Alhazen.

Yet before there could be any real progress in making sense of how the eye worked anatomically and physiologically, it was first necessary to ground "laboratory"-based experimental optics on a much firmer footing. When Alhazen, followed by his European Christian disciples, began to experiment with glass cylinders, simple lenses, and convex and concave mirrors, and to elucidate

60

the mathematical geometry of the rainbow and shadows and why, when the sun is below the horizon at twilight, the sky can remain bright, significant breakthroughs became possible. The study of light and vision provides us with the first great instance in which a non-medical science, such as laboratory optics, would help bring about fundamental change in medical understanding. And as we shall see in the following chapters, there would be many more.

Rainbows, colours, and perspective: Medieval Europe's new key to physics

Alhazen's *Kitab al-Manazir*, or *Optical Thesaurus*, was an inspiration to European thinkers, and to none more than scientifically minded monks and churchmen, who read the ingenious Muslim doctor with delight. Born in 965, Ibn al-Haytham, or Alhazen as he was known in Europe, spent most of his life practising medicine and doing ingenious things in Cairo. Drawing on Ptolemy, Aristotle, and other classical writers whose works were already available in Arabic as his starting point, Alhazen went on to become, in many respects, the founding father of optical physics.

What has always fascinated me about Alhazen is his sheer brilliance and ingenuity as an experimental scientist: a brilliance also exhibited by his disciple Kamal al-Din al-Farisi. Working during this "golden age" of Arabic civilization, before the formal codification of Islam and before a strict fundamentalist interpretation had taken hold, Alhazen's scientific ingenuity enjoyed free play, without his activities being regarded in the least bit "blasphemous". And what so captivated the devout Alhazen was the intimate relationship between his science and his monotheistic faith, for "God is the light of the heavens and the earth". A statement with which no Christian or Jew would have disagreed.

Although Alhazen, like other Arabic doctors, was interested in the eye, its structures, its perceived modes of operation, and its afflictions, what especially preoccupied him was the behaviour of light in the heavens, in the air, and on the earth, before it encountered the physiology of the eyeball: optical physics. This external, universe-generated model of light would enable him to make observations of natural optical phenomena and conduct ingeniously thought out experiments to test and mathematically model the behaviour of light rays from an external, objective standpoint. Alhazen, when translated into Latin, would become the original primary source for medieval European optics, while his advocacy of the intromission theory would help to form the thinking of a succession of European optical researchers, extending from Bishop Robert Grosseteste down to Sir Isaac Newton.

Alhazen and al-Farisi undertook two particularly significant lines of research in physical optics, both to do with the nature of refraction. One of these involved the causes of twilight, and the other of colours in the sky. Why is the sky capable of being bright long before, or long after, the sun has risen or set? The ancients had various ideas about the air bending sunlight, but Alhazen began to investigate it mathematically: almost certainly using an astrolabe, which was an astronomical angle-measuring instrument devised by Claudius Ptolemy in *c.* AD 180, but brought to practical and enduring perfection by the early medieval Arabs.

Upon noting when the first hint of dawn appeared, Alhazen measured the altitudes and the positions of certain bright stars in the eastern sky. Then, just before their light was drowned out by the imminently rising sun, he took the same stars' positions for the last time. Exactly the same could be done, in reverse, at dusk, to time the period between geometrical sunset, when the sun touches the western horizon, and complete darkness. From these astronomical sightings, Alhazen was able to calculate the duration of twilight, and from it how far the sun had been below the horizon when the first and last glimmerings of daylight had been visible. From this data he computed that the sun was 19° of arc below the horizon at first and last light.

So what made the sky light up? Very clearly, the earth's atmosphere was bending, or refracting, the light around the curvature of the globe, for both Muslims and Christians were in full agreement with the ancient Greeks about the spherical character of the earth.

The atmosphere acted in the same way as what the Greeks called *krustallos* and what, in Latin, would be called *lens* (from the similarity in shape to the original meaning, "lentil") – a lens.

Atmospheric refraction was by no means a new study in *c.* AD 1000, and Aristotle, in his *Problemata* ("Problems") of *c.* 350 BC,[1] had discussed the formation of solar and other "pinhole" images cast into darkened interiors through narrow cracks, such as those in basketwork screens: "pinhole" cameras. But it was Alhazen, using the brilliant, blinding light of North Africa, who studied the upside-down and right-to-left full colour projections in an effort to understand the refraction of light and colour.

We will be looking at early theories of how we were thought to *perceive* coloured light in more detail in the following section, for the *Optical Thesaurus* details Alhazen's "laboratory" experiments with mirrors, plain and curved, as well as with simple transparent pieces of quartz figured into cylinders, spheres, and other shapes: with early *lenses*.

Why did a quartz cylinder, or a concave metal mirror, produce such a distorted view of the world when its projected images were carefully examined? And when al-Farisi was building on Alhazen's work, how could he create a kind of artificial rainbow on demand, by directing a narrow shaft of sunlight into a suspended quartz sphere? Natural rainbows were rare occurrences in the hot, dry climate of northern Egypt. Alhazen showed that to understand the physiology of vision, one first had to understand the physical behaviour of light itself.

Alhazen was to leave a remarkably detailed account of the basic anatomy of the eye, which clearly suggests a "hands-on" familiarity with ocular dissection: probably of the eyes of cattle. He describes the aqueous and vitreous humours, the conjunctiva, iris, and lens. And, in particular, Alhazen traces the route followed by the optic nerve from the retina of each eye to the brain. While these basic structures were already familiar to medical men by *c.* AD 1000, he provides an especially lucid account which, both in Arabic and in Latin translation, would pass down the centuries and be regarded as authoritative by Western European writers.

Alhazen had, however, only an imperfect understanding of how the eye functioned physiologically, and how all the parts within the eyeball worked together to enable a person to finally "see" and to make sense of the outside world. This would require the work of physical and ophthalmic researchers in the centuries that lay ahead.

UNRAVELLING THE COLOURS OF THE RAINBOW: MEDIEVAL EUROPE'S GREAT DISCOVERY

More than any other source, it was the Latin translation of Alhazen's *Optical Thesaurus* in the late twelfth century that really initiated serious optical study in the West. Central to that saga was the wonder of coloured light. What *was* colour? Why did sky colours change during the course of the day? Why did colour disappear at night? Once again, one encounters a coalescence of puzzles and paradoxes which resonated through physics, physiology, the psychology of perception, and theology. Why had God made colour so beautiful on the one hand, yet so elusive on the other?

Medieval and Renaissance European and Arabic optical writers seemed happy to accept the classical explanation for colour: it was caused by the pure, white divine light of heaven being somehow "corrupted" into colours by contact with common earthly, degenerate, or fallen terrestrial matter, such as air, water, rock crystal, or human eyes. Advocates of both the extramission and the intromission theories accepted this idea in one form or another. Were not colours "mutable" and subject to change, as pink skies slowly turned into flaming red skies, then to darkness, and all colours could be bright or muddy, depending on the circumstances under which they were seen? This "white light degradation" idea of colour would not be significantly challenged until 1665, when Robert Hooke argued from experiments that red and blue light were fundamentals and all the rest were intermediary tints, and then, after 1672, when Newton proposed his prism-based six-colour spectrum.

Most baffling of all was the rainbow, especially when (as medieval scholars would discover) the right meteorological conditions produced a pair of concentric rainbow arches, one inside the other, but with their colours reversed. Alhazen and al-Farisi had produced "rainbow" colours by shining light into a suspended quartz sphere, but it was in the much damper air of England, Germany, and Eastern Europe, where rainbows are much more common, that the study of this phenomenon really began to bear fruit.

Europe's true founding father of optics began as a poor farm boy from rural Suffolk, who, through ecclesiastical educational patronage, would first study in the new University of Paris, then become a lecturer to the Oxford Franciscan friars, and finally rise to the dignity of chancellor of the new University of Oxford and Bishop of Lincoln. This was Robert Grosseteste (c. 1175–1253). While there had been earlier medieval writers on science, such as the Venerable Bede of Jarrow, England, and Gerbert of Aurillac in France and Spain, Bishop Robert not only conducted optical researches, but also wrote on what would later come to be called the scientific method: how we should go about conducting scientific, experimental, and mathematical researches into the secrets of God's creation. Grosseteste's treatises De Luce ("On Light") and De Iride ("On the Rainbow"), written in the first part of the thirteenth century, would prove inspirational for the age to come. To Grosseteste, as to Alhazen, light, experimental optics, and both the physiological and the psychological aspects of seeing were understood in the context of divine wonder.

To Grosseteste, light signified God's presence throughout the universe and the terrestrial realm, for was not light God's "first" or primary form of creation? By the time of Bishop Robert's death, in his seventies, in 1253, European science had been bequeathed a legacy which still enriches Western civilization today. His foremost disciple was to be another young Franciscan friar, Roger Bacon (c. 1214–92) of Oxford and Paris.

Bacon was a true pioneer of what we now call physics. Not only was he fascinated by light, but by other forms of what we now call energy – such as heat, and why falling objects are attracted to the earth – and also by perception. How do we see? All of this chimed in elegantly with his love of the truths of geometry and mathematics.

Somewhere around 1260 (I like to think at the Franciscan House in west Oxford, though we cannot be sure) he developed a model for the rainbow based upon careful mathematical measurement. Like all great discoveries, it was built upon the insights of former researchers, such as Aristotle, Ptolemy, Alhazen, and Grosseteste. Bacon suggested that a rainbow formed a section of an optical cone. The sun was at the apex of the cone, while part of the cone's opposite circular end (the coloured bow) was in the sky and the mathematically postulated lower part of the cone was below the horizon. When the sun was low in the sky, one had a large rainbow, for then the optical cone of the bow ran almost parallel with the surface of the earth.

Bacon realized that a rainbow required three components: the sun at the apex of the cone; a facing body of cloud, 180° across the sky; and a human observer on a geometrical axis connecting the two. Then, crucially, he discovered a fact which one might consider as the "first" law of geometrical optics. For when Bacon, like Alhazen, used an astrolabe to measure first the height of the sun and then the height of the facing rainbow, he found that the coloured light coming down from the rainbow to the observer always did so at an angle of 42° to the ray of sunlight hitting the cloud.

Irrespective of how high or how low the sun and the rainbow were in the sky, the coloured light's exit angle was *always* 42°. Bacon realized that the vapours in the cloud somehow both reflected *and* refracted the incoming sunlight, to beam it down to earth in a fixed angle. One could even get a rainbow effect from the spray of a waterwheel, when the sun and the observer's eyes were in the right alignment. Geometry, colour, beauty, vision, and reason all seemed bound together in an exquisite repeat of the act of creation, and of the rainbow which God had set in the heavens as a sign to Noah after the flood.

Another friar would build upon Bacon's work, around 1290, although Theodoric of Freiburg belonged to the Order of St Dominic rather than Bacon's Order of St Francis. Like al-Farisi, Theodoric experimented with quartz or glass spheres as well as with glass vessels full of water, and obtained rainbow colours projected from the sphere. Theodoric went further, however, in his attempts to "ray-trace" the sunbeam's path within the sphere. He discovered that multiple refractions and reflections were taking place inside the sphere before the ray exited at 42°. These compound refractions and reflections within the sphere seemed somehow to be generating the colours of the rainbow.

From these and other experiments and measurements made with an astrolabe, Theodoric realized that what generated the colours of the rainbow was not sunlight entering a single, amorphous *cloud*, but the millions of individual droplets which the cloud contained, each of which acted as a complex internally reflective and refractive sphere, sending out millions of separate spectra. Then our eyes, minds, and souls somehow assembled them to form the colours and shape of the rainbow.

Bacon and Theodoric not only made seeing colours innately mathematical, but Theodoric did something momentous for the future development of science: he showed that one could reconstruct God's creation in the laboratory, by experiment, whenever one chose – or had clear sunshine.

Yet thirteenth- and fourteenth-century European science did something else equally momentous in its studies in optics and seeing: it invented spectacles to aid weak eyesight!

Spectacles: The invention
that changed the world

Medieval Europe invented a whole range of clever devices which came to transform the world and give Western civilization its science- and ingenuity-based cultural edge. These included mechanical clocks, firearms, big musical pipe organs, windmills and watermills to do hard work, more efficient agricultural tools, the perfecting of the Chinese magnetic compass – and the humble pair of glasses. Spectacles represent the first application of experimental physics to medicine, thereby initiating a tradition of cross-fertilization between the sciences which we see in its most modern forms in CAT (Computer-Assisted Tomography) and MRI

(Magnetic Resonance Imaging) body scans and computer images.

Spectacles were not the first prosthetic devices used in medicine – goodness knows how far back in time crutches and wooden legs go – but unlike these simple devices, spectacles predicated some knowledge of how light behaves, and how curved pieces of glass can function in relationship with soft, complex organs such as eyes, to improve a person's quality of life.

Roger Bacon's *Opus Majus* ("Greater Work") of 1267 makes the first explicit reference to lenses being used to aid weak vision, in the form of the magnifying glass. In Robert Smith's translation of a passage from the *Opus Majus* in 1733, "If the letters of a book... be viewed through a lesser segment of a sphere of glass or crystal, whose plane [flat] base is laid upon them, they will appear far better and larger."[2] In other words, with a "plane-convex" magnifier, one face of the glass being flat and the other curved.

Various early references survive to lenses being held before the eyes to produce spectacles, some rather ambiguous ones from as early as 1285, although it is said that a Florentine tombstone of 1317 mentions one Salvino D'Armate as the "Inventor of Spectacles", along with his Dominican friar friend Alessandro della Spina. Either way, the device – a pair of lenses held in a frame of metal or leather and fastened round the head or perched on the nose – caught on, and spread rapidly across Europe. Late medieval and Renaissance art depicts a growing number of Christian saints and scholars wearing glasses, and they soon became the hallmark of the scholar. By the fifteenth century, even portrayals of scholar-saints such as Augustine and Jerome would often include spectacles.

Early spectacles were mainly intended to improve eyesight for reading, and these were the convex, magnifying lenses mentioned by Bacon. By 1600, however, spectacles with concave, or "diminishing" lenses were being used to help rectify long or short sight, and to facilitate seeing at a distance. Proper ophthalmic tests and prescription spectacles would not come for centuries, for that would demand new discoveries in the physiology of the eye. Earlier spectacles were chosen entirely on empirical grounds, the patient visiting a lens-grinder's shop, trying various glasses until they found the ones that worked best, and then having them put into a frame.

The proliferation of spectacles by the sixteenth century led to the spread of the art of lens-grinding across the length and breadth of Europe by 1600. Those skills which made it possible for those with poor sight to see clearly would pave the way for the invention and development of that quintessential medical research instrument: the microscope. We will return to that optical wonder in Chapter 22.

Couching for Cataract: Albucasis and Medieval Eye Surgery

It may seem incredible that an organ of such delicacy as the human eye, which was not properly understood, anatomically and physiologically, until recent times, should have been the subject of a surgical technique that extended back into remote antiquity. Yet "couching for cataract" is described in the surgical writings of Sushruta (pre-500 BC), Hippocrates, Celsus, Galen, and the medieval Arab writers.

Amazing as it may seem, this over 2,600-year-old procedure is still being practised today in village India and elsewhere. Some years ago, I saw a modern medical film showing the operation being performed by an indigenous operator in India. Then in the summer of 2014, I had a visiting medical student in Oxford – herself of Indian ancestry and the daughter of a surgeon now practising in the USA – who used to return to India with her father to teach basic Western medical skills to local village healers. This young woman not only told me that she had assisted a local healer in a cataract operation in a remote northern Indian village, but she provided me with a detailed written account of the operation, complete with drawings of the couching needles and other basic equipment used by the village surgeon, who even appeared to be familiar with the ancient Sanskrit writings of Sushruta![3] (See Appendix 1).

In this time-honoured procedure, the patient – who needed nerves of steel – lay down on the ground looking directly upwards, with his or her head held firmly between the knees of the kneeling operator. It was crucial that the patient did not flinch in the least as the needles were inserted into the eyeball, otherwise they would be blinded. Given a courageous patient, a well-practised surgeon with a steady hand, and basic post-operative care, if infection did not set in, the patient stood a decent chance of having some vision restored.

What is cataract? The Greek word *katarraktes* means a "swooping down" or "waterfall". Celsus called it *suffusion* ("covering over"), and the Arabs, in Latin translation, *gutta opaca* ("opaque drop"), though all basically describe the same thing: a progressive, invading opacity of the lens. One of the classic descriptions of the "couching" or treatment procedure for cataract is to be found in the *Kitab al-Tasrif* ("On Surgery") of Albucasis, of *c.* AD 980.

Albucasis preferred his patient not to lie down, but to kneel, facing the operator. The following couching procedure was simple and quick, albeit spine-chilling. A fine, sharp, brass needle is inserted at a flat angle into the white of the eyeball, at the base of the aqueous humour, from either the left- or the right-hand side, depending upon which eye is being treated. The operator must keep turning the needle in their fingers to prevent adhesion and ensure an easy sliding passage. The needle must be pushed until the operator feels "something empty", at which point they will see the needle through the cornea with relation to the pupil and the iris. The needle is now gently forced downwards, until the lens snaps free from its muscular anchoring. The point of the needle is then levered backwards to force the loosened, cloudy lens down into the transparent mass of the vitreous humour, where, hopefully, it will sink. At this point, light now floods in through the hole, and the patient can suddenly see.[4]

Sometimes, however, the detached lens can float upwards in the vitreous humour, to impede vision once more, although Albucasis tells us that he had heard of an Iraqi/Persian surgeon who had developed a new technique to remove the clouded lens altogether. In this new technique, a cut would be made in the eyeball with a sharp scalpel, and the clouded lens detached as above. At this point, however, a fine hollow tube was inserted, and by means of a syringe or other device, the lens was sucked out of the eye altogether: a very early example of a non-urinary catheter device in surgery.[5] Albucasis's unnamed colleague was almost certainly Ammar ibn Ali al-Mawsili, a tenth-century Persian, who described his use of such a needle and syringe technique upon a patient in Egypt. Al-Mawsili's technique would be used, with increasing safety and refinement, down to modern times.

Albucasis goes into a lot of detail regarding the aftercare of the patient. Following surgery, the eyes had to be bound up and the patient laid upon their back, looking upwards, for three days. On the third day, the bandages were removed, and the patient's vision would be tested, after which the eyes would be re-bound until the seventh day. Salves might also be applied. The big problem was infection, which could blind the patient permanently or even kill them if it spread. But all being well, the patient should be able to see again.[6]

Yet how could a person see without a lens? Not brilliantly, but much better than with a cataract: the lens focuses, refines, and sharpens about 25 per cent of our vision, though the optical properties of the cornea do about 75 per cent of the general focusing, enabling one to see shapes, colours, and the outlines of objects. To go from being blind to having blurred vision was no mean improvement.

Cataract operations were widely practised over the ensuing centuries, using the lens-depressing couching, or the more sophisticated lens removal procedure. Then almost 900 years after Albucasis and Al-Mawsili, the 70-year-old Reverend Patrick Brontë, father of the Brontë sister novelists, was operated on in Manchester by Mr William James Wilson FRCS, in August 1846. Charlotte Brontë, who accompanied her father to Manchester, left us a remarkable account. Mr Wilson, Patrick later recorded, first administered belladonna to expand the pupil and then performed a lens extraction procedure. The patient bore the 15-minute unanaesthetized operation with "extraordinary patience and firmness", so Charlotte recorded. The pain, Patrick remarked, was "of a burning nature – but not intolerable". After surgery, Brontë's eyes were bound, and he spent a month lying upon his back, apparently healing without significant complications. Then strong spectacles – unknown in Albucasis's time – enabled him to see clearly once more.[7]

THE EYE AS AN OPTICAL PROJECTOR

When did scientific men begin to understand the optical functioning of the eye? Ironically, that breakthrough came not from a medical man, but from the astronomer Johannes Kepler, whose *Dioptrica Nova* ("New Optics") of 1604 not only recorded carefully conducted studies of dissected eyes, but also laid the foundations of our modern knowledge of the eye as a projection system. Kepler was fascinated not only by the refractive properties of the lens and humours, but also by the retina and optic nerve. While being familiar with the classical Greek optical writers, and Alhazen, he developed a model of the eye as a sort of projector, in which the lens projected an inverted image upon the retina, which was then somehow conveyed to the brain by the optic nerve, then turned the right way up, to give a view of the outside world.

Ophthalmology was the first branch of medicine to "cross-fertilize" with a non-medical science: physical and experimental optics. Yet while by 1650 we had Kepler's, René Descartes', and a few other models of eye–brain function to weigh into the equation, the puzzle of how we actually see, perceive, and understand remained. And in many ways it still does today.

But now let us look at what was taking place in medicine within medieval Europe.

CHAPTER 5

Rahere the Jester Meets
St Bartholomew

*J*mplicit within the folklore, or received wisdom, of the post-Renaissance and Romantic eras is the belief that there was no rational science in those Dark Ages that fell between the end of the Roman Empire and the late fifteenth century. While the Arabs kept the torch of classical civilization burning in their own lands, Europe was a barbarous backwater, dominated by ignorant churchmen, witches, and superstition. Yet as I have shown in other books, astronomy, mathematics, and geometry – as well as architecture, civil engineering, theology, philosophy, and a whole spectrum of literary disciplines – *flourished* in Christian Europe, especially after *c.* AD 1000. And so did medicine.[1]

Medicine, however, lacked the externally verifiable criteria of proof and demonstration which were present in astronomy and mathematics. Medicine had nothing like the self-contained system of the 360° scale into which circles could be divided, or the unfalsifiable truths of Euclidean geometry, the rules of proportionate triangles, or arithmetic sequences of pure numbers. Medicine lacked certainties. Pain, fever, mental illness, or a mysterious lump seemed to come from nowhere. They sometimes killed, and they sometimes went away as mysteriously as they came, and trying to make sense of them baffled the wisest observers. So while medieval people might be able to build a spectacular cathedral, predict an eclipse, or devise a self-acting machine such as a clock, they were largely at sea when it came to facing illness. All the wisdom of Hippocrates, Aristotle, or Avicenna still left one wringing one's hands in bafflement when faced with human suffering. Even though doctors had made enormous progress with clinical observation and the taxonomy of illnesses from the fifth century BC onwards, and anatomy, practical surgery, and herbal therapeutics had laid key foundations for the future, disease remained intractable. It would continue to do so until the last couple of centuries before our own time, as we shall see in our later chapters. But what was on offer to a sick person in medieval Europe?

EARLY MEDIEVAL CARE:
LEECH BOOKS AND HERBALS

One great treasure, now preserved in the British Library, is that vellum manuscript *Laeceboc*, or *Leechbook*, compiled by one Cild for a patron named Bald, and known as the *Leechbook of Bald*. The word "leech", does not derive especially from the use of leeches, but rather from Old English *Läéce*, one who practised the craft of a healer.

It is an Anglo-Saxon text, with some Latin additions, probably compiled during the time of England's King Alfred the Great (849–901). In addition to unifying England under his benign rule, he was England's first great patron of learning. The *Leechbook* was translated into English and edited in 1865 by the Reverend Thomas Oswald Cockayne. Cockayne's meticulously titled, ground-breaking piece of scholarship *Leechdoms, Wortcunning and Starcraft in Early England*[2] contains three other Anglo-Saxon medical texts in addition to the *Leechbook*. These include the *Lacnunga*, another medical recipe collection; the Anglo-Saxon *Herbarium* ("Herbal"); and *Medicina de Quadrupedibus*, dealing with veterinary medicine, the last two of these also being reprinted in the *Old English Herbarium*.[3]

Essentially, the *Leechbook of Bald* is a collection of recipes rather than a systematic medical treatise, and its contents appear to be north European, Anglo-Saxon, and Nordic. It is *not* in the Mediterranean tradition of Hippocrates and Galen, and at the time of its compilation, Arabic medicine was still in its infancy

74

and not known in the West. It contains no coherent theories of disease, and there are no attempts at anatomical or physiological explanation: just a collection of aches, pains, and injuries, along with the botanical and natural substances that may be used in their amelioration. Treatments for headache, shingles, impotence, sore feet, and a collection of other afflictions are listed, along with some surgical items, such as an attempted treatment for hare lip.

Bald's *Leechbook* must not be thought of in isolation, however; for there were the other classes of works which formed what one might think of as companion pieces to it: the *Herbals*, or books of plants, and the *Bestiaries*, or books of beasts, all of which together provide context for both the medical and natural historical worlds of early medieval people.

More obvious, in a medical context, were the *Herbals*. These books of plants drew on a long lineage, such as the *Materia Medica* ("Medical Pharmaceuticals") of Dioscorides of *c.* AD 80 and the lists of plants included in Pliny's *Historia Naturalis*, also of the first century AD. Herbals in their various traditions were also known in China and the Arab world. But it is Albertus Magnus (Albert the Great), in the mid thirteenth century, whose *De Vegetabilibus* ("On Plants") brought together classical, Arabic, and later European knowledge. In addition to the botanical descriptions, he naturally included sections on the "vertues" or healing properties of plants. In the edited translation *The Boke of Secretes of Albertus Magnus, of the Vertues of Herbes, Stones and Certayne Beastes*, of 1560 (in the Bodleian Library, Oxford), Tudor English readers were informed of the medicinal and other properties of a variety of plants, many common to England. These included vervain (or verbena), marigold, and henbane. Vervain, for example, when combined with peony seeds and gathered when the sun was in the astrological sign of Aries, in March and April, was said to be good for the "fallyng sykenes", or epilepsy.[4] Vervain also imparted "greate strength in venereal pastymes" producing children "very apte to learne".[5]

Bestiaries, which Albertus Magnus also touched upon, were essentially early Christian compilations of true and mythical animals, and their magical properties, with roots extending back to Aristotle,

Pliny, the anonymous *Physiologus*, and other classical authors. They were not explicitly medical works, so much as zoological, perhaps dating back to the later Greek *Physiologus*.[6] But Isidore of Seville, in his famous encyclopedic *Etymologiae* (c. AD 610), Book 12, along with subsequent writers, expanded the "Book of Beasts" genre, using specific real or imaginary animals to exemplify certain Christian virtues, such as temperance, and indicate how they could conduce to health. This was an approach to health rooted not necessarily in surgery or chemical medication, but rather concerning itself with an individual's wider moral, spiritual, or psychological state.

In their world, the physical aspect of a remedy was seen to operate in a spiritual context and was accompanied by prayers to saints and the performance of rituals. In many respects, prayer would have been viewed as the most important part of the treatment: for if the cause of a physical affliction lay in general or specific sins, then unless a spiritual ally could be enlisted, the purely medical components of the treatment would be of little worth.

This, I would argue, is not being superstitious, but downright practical and realistic, for Anglo-Saxon people had a wholly different concept of disease from that which we have today. In their way of thinking, the body was not a complex piece of biological mechanism to be understood in the light of purely physical, chemical, and biochemical criteria, so much as the transient wrapping enfolding an immortal soul. It was the soul, and its long-term health and salvation, that really mattered, for at the end of life that soul would slough off its bodily wrappings and return to God. One might think of it as analogous to wearing a warm overcoat in cold weather. During winter, we make sure we take care of the coat as best we can, but we know that when summer comes, we will cast it off, and enjoy the warm days without need of its comfort. Likewise, the soul would no longer need the comforts of a physical body once it had left this world and gone on to enjoy (all being well) the summer of heaven.

This wider theological context of the relationship of body, soul, and immortality was not the exclusive worldview of the Anglo-Saxons, but was shared by all Christians, from the days of St Luke the "beloved physician" onwards. It stood to reason that it was the owner's duty to look after the body and keep it in the best possible running order as long as the soul required it for a place of earthly residence. And it is here that the leeches, herbalists, and healers had their function, much as one might, from time to time, require the services of a tailor to repair the rents and tears in a favourite coat. And in the same way that "health" was seen in a spiritual as well as in a physical context, so were cleanliness and hygiene.

SALERNO, NEAR NAPLES: EUROPE'S FIRST HOSPITAL
AND MEDICAL SCHOOL

Southern Italy was a place of exotic cultural cross-currents in the eleventh century. This Christian peninsula and the adjacent island of Sicily had been invaded by the Muslims in AD 827, though by 1050 the Normans under Robert Guiscard and the Emperor of Byzantium had driven them out of Sicily, thereby restoring Italy to Christian control. Within the territories of the kingdom of Naples, which included most of southern Italy along with the island of Sicily, was a fascinating mixture of peoples and cultures. There were medieval Italians, proud of their Christian and ancient Roman ancestry; Byzantine Christian Greeks in the Greek Orthodox tradition; Jews; and the remnants of the old Islamic occupation. Latin and Greek, with pockets of Arabic and Hebrew language and culture, were all in evidence. An integral part of this culture was science, medicine, and the study of how the natural world and living things worked.

Central to this cultural tradition was the great monastery of Monte Cassino, about 60 miles north of Naples on the way to Rome, founded by St Benedict in AD 529. While the great monastery was successively destroyed and rebuilt by the Lombards, the Saracens and others, it came to its greatest fame and prosperity after 1050. In particular, Monte Cassino became a centre of scholarship, as men such as Alfanus, subsequently Benedictine Abbot of Monte Cassino and then Archbishop of Salerno, began turning major Greek and Arabic texts into Latin. Alfanus's translation of Bishop

Nemesius of Emesa's *Premnon Physicon* (*De Natura Hominis*, "On the Nature of Man") inaugurated Western, Christianized Galenic medicine. Then there was Constantinus Africanus (Constantine the African), *c.* 1020–87, a native of Carthage and probably a Muslim convert to Christianity, who worked in Salerno and also became a monk of Monte Cassino. Far from being a "dead" language, Latin was the *lingua franca* of learned Europe: of the church, the law, and of diplomacy. It was also the international language of medicine and science – a primacy that it would retain, through its various permutations and changes of style, down to the seventeenth century and beyond.

In particular, Constantinus Africanus is credited with the Latin translation of the *Pantegni* ("Whole Art" [of Medicine]) of the tenth-century Persian medical encyclopedist, known in the Latin as Haly Abbas. To Constantinus is also ascribed the translation of some of Hippocrates' and Galen's works into Latin, along with the Nestorian Christian Hunayn ibn Ishaq's (Johannitius') textbook *Liber Ysagogarum* (or *Isagogarum*), "Introduction" to medicine, of the mid ninth century. These and other works produced in Monte Cassino or nearby in the kingdom of Naples in the eleventh century constituted the rebirth of the classical Greek medicine of Hippocrates and Galen in post-classical Europe; a medical Renaissance.

Immediately south of Monte Cassino and the city of Naples standing on its beautiful bay is another great bay, or gulf, on the south-western side of the Italian boot. This was the Bay of Salerno, with its principal city, Salerno, standing on the northern shore, some 40 miles south of Naples and about 100 miles south of Monte Cassino. Legend, custom, and historical evidence all point to Salerno becoming the birthplace of systematic European medicine, based not on folk-cures but upon Latin translations of classical Greek texts. This was primarily under the protection and patronage of the Norman rulers of Naples and Sicily, with Constantinus Africanus becoming secretary to Robert Guiscard in addition to being a medical translator.

Legend – and it is frankly impossible to confirm the story – maintains that the medical school that came into being in Salerno was the work of four men: a Roman Catholic and a Greek Orthodox Christian, a Jew, and a Muslim. This is just one of many stories we have about Salerno. Another is that it had as a member of its "faculty" a female doctor named Trotula: a lady who, whether she actually existed or not, would be immortalized in legend. Chaucer makes passing reference to her in "The Wife of Bath",[7] while in the medieval English medical tradition she is the famous "Dame Trot".

In many respects, the "School of Salerno" presents something of a puzzle for historians, for its history had entered into legend even by the time of Geoffrey Chaucer in the late fourteenth century. But the four reputed founders and Dame Trot apart, there are certainly a number of significant details about which we can be confident. For one thing, Salerno, Monte Cassino, and Naples, especially after the Emperor Frederick II founded the University of Naples in 1224, did become a hub of translation from Greek and Arabic, and works like the *Articella* ("Little Art"), *Pantegni*, and *Liber Ysagogarum* certainly were appearing in the eleventh and early twelfth centuries.

Around the same time that this was happening in southern Italy, a parallel movement was taking place in Spain. In the wake of the "Reconquista", when military leaders such as Rodrigo Diaz, known to history as "El Cid", were driving the Muslims into the southernmost parts of Spain and reclaiming the country for Christendom, scholars from various European countries were mastering Arabic and translating the great store of Arabic books found in cities like Toledo, which the Christian knights took in 1085. Classical Greek texts, having passed in late antiquity from Greek to Syriac and then into Arabic, were now being made available to Latin readers. What is more, original Arabic medical and scientific texts were being translated as well. By 1140, for example, the Italian Gerard of Cremona had translated both Rhazes' medical and Albucasis's surgical writings into Latin. And these works, along with dozens of others, would be copied and recopied, and gradually spread around Europe.

Many of these works came to be fed into the Salerno story, bestowing upon Salerno the reputation of being the birthplace of

Europe's medieval medical "Renaissance". And nothing epitomized the subsequent enduring fame of Salerno more than a composition probably made by the late thirteenth-century Spaniard Arnaldus de Villanova, or as he is known in English translation, Arnold of the New Town. This was the *Regimen Sanitatis Salernitanum* ("The Salernitan Rule of Health"): a versified body of rules and mnemonics which aimed to establish a regimen of health.

In its basic assumptions, the *Regimen* was Hippocratic and Galenic, focusing upon diet, moderation in all things, exercise, and the cultivation of a quiet mind. "Doctor Diet" and "Doctor Quiet" in the Salerno rhymes epitomized the whole system, for only through a balanced diet, an exercised body, and a mind at peace with itself could health be found. Indeed, through the whole Salernian system health was envisaged as akin to wisdom in the Old Testament. It was a pearl beyond price that had to be sought, for it had a tendency to hide from the greedy and the self-indulgent, and needed a special frame of body and mind to discover it. In many ways, this health-wisdom was similar to the Philosopher's Stone of the alchemists, which only a devout and balanced chemist could discover.

The advice offered in the *Regimen* was sound and practical for the time. It contains, for example, little mention of the use of astrology in diagnostics (astrology being first engrafted on to classical medicine by the Arabic writers), although the employment of celestial criteria would become commonplace in sixteenth-century medical practice. Thomas Paynel's English translation of the *Regimen*, published in London in 1541, is remarkably pragmatic. In its description of how and when to let blood, for example, it confines itself to how best to open a vein and drain away bad humours, and says nothing about the best lunar phase in which to do so. Similarly, its treatment of melancholy makes no reference to Saturn (generally seen as the patron planet of melancholy) and deals only with the then understood causes of the disease, such as excessive dryness or being too solitary, and how to counter them with medicinal or lifestyle changes.

Lying at the heart of the whole classical and Arabic medical tradition was the idea that in most cases the patient brought disease on themselves: by being greedy, angry, selfish, lazy, drunken, or silly. All these modes of behaviour hindered the correct flow of the humours and other vital life forces within the body, leading to blockages, tumours, overheatings, fevers, or madness. The way to cure them came through the adoption of a "temperate" lifestyle. It was a portmanteau approach to health and disease which ran from the ancient Greeks to the Victorians, and whose common-sense home truths still apply today, lying at the heart of many modern "alternative" medical regimes.

The *Regimen Sanitatis Salernitanum* was not the only "guide to health" book on offer. There was also, for example, the Taqwim al-Sihha ("The Maintenance of Health") or *Tacuinum Sanitatis* as it is called in various Latin versions, by the eleventh-century Baghdad Nestorian Christian physician and philosopher Ibn Butlan (also known as Ellbochasim de Baldach, or Baldac). Several beautifully illustrated mainly fourteenth-century copies of this work survive, most notably in Vienna, Paris, the Roman Casanatense Library, Rouen, and Liège. The Rouen copy states that the book is about six things: the air and the heart; the right use of food and drink; correct movement and rest; sleep and wakefulness; retaining and expelling humours; and states of mind. These six aspects of health are paralleled elsewhere in Arab medicine, such as in Hunayn ibn Ishaq's *Liber Ysagogarum*, and possess clear resonances with Hippocrates. As one would expect, the whole body of advice is firmly placed within a religious context, for such wisdom and advice comes via the grace of God.

Each beautifully detailed painted and illuminated manuscript page – displaying fascinating details of dress, trades, buildings, and agricultural techniques – carries an account of the "vertues" or properties of the plant being discussed. Spinach, for example, is said to be cold and humid, good for coughs and chest complaints, though capable of disturbing digestion, although this could be mitigated by frying the plant with salt water and vinegar. Sage is good for paralysis, rue sharpens the eyesight, rye breaks down congested humours, south winds are good for the chest, and roses benefit "inflamed brains". One can sample both the medical and the artistic content of the *Tacuinum* manuscripts in a beautiful fine-art reproduction modern scholarly edition.[8]

Though Arabic in its source, the *Tacuinum* nonetheless drew from earlier sources which had themselves been absorbed into medieval academic Arabic medicine which, as we saw above, was still very much Hippocratic, Dioscoridean, and Galenic in its style of thinking, with the achievement of balance being an essential prerequisite of finding one's health.

While the *Regimen*, *Tacuinum*, and other writings dealt with what might be called the ideal or theoretical conditions under which an individual found and retained health, how were the sick treated and managed on a practical level? What happened to the sick poor? This leads us to the rise of that great therapeutic institution – one of the enduring legacies of medieval medicine to the world – the hospital.

THE FOUNDING OF ST BARTHOLOMEW'S HOSPITAL
IN TWELFTH-CENTURY LONDON

According to legend, one of Europe's great hospitals was founded by the court jester to King Henry I. Rahere, Rayer, or Raherius (spellings vary) certainly appears to have been a real historical figure, who died in London in 1144. Whether he really was a jester or a monk and a priest remains a moot point. There is no reason why he should not have been all three, for monks and priests came in all types. Some worked out their faith in solitary contemplation, whereas others were witty and sociable. It was part of St Benedict of Nursia's vision, from the early sixth century, that those men who did not choose to seek God as solitary hermits might find him instead in disciplined communal living with other monks, in monasteries, where, among other things, St Benedict's "Rule" instructed them to care for the sick.

If Rahere was, or had been, a priest-jester, he would not be unique, as history contains well-documented accounts of churchmen who won notoriety as entertainers. Three centuries after Rahere, the outrageous and often reputedly bawdy Oxford graduate John Scoggin was said to have raised laughs at the court of King Edward IV with his pranks, before dying, it was sometimes claimed, as a resident in the monastery at Westminster Abbey. It was even reported that Scoggin died cracking jokes, for he requested to be interred beneath one of the waterspouts of the abbey roof, saying that he had always enjoyed "good drink", such as that which flowed from an abbey roof! His gags and capers, true or apocryphal, were later immortalized in print as *Scoggin's Jests* (1626) and in other early joke books. The medieval Benedictine community of Westminster Abbey had a *hospitium* or *infirmarium*, where the sick were cared for, and whether monk, priest, scholar, or secular fool, Scoggin would probably have been in it when requesting his valedictory "good drink".[9]

6. Rahere the Jester, represented on his tombstone in St Bartholomew the Great, London, as a genial monk or priest. (Henry Morley, *Memoirs of Saint Bartholomew Fair* (1859). A. Chapman collection.)

How did a twelfth-century monk-jester come to found a hospital? The legend further tells us that Rahere, who also seems to have been a court favourite of King Henry I, became seriously ill, either in England or on a pilgrimage to Rome. In his illness, he had a vision of St Bartholomew the apostle, patron saint of tanners, ironically, with many healing miracles to his credit. Rahere promised the saint that, should he recover, he would found a house, or monastery, dedicated to St Bartholomew in London.[10]

As one of the functions of a religious house was – in the tradition of the Good Samaritan – to care for the sick, the wayfarer, and the destitute, it would also provide *hospitality*. This essential feature of a monastery takes its linguistic root from the Latin words *hospes* ("host") and *hospitium* ("place of reception" or "guest room"), from which we derive the word *hospital*. In particular, Rahere's St Bartholomew's would have a particular institutional responsibility for the sick and destitute. It would even bring up and look after orphaned children born to mothers who died in the hospital.

Exactly how Rahere obtained the funds to build and endow his religio-medical community in the heart of the city of London adjacent to Smithfield Market is not clear, but he did so, around 1123, and by the time of his death 19 years later, he was prior, or senior monk.

I have always found Rahere a captivating figure. My own suspicion is that he was much more than a plain working entertainer, especially as one story tells us that as early as 1115 he had been a canon of the nearby St Paul's Cathedral. Rahere was quite likely a well-born Norman gentleman who became a priest or monk and whose social status enabled him to move in royal courtly circles: a worldly priest with a gift for good humour and an ability to crack jokes and perform entertaining capers. But then, in the throes of fever in a foreign land, he had a life-changing religious experience. He spent the rest of his life in the Christian service of the sick and the poor, yet knew how to use his social clout to found a monastic hospital. Maybe Rahere also eased the sufferings of the patients by making them laugh.

A medieval monastic *hospitium*, however, should not be thought of as similar in its goals and intentions to a modern hospital. Its purpose was not to take in the sick, diagnose their infirmities, apply some kind of rational or scientific cure, and then discharge them fit and well. St Bartholomew's – or Bart's as it would come to be known – was primarily, like other monastic hospitals, a place of *care*, and the services it dispensed were probably closer to modern-day nursing than to "medical science" as such. Its clientele would have been drawn almost entirely from the poor, whether indigenous Londoners, visitors, pilgrims, or travellers. It would also have attracted sick persons hoping to gain the attention of St Bartholomew himself, who would, perhaps, bring them back from the brink of the grave, as he had done Rahere. This lively sense of saintly intervention was central to the whole rationale of medieval Christendom, and still is, for many Christians worldwide. Bart's great "rival" hospital across the river Thames, St Thomas's, dedicated to the memory of Thomas Becket – "Tommy's", as it would be nicknamed – was founded in Southwark in 1215 by a community of Augustinian canons: St Thomas was also famous for his healing miracles.

To understand the function and role of a place like St Bartholomew's in twelfth-century society, one must think in terms not of Hippocrates, Celsus, and Galen, but of Bald's *Leechbook* and the herbals and books of beasts discussed above. It was a place where herbs, natural animal products such as fat and honey, potent words, divine and saintly relics, natural magic, and prayer combined with the practical resetting of a broken bone. And of all these agents, prayer, the saying of the Mass, the anointing of the sick with holy oil, the Last Rites upon passing over, proper Christian burial, and prayers and candles to see the departed soul safely to God, would have been by far the most important. For many people would have left Bart's wearing nothing more than their burial shrouds.

This was in no way an indictment of the competence or care of the hospital staff, so much as a stark fact of life at that time. In Anglo-Saxon and Norman times, and several medieval centuries later, the overwhelming number of diseases doing the rounds were simply incurable. Most of those maladies, however, would not have been diseases like our modern-day killers, cancer, stroke, and heart disease, but rather person-to-person infections that were rife in congested and often, at least for the poor, unwashed communities.

85

While descriptions of what we now know as cancer, stroke, and heart disease exist in surviving medico-religious literature, they would have been relatively rare, for these diseases occur most frequently in ageing populations. And while very little in the way of systematic age-at-death records survive from early medieval times, at least for the vast generality of ordinary people, we know from other evidence that life was short. The rapidly advancing science of palaeopathology (the careful clinical study of exhumed human remains from accidentally uncovered medieval and other graveyards), tells us how short life actually was. A skilled pathologist, by examining the long bones, joints, teeth, and skull sutures, can place a dead person quite accurately within an age group, be it early or late teens, twenties, thirties, and so on, as well as determining the sex of the deceased. While childhood mortalities were, not surprisingly, very high, palaeopathological evidence indicates that death rates were alarmingly high throughout all age groups. A medieval person who reached 35 or 40 was doing well, even for a high-status person, and to reach the biblical three score years and ten was an achievement.

In addition to the wider social care problems of acute illness, such as a fever epidemic, organ failure, or accident, there were the chronic problems occasioned by physical disabilities, mental illness, and old age. Very few medieval people in any generation would be likely to reach the age of 60, but there were enough to necessitate the founding and endowing of almshouses, or care homes, such as those of Ewelme in Oxfordshire, Winchester, and elsewhere across Christian Europe. Rahere's St Bartholomew's in London would have shared some of these functions, along with hundreds of less famous monastic *hospitia* across Europe.

One early patient of St Bartholomew's, who would later become an English libertarian icon, was Wat Tyler, the charismatic political leader of the Peasants' Revolt, when people across several south-eastern English counties, in protest against an iniquitous double-taxation imposition, rose up and occupied London in June 1381. After he had been duped by several of the boy-king Richard II's advisors into attending a "peaceful" meeting in Smithfield,

London, Sir William Walworth, Lord Mayor of London, laid into the unarmed Tyler with a sword. The badly wounded Tyler was, however, able to ride away to re-join John Ball and several other of the revolt's leaders before dropping from his horse. The monk doctors of the nearby "Bart's" kindly conveyed him to their *infirmarium*, where he was cared for.

In a barbaric deed, a clear breach of monastic sanctuary, some of the king's men, on hearing that Tyler was still alive and being tended, proceeded to storm into the monastic hospital, drag Tyler outside on to Smithfield market ground, and there behead him. John Ball, the monk whose profoundly egalitarian Christian sermons and rhymes had helped to inspire the revolt in the first place, was arrested soon after, and brutally executed in St Albans.

Bart's, Tommy's, and numerous other monastic hospitals around Great Britain and across Europe may have been very limited when it came to curing physical illness as we now think of it, but as we saw above, that was not the sole goal of healing in medieval times. In the twelfth century, and for the next 700 years to come, and certainly up to the Reformation in the sixteenth century, healing involved a much bigger package than simply walking away fit and well. And it still does today, although modern emphases may be different.

In Rahere's day, Europe was Christendom, in which a human being, both prince and peasant, needed spiritual as well as physical care. The vast majority of sick people, from emperors to beggars, would have seen their lives as part of a bigger Christian redemptive drama. Today we use drugs, counsellors, psychotherapists, and social workers to ease the emotional distress of a patient facing trauma or death. In medieval times, one had the full apparatus of the church and the hope of redemption and eternal life to give comfort and assurance. And while the sick may not have seen modern-style doctors, nurses, and sophisticated pieces of medical technology moving around the ward, they would have seen priests and crucifixes, and witnessed the Host, or body of Christ, being elevated at Mass, either in chapel or in the ward. And they would have heard comforting prayers, proclaimed publicly and privately, at the ward's own altar, and at the individual's bedside. When a fellow-patient lay close to death, one would have heard the reassuring words of the Last Rites, and the dying person would have felt the touch of holy anointing oil, to defend the newly released soul from the clutches of marauding devils, and ensuring its safe passage to Christ.

The Protestant Reformers, like many people today, might have dismissed all this as superstition; yet in a culture that saw life, death, and eternity within the context of a greater spiritual and cosmological drama, it was powerful stuff. This is one of the reasons why medieval people so loved the monasteries and their hospitals, and why the Reformation and the abolition of the English, Welsh, and Scottish monasteries in the sixteenth century left people bereft of so many comforts. This aspect of medieval life has been re-examined and re-evaluated by a growing number of modern historical scholars, such as Eamon Duffy.[11]

By the time of the Reformation, London, with its *c.* 40,000 population, contained numerous monastic institutions which provided vital medical, psychological, and social care for the poor and the sick. All of them fell foul of Thomas Cromwell's zealous commissioners and were dissolved – shut down and their assets seized by the Crown and its agents. This dissolution would provoke a popular reaction not only across London, but across England, especially the north in what became known as the Pilgrimage of Grace. It would be put down with brutal thoroughness.

By 1540, therefore, England had suddenly lost its whole infrastructure for dealing with the poor, the homeless, the sick, the mentally ill, and the dysfunctional. It had taken centuries to build up that infrastructure, and it would take a long time to even begin to replace it. Fortunately, however, St Bartholomew's survived, for after Rahere's monks had been evicted, King Henry VIII was prevailed upon to re-found it over 1546–47 as a secular hospital, to be managed by the City of London Corporation.

Similarly, in 1551 the City of London prevailed upon the advisors of the boy-king Edward VI to reopen St Thomas's, which, as a monastic foundation, had likewise been dissolved in 1539.

The 1247 Priory of St Mary of Bethlehem – popularly known as Bedlam – was also re-founded as a secular hospital to continue in its medieval role of providing sanctuary for the mentally ill.

CURE OF BODY AND CURE OF SOUL: HOW CLEAN WERE MEDIEVAL PEOPLE?

We still labour under the popular post-Enlightenment myth that medieval people – especially in Western Christendom – were filthy and verminous. It is believed that people were cleaner in the Arab world, due to their wiser doctors. The Western filthiness myth was seemingly confirmed by Andrew Dickson White in his *A History of the Warfare of Science and Theology in Christendom* (1896), where he cited cases of medieval monks who positively revelled in their supposedly godly filthiness, on the grounds that washing was "luxury" and consequently monks should have nothing to do with it.[12]

As all too often with Dickson White, there is nothing wrong with the individual cases he so meticulously cites, but he gives a distorted perspective by omitting the evidence to the contrary. Cleanliness, however, has four important prerequisites: regular washing of the human body; the washing of clothes; the provision of a clean and unpolluted water supply; and the efficient removal of sewage. In north European countries such as Great Britain, northern France, Holland, Germany, and Scandinavia, with their abundant rainfall and fast-flowing mountain-to-sea rivers, the two latter requirements – clean water and sewage removal – could be supplied with relative ease. Regular body and clothes washing could be more problematic, especially in winter.

Just imagine, if you lived in a draughty stone or wooden building, with one central fire and a hole in the roof to let the wood smoke out, you might feel strongly disinclined to strip off and plunge into a tub of water, let alone duck yourself in a nearby river. It was much more appealing to keep the same shirt and close-to-the-body garments on all winter, from October to May, for keeping warm would take priority over being clean. In practical terms, keeping warm would doubtless be much better for your health than being clean.

Washing clothes in winter would have posed a problem, while drying them afterwards could have presented an even greater one. Wool was the principal fabric of the north, and in an age without chemical detergents, washing woollen fabrics involved long soaks, beatings, and scrubbings in lye, a primitive alkaline soap solution consisting primarily of stewed wood ash. Urine, cheap and easy to come by, was also used for washing clothes, as the alkalinity of one chemical and the acidity of the other could act as a solvent to the essentially greasy body substances that dirty clothes: provided, that is, one had sufficient fuel to boil them for a period.

In addition to wool, the better-off classes started to wear linen next to the skin by the twelfth or thirteenth centuries, but linen also requires long soaking in the absence of post *c.* 1800 bleaching agents. And while cotton fabrics are lighter and easier to wash than wool and linen, this Egyptian plant would scarcely have been known north of the Mediterranean for much of the Middle Ages. So being washed thoroughly, with frequent changes of garments, posed problems when the rain or the snow was falling for weeks on end.

Yet with the abundance of water came other hygiene advantages. Springs and fast-flowing streams could ensure a relatively clean water supply, especially if one were not living in a cramped and congested urban space; and even there, hollowed-out tree-trunk water pipes might be used to bring in cleaner water than that available from the river. Archaeologists have found the remains of such wooden pipes in London and elsewhere, while the Roman "plumbers" (or workers in *plumbum*, lead) had left many examples of their lead pipe-making skills across Europe.

It is in the great monasteries, however, that one encounters medieval sanitary engineering at its best, and nowhere more than in the remains of the great Cistercian abbeys of the twelfth and later centuries. Although not the founder of the Order, it had been St Bernard of Clairvaux after 1113 who became its driving force. St Bernard's monks were enjoined to found new communities in wild and remote places, and some 2,000 Cistercian houses (from the original house at Cîteaux – *Cistercium* – near Dijon) were established across Europe. In addition to their uniquely beautiful, soaring abbeys in the new Gothic style, Cistercian monasteries came to embody the most sophisticated sanitary engineering of the entire medieval period.

All monasteries and nunneries, starting with those communities first inspired by St Benedict after AD 590, being disciplined, carefully regulated religious communities, had always pioneered the highest hygiene standards of the day. Monasteries were social communities, with their *lavatoria*, or places to wash before and after meals, bath-houses, and laundries. They had spearheaded the post-Roman cleanliness movement. But in many ways, the Cistercians made refinements to the basic Benedictine pattern. Often built in remote places, Cistercian abbeys could, in some ways, engineer the open landscape around them: not only for the well-managed hillside sheep farming on which the Order largely depended for its income, but also for the design of the valley-bottom abbeys themselves. One has only to visit the ruins of an English Cistercian abbey to see this even today.

Take Tintern Abbey, in Monmouthshire, on the Welsh borders, founded by French missionary monks in 1131. Even in ruins, it is magnificent, with its roofless church and central tower still standing and dominating the beautiful valley. One notices that the monks had not only laid out their twelfth-century abbey close to the River Wye (the Cistercians invariably built near to a fast-flowing river), but that a special stone-lined channel had been dug to divert a constant stream of fresh water through the monastic site. Partly culverted in, but being open in a number of places for easy access, it would have been used upstream – to draw off clean, cold drinking and cooking water to serve the kitchens and dining hall, while downstream, prior to discharging back into the river, it would have served the *lavatoria*: washing areas and lavatories.

The monastery would have discharged raw sewerage and kitchen and other waste directly back into the Wye, but in a region that was only thinly populated in the twelfth century, that waste would have biodegraded fairly quickly. It must have been discharging bacteria into the river, but in an age still ignorant of germs, and where even the most learned physicians attributed fevers and infections to bad smells and wind-blown effluvia, one can hardly blame the Tintern and countless other monks across Europe for that. The monks were doing their very best, engineering wise, to keep their persons and environment clean and fresh, revealing a radically different approach to hygiene from that of those monks who supposedly revelled in godly filth. And no matter what state of unkemptness a solitary hermit living in his desert cave might have chosen, it was recognized by St Benedict of Nursia in the sixth century AD, then by St Bernard and others, that when people lived alongside each other within the walls of a monastic *community*, filthiness was not a viable option, and strategies had to be adopted to keep the place clean and sweet. So, in the 1,400 years between the civil engineers of ancient Rome and those of Industrial Revolution Britain, it was the monks and nuns of Europe who were the trail-blazers when it came to public health precautions and sanitary engineering.

When did lay people start to wash their bodies on a regular basis in the period following the collapse of the Roman world in the fifth and sixth centuries? It is hard to be sure, but by the fourteenth century, if not before, commercially operated bath-houses were appearing in many of Europe's major – and minor – cities. They were places where people could go and soak in a large wooden tub of hot water, usually, judging from surviving illustrations, sitting in an upright posture. These "stews", as they came to be called, may have been conducive to the physical cleanliness of their clients, but they also acquired the reputation of doing the exact opposite for their moral cleanliness. The male bathers were generally waited upon by young women who, in many cases, provided services above and beyond those of handing out jugs of hot water, pieces of early "castile" soap, and towels.

We must not allow these "stews" to distort our view of medieval cleanliness. There were also bath-houses that dealt with a female clientele, and not all secular bath-houses were brothels. The Florentines, the Londoners, the Parisians, and many other inhabitants of the cities of Europe had come to realize that being clean was more pleasant than being filthy, and it was becoming progressively easier. As the overall European economy began to accelerate after *c.* 1050, merchant wealth began to abound, private houses were better constructed, glass windows allowed in light without draughts, and architecturally designed fireplaces replaced smoky central hearths. All these factors helped to make the homes of at least the better-off classes more comfortable and cleaner.

CHAPTER 6

Spiritual Inspiration, Miracle, Possession, Mental Illness, and the Brain

*O*f all the branches of medicine, none has been more elusive than that of the mind. Fevers, tumours, and being "struck" by a paralysis were not without their explanatory puzzles – and inherent problems in attempting to cure them – but the mind seemed a different terrain altogether. It struck at the heart of personal identity, the soul, and the ability to maintain coherent relations with family, friends, and society as a whole. Mental doctoring was always, and still is, replete with philosophical and theological problems, and poses perennial questions about how matter, spirit, meaning, reason, and un-reason interrelate.

DISCERNING CLINICAL ILLNESS FROM SPIRITUAL STATES

In a deeply devout society such as that of medieval Europe, mental illness posed particular questions regarding an individual's relationship with God. Was a person who went into trances, saw and conversed with beings invisible to the sane, and uttered prophecies that appeared to come true, clinically mad or touched by God? Were they God's messengers to a fallen world? What about mentally disturbed people who uttered vile curses and blasphemies, and, if not restrained, might seriously damage themselves and others, and even go on to commit the sin of suicide: the rejection of that life which was each individual's personal gift from God? Were such people possessed by the devil?

The Bible gives examples of both types. Some believed Elijah, Jeremiah, and even St Paul to be mad: not only did his Christian conversion come about via a traumatic incident on the road to Damascus, but when he was invited to give an account of his beliefs before King Agrippa and Porcius Festus, the Roman governor, his impassioned preaching caused Festus to exclaim "Paul, thou art beside thyself: much learning doth make thee mad."[1] Deep study was one of the acknowledged causes of mental derangement.

But Paul, along with Elijah, Jeremiah, and a host of other prophets, came to be seen both by their contemporaries and by later centuries as genuinely *inspired* by God, since their prophecies were deemed to have been wise, subsequently came to pass, or – in the case of Paul – had a transformative effect upon world history.

Nor is the Bible short of examples of the genuinely deluded and the spirit-possessed. The Babylonian king Nebuchadnezzar is suddenly struck down by God at the height of his power and becomes so deranged that he ends up eating grass like a wild animal.[2] Then King Saul, when God withdraws his favour due to his disobedience, is afflicted by an episodic condition which looks rather like violent paranoid schizophrenia, causing him to suddenly throw javelins at the young David when David's harp-playing failed to pacify him. (This last incident, it has been suggested, makes King Saul history's first documented music critic.)

The New Testament contains an abundance of persons regarded as "lunatic": from the epileptic youth whose dramatic release from demons left him looking like the dead until Christ revived him, to the psychotic "Legion" who harboured enough devils to cause a whole herd of pigs to run berserk and drown themselves. There were many more. All these cases were attributed to spirit or demonic possession in one form or another, and in some instances the clinical details are so good that one might even risk making a modern-day psychiatric diagnosis. Yet, we are told, *all* of them were miraculously cured, leaving a powerful body of precedents for the

95

understanding of mental illness in the centuries to come.

Discerning these agencies, divine and diabolic, was an inherent part of dealing with mental illness during the medieval period, as well as for several centuries afterwards. Medieval priests and physicians also recognized clinical mental illness as a *disease* possessing an organic basis, as witnessed in the founding of the Hospital of St Mary of Bethlehem – Bedlam – London, in 1247, for "poor men deprived of reason", of which more will be said in the last section of the present chapter. There is no shortage of instances in which medieval bishops, clergy, and others, when examining cases of supposed visionaries and prophetic mystics, knew that it was imperative to discern the difference between a person genuinely "touched by God" and someone who was simply deluded or mad. Such cautions were exercised by those persons who had to evaluate the visions of Margery Kempe, as we will see below.

Central to understanding medieval ideas of divine inspiration and mental illness was the nature and status of the miraculous. For miracles, and discerning the hand of God in the workings of the world, were an integral part of medieval reality. This was in no way superstitious, but stood upon impeccable precedents in the Gospels and the church fathers. There were also the healing miracles of the Old Testament, such as that of Naaman the Syrian leper,[3] but much more significantly, there were those of Christ, and even of St Peter, in the New: an impressive catalogue of cases, embracing the lunatic and frenzied, the lame, the blind, the fever-racked, and even the dead – all cured miraculously by the touch of God.

The intellectual foundation underpinning the miraculous derived from the works of the early Christian theologians, most notably St Augustine. To Augustine, the whole existence and continuation of the created world was miraculous, for without divine miracle nothing would have existed at all. Human beings could observe this ongoing miracle on an everyday level, in the sheer beauty and constant replenishment of nature by the rain, sunshine, warmth, light, and the seasons. Yet to stop humanity becoming blasé about these everyday miracles, God created occasional miracles that seemed to differ from the daily miracles: these could include miraculous cures

96

and deliverances, the restoration of the mentally deranged to sanity, and visionary encounters that led to a fundamental change of life for the recipient – such as Rahere the jester's visionary encounter with St Bartholomew.

Irrespective of what a twenty-first century person might think of these miracle narratives, they cannot be excluded from any modern evaluation of medieval healing, nor can they be simply dismissed as "superstition". Nor, indeed, can "demonic possession". For while we may now use terms such as "psychosis", "paranoid schizophrenia", and "bipolar disorder" to classify and hopefully treat the symptoms manifested by the mentally disturbed, the scenarios described by these people often have remarkable parallels to those found in medieval and even biblical narratives. While we may no longer associate mental trauma with cloven-hooved grinning fiends, the old terrors nonetheless manifest themselves in a different garb, and the sense of menace felt by the sufferer remains the same.

I have met schizophrenics who have told me that when they became aware that an acute episode was developing, they would try their very best to resist it, for they knew that it would be very frightening. Yet try as hard as they might, the "thing" invariably takes possession of them and drags them down into a hellish trauma that may last for weeks. It is just like being "possessed" and taken over by a predatory, external, alien force, as it has been described to me. Whether we choose to call this force "the devil", "evil", or an imbalance in the sufferer's neurological chemistry, it is all the same for the victim: it is a season in hell, be that a religious or a secular hell, depending upon one's wider beliefs.

Then there are those people whose unusual psychological episodes are positive and lead to a new, more generous, and optimistic view of the human condition. Suspending judgment for the time being on the current clinical explanation for "out-of-the-body" experiences, a number of people whom I have met and whose writings I have read claim to have journeyed to non-physical realms and frequently (but not always) to have undergone some kind of subsequent life-change. Not infrequently this life-change might incline them to become more generous, kindly, and charitable than they have been before, and it is often spoken of as resulting from visions of a greater truth which they might link with heaven or with God, in one form of religious tradition or another.

While not being mystically inclined myself, I have talked with quite a few modern mystics and have been struck by parallels with individuals in medieval literature. Such people are wholly sane and well-balanced in their day-to-day lives, yet claim to have had experiences that might be described as ecstatic, beatific, glorious, and deeply enriching: a state reached by prayer, contemplation, and spiritual discipline. This spiritual state, one must emphasize, is in no way confined to people in the Judeo-Christian religious and cultural traditions. While I lack the sensitivity to share in such visions, I respect them, and accept their genuineness.

This brings us back to the medieval worldview in which visions and heightened mental and emotional states had to be sifted and tested for authenticity. For while a modern-day psychiatrist may not see a patient's condition in a religious context, as would a medieval physician, we still face the problem of how best we might try to understand and to help the mentally distressed, and hopefully, give them a meaningful role in society. Yet in addition to the medieval view of disordered mind-states as being spirit-based, we must not forget that the classical legacy, known and studied in Latin translation in Europe's medieval universities, also contained a physiological approach to the mind–body dilemma.

EPILEPSY AND THE HIPPOCRATIC TRADITION IN MEDIEVAL EUROPE

In the thirteenth and fourteenth centuries, when the Hippocratic and Galenic writings were being studied in Latin translation in the medical schools of Europe's universities, a physical explanation for mental action and for psychological disturbance was also becoming available. The Hippocratic *On the Sacred Disease* (epilepsy) and *Dreams* both display an essentially pragmatic approach to mental illness, visions, and delusions, rooted in an early usage of Hippocrates' model of humoral pathology (discussed in Chapter 1). And whether these writings were the work of one man or a "school" of Hippocrates' followers based on the Greek island of Cos, they exerted a profound influence and were essentially absorbed into the physiological system of Galen. From Galen, they passed first into the Islamic Arab world, and then into Christian Europe, via Greek, sometimes Syriac, Arabic, and then Latin.

Whoever was the author of Hippocrates' *On the Sacred Disease*, he appears to write as an individual rather than a "school", putting his cards on the table from the first sentence, and affirming "I do not believe that the 'Sacred Disease' is any more divine or sacred than any other disease."[4] He also argued for the primacy of the brain in trying to explain mental states; as the "interpreter of comprehension".[5]

This disease, epilepsy, had long since acquired a divine mystique, not only from the fact that it sometimes affected royal personages, but also from the suddenness with which it seemed to strike its victims. And the symptoms could be frightening, with paroxysmal fits, paralysis, shrieking, foaming at the mouth, and sometimes a corpse-like appearance at the conclusion of the attack. Yet Hippocrates argued that there was no real reason to consider epilepsy as any more sacred than, let us say, an episodic fever. Indeed, he considered that such divine associations were tantamount to blasphemy against the pagan gods, for the disease was *not* caused by them. Epilepsy, rather, was caused by natural factors similar to those that caused other diseases: namely, the patient's temperament, a particular humoral imbalance, and even circumstantial factors, such as wind direction at the time of attack. And most important of all, epilepsy was rooted in the brain – the earliest association of a particular pathology with that organ – though Hippocrates' explanation of the causes of epilepsy is not the one we would use today.

Because the writings of Hippocrates and Galen were so influential upon medieval medical thinking, however, many medieval doctors too correctly associated epilepsy with the brain: a view that was not infrequently held in tandem with demonic possession or other spiritual explanations.

Hippocrates tells us that those most susceptible to epilepsy are possessed of a *phlegmatic* temperament: phlegmatic meaning those whose humoral composition was predominantly governed by Phlegm. "Bilious" temperament types, governed by Yellow or Black Bile, and "sanguine" temperament types, governed by Blood, were less at risk. And the reason for this is straightforward: Phlegm had a particular association with the brain. Phlegm would most commonly manifest itself as nasal mucus, tears, and saliva, being cold and moist, and, in a more refined form, was equivalent to the "animal spirit" or life-force-giving agent of Galenic physiology.

If, for whatever reason, Phlegm began to move down from its natural seat in the "cool" brain to mingle with the warm blood in the heart, lungs, or liver, the blood would be unnaturally cooled, thereby causing these organs to jump violently as the cold Phlegm dissipated in the bloodstream. And should the vessels leading the Phlegm downwards be obstructed, they could affect respiration, causing choking, gasping, asthma, and the utterance of groans and other distressful noises. In a person possessing the phlegmatic temperament, even *fright* could trigger an attack, for shock disturbed the fluid flow.[6]

Then, by way of comparison, Hippocrates tells us that some animals, most notably goats, are also prone to epilepsy, one reason being the moist and watery membranes inside a goat's head, and the apparent sloppiness of its brain.[7] Epileptic humans, he says, have also been found to have especially wet brains on the occasion of their heads being opened. But in addition to the "phlegmatic" humoral make-up, other factors were seen as playing a part in inclining an individual to epilepsy: as they did with most diseases, according to the Hippocratic tradition. These included heredity, age, location, and winds.

As an astute clinician and the inventor of the patient "case history", Hippocrates was aware that many medical conditions ran in families. Age too played a significant part. Infants were especially at risk, as "their little blood vessels are too narrow to absorb a large quantity of insipissated phlegm", such as an epileptic child's brain might generate.[8] Yet few people developed epilepsy after the age of 20 if they had never had it previously, while epilepsy in the old was only generally found in people with an established history of the disease. Elderly epileptics were especially susceptible in winter, after getting their brains overheated by sitting over a roaring fire and then going out into the cold, causing the Phlegm to coagulate. Warm south winds were especially dangerous for epileptics, tending to make the brain "flabby" and to upset the humoral flow, thereby precipitating a fit.

Two other types of brain-based mental disorder were also recognized. Phlegm, Hippocrates argued, was the cause of quiet madness, with no shouting or violent behaviour (depression). Bile, on the contrary, produced noisy, physically active, troublesome behaviour (psychosis). Both types were seen as caused by what we might today identify as the effects of chemical action in the cortex.[9] And just as the Hippocratic physicians firmly identified epilepsy as a physical and not a spiritual disease, so they judged dreaming and the images generated in sleep to be equally occasioned by diet, lifestyle, and memories of our waking hours (*Regimen* IV, "Of Dreams"). (Indeed, without specifying Hippocrates, this natural, "physical" cause of dreams features in Chaucer's "The Nun's Priest's Tale", where Pertelote the hen chastises Chanticleer the cock for being frightened by a dream: dreams merely being "engendered in the too-replete / From vapours in the belly".[10]

The Hippocratic treatise *On the Sacred Disease* presents us with a fascinating mixture of data. The wider humoral physiology in which its explanatory rationale is set would have a profound effect upon the history of medicine over the next two millennia, especially when absorbed into the anatomical and physiological system of Galen, and taught not only in the medical schools of medieval Europe, but in those of the Arab world as well.

Explaining a disease was one thing: effectively treating it was quite another matter. And when it came to treatment – as was the case with most illnesses before the nineteenth century – the healer would try anything that had a chance of working. Chaucer's fourteenth-century doctor "contemporary" John of Gaddesden (*c.* 1280–1361), for example, recommended in his *Rosa Anglica Medicinae*

("The English Rose of Medicine")[11] a portmanteau treatment of Bible readings, prayers, amulets, and herbs for epilepsy. In addition to being an Oxford medical graduate, John was a priest or deacon holding a canon's stall in St Paul's Cathedral, and would have seen no incongruity whatsoever in enlisting both physical and spiritual powers in the fight against human suffering.

The thirteenth-century Catalan philosopher and physician Arnald of Villanova, who, as we saw in Chapter 5, has been suggested as the author of the *Regimen Sanitatis Salernitanum* rule of health, advocated the Galenic remedy of bleeding the patient to stabilize the humours and control epileptic episodes. Even the greatest surgical writer of the entire medieval age, Guy de Chauliac in the fourteenth century, saw no incongruity in recommending charms and prayer in the treatment of epilepsy.

Yet while medieval doctors inherited the Hippocratic and Galenic brain-centred pathology as a way of explaining epilepsy, how did they think the brain was constituted anatomically and physiologically, and how did its operations relate to health, disease, thinking, feeling, and a person's ability to function? Medieval neurological ideas were far from superstitious, as we shall see.

CELLS, CHAMBERS, AND FLUID FLOWS: THE
MEDIEVAL EXPLANATION FOR BRAIN FUNCTION

Not only did medieval doctors accept the ideas of Hippocrates and Galen about the brain being a causal factor in epilepsy, but they were also fascinated by the brain's relationship with sensation, the soul, and wider mental states, and how it was connected to the nerves, muscles, and organs of the body: with what one might now call neuroanatomy and neurophysiology.

But was the brain the centre of *all* cognition? One ambiguity which ran from Greek antiquity into the sixteenth century concerned the roles played by the heart and the brain. Intellect and thinking might well reside in the "cool" head, but where did the feelings and passions lie? Some traditions in Greek thought had said that they lay in the "warm" heart, just as in common parlance today clever, level-headed people are still described as "cool-headed" and kindly, emotional people as "warm-hearted". Indeed, Aristotle had placed the soul, and hence "feeling", in the heart because, as the organ of motion and heat, the heart was the obvious seat for the vibrant, active soul rather than the cold, static brain. This dichotomy of function lay at the core of Hippocrates' cold Phlegm and hot Blood pathology for epilepsy that we saw above.

Shakespeare even used aspects of this pathology to explain human action in many of his plays. In *The Merchant of Venice*, as Bassanio ponders upon which casket to choose for the hand of Portia, he asks:

Tell me where is fancy bred
Or in the heart or in the head?
How begot, how nourished?[12]

"Fancy" in Tudor parlance encapsulates the feelings, hunches, and hopes that lie behind the making of a random choice, and Bassanio is musing on whether such choice is governed by the cool brain and the intellect, or by the warm, emotional heart. Likewise, in *The Winter's Tale* King Leontes, in a fit of jealous anxiety, exclaims:

I have tremor cordis on me, my heart dances;
But not for joy; not joy.[13]

A *tremor cordis* (a heart flutter or palpitation), in classical medical parlance, was seen as being induced by a disturbed emotional state.

So how did medieval doctors see the "cool" brain as operating? As in most things, their ideas were guided by the ancients: Latin translations of Galen (obtained through Byzantine Greek or Arabic intermediaries), of Hippocrates, and of the fragments, preserved by Galen, of Herophilus, Erasistratus, and the third century BC Alexandrian school of anatomy – the first anatomists to open up the human body and attempt to work out the functional relationships between the organs.

Lying at the heart of their endeavour was an attempt to define and classify those seemingly intangible, semi-mysterious processes that activated living things: the spirits and souls. Galen (as we saw in Chapter 2) had seen "natural spirit" as that nourishing force present in the liver, which turned *chyle*, or food juices from the stomach, into life-giving blood. Likewise, in the warm heart, the blood had been further endowed with a "vital spirit" which imparted a power of action beyond the simple business of nutrition. Then finally, that small quantity of blood that was directed into the brain was transformed into "animal spirit" (not "animal" in the sense of beasts, but of the Latin word *anima*: "soul", or "life force").

These forces had parallels in the "vegetable" soul, or mere power of growth, as found in plants; the "animal" soul, possessing not only a power of growth, but also of motion and instinct; and the "rational" soul. The rational or immortal soul resided in the head, enabling its human possessor to think, deduce, use logic and language, and have self-awareness and discernment that transcended and complemented the two lower souls. Especially in the Christian tradition, this rational soul also possessed "mind", a power of eternal life, and a capacity for union with God.

Neuroanatomy began around 300 BC, when Herophilus of Alexandria and his pupils and successors began to open up human and animal heads, and inspect the brains that lay therein. Brains were found to consist of two halves, or hemispheres, seemingly quite separate from each other, but nonetheless possessing linkages of nerves and blood vessels.

When the two "hemispheres" of a brain are sliced in two, however, each hemisphere is found to have two sets of empty chambers within it, running front to back. These are the "ventricles", and as the ancient anatomists discovered, they contained a fluid which was salty to the taste. In that integrated system of anatomy and physiology which Galen bequeathed to the world in his *De Usu Partium* ("On the Function of the Parts"), *De Anatomicis Administrationibus* ("Anatomical Procedures" or "Methods"), and other works, he came to see the ventricles as holding the "animal spirit": a refinement of Herophilus's idea of locating the soul in the fourth ventricle.

By the time of Galen in the second century AD, a succession of Greek anatomists and physiologists, extending from Herophilus in *c.* 300 BC to Erasistratus and Aristotle onwards, had concluded three things about the brain. Firstly, the soul, the person's vital and immortal essence, resided in the brain. Secondly, there was a complex network of blood vessels at the base of the brain that seemed essential to the generation of animal spirits: the *rete mirabile*, or "wonderful net" – though, as Thomas Willis would discover in 1664, such a *rete* was only to be found in certain types of animals, whereas the two hemispheres of the human brain obtained their blood supply from one large circular artery. (Willis's name would be immortalized in this artery: "the Circle of Willis".) Thirdly, the brain was the seat of all "motor" functions, with nerves extending from the cortex to the organs of touch, taste, smell, and so on.

Galen's work established many of the assumed functions of the brain that would endure for the next 1,500 years. For example, he traced the motor nerves (which activated body movements) to the cerebellum region and sensory nerves (responsible for sensation and feeling) to the cerebellum/cerebrum.[14] While many of Galen's specific connections and explanations are now known to be either oversimplified or else incorrect, he grasped the big picture, in so far as all the nerves relate to specific brain structures and different parts of the cortex do in fact control different systems of nerves and bodily functions.

All of this passed into medieval Europe and formed the stuff of medical teaching in Paris, Montpellier, Bologna, and elsewhere. It also lay at the heart of that late classical and medieval model for the human brain known as the "cell doctrine". Building on Galen and Herophilus, this cell model of the brain was brought to its enduring form under the aegis of two eminent churchmen: Nemesius, Bishop of Emesa, and the great St Augustine of Hippo himself, around AD 390–400.

In this cell doctrine, the four lateral, or longitudinal, ventricles, two in each brain hemisphere, were associated with specific mental functions. In its earlier form, in the early Christian period, there were three cells. The first of these cells was *sensus communis*, the

105

location of our basic sense inputs of sound, taste, smell, touch, and vision. In a rather crude sketch of *c.* 1500,[15] sound, smell, impressions entering the eye, and the sensations of heat from a fire, a snakebite, and food or drink upon the tongue are all connected by lines to the same *sensus* cell, or chamber of the brain.

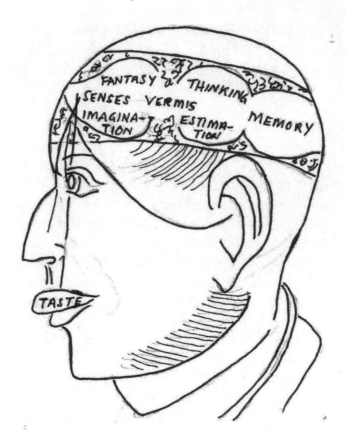

7. The medieval three-cell model of the brain, *c.* 1525. Though schematically depicted, the cells approximate to the real (four) ventricles of the human brain. Note how taste, sight, and smell all enter the frontal or "sensitive" cell before passing on to produce "Imagination", "Estimation", "Memory", and other thinking processes, via the Vermis ("worm") or connecting passage. (Drawn by A. Chapman from Plate 52 in E. Clarke and K. Dewhurst, *An Illustrated History of Brain Function* (1970), p. 36, with Latin text translated into English.)

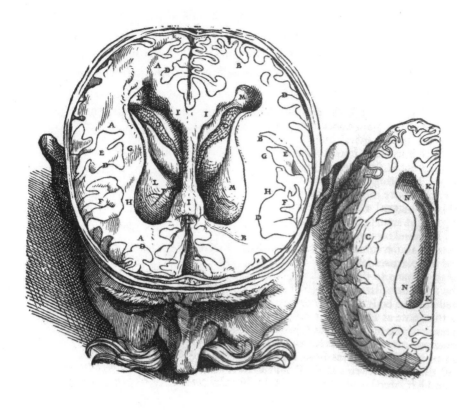

8. Dissected human head with the skull dome sawn away, displaying the left
    and right lateral ventricles of the brain, as demonstrated by the Renaissance
    anatomist Andreas Vesalius. Vesalius, who gives the first accurate drawing of
    the human brain, doubted the cell theory, and side-stepping philosophical
    considerations, examined and depicted the brain ventricles purely as
    anatomical structures holding the "animal spirit" (or cerebrospinal fluid).
    (Andreas Vesalius, *De Fabrica Humani Corporis* (1543), Book VII. Christ
    Church Library, © Governing Body of Christ Church, Oxford.)

The second cell was ascribed to reason and cognition, or
intellectual processes, while the third dealt with the function of
memory. This threefold division of mental processes – sensation,
thinking, and memory – was not evenly distributed between the
ventricles, and in some later versions of the cell model there were
more than three chambers and functions. Some later writers added
a fourth chamber-function to deal with *phantasia* or imagination,

while Avicenna divided mental processes across five cells. The cells, it was posited, were connected by a *vermis*, or worm-like aperture, thereby enabling a sort of internal discourse within the brain, as sensation, intellect, memory, and imagination interacted to explain all aspects of mental life.

The "cell" model, based on the best anatomical knowledge available by the fifth century AD, tried to create an integrated physiological, neurological, intellectual, and spiritual model of the human being. Endorsed by churchmen and taught across the universities of Europe, it epitomized that quest which had eluded the ancients, fascinated medieval Christians, Jews, and Muslims, and still occupies us today: How do our conscious minds, with all their teeming powers of imagination, conceptualization, and creativity, relate to the brute biochemistry of our brains? How do our souls connect with our bodies?

MARGERY KEMPE (NÉE BURNHAM OR BRUNHAM)
AND RELIGIOUS VISIONARIES

We saw above that in the Middle Ages many aspects of behaviour which a modern secular person might dismiss as deranged could also be assessed from a spiritual perspective. In a culture where good and bad spirits were believed to be at work, and a contest was perceived to be raging between good and evil forces, it made eminent sense to "test the spirits" in individual cases. Yet as we saw with epilepsy, and with medieval ideas of brain function, it is clear that medieval Europe possessed a remarkably well-thought-out and pragmatic concept of organic mental illness as well. Could not mental derangement occur when the cells within the brain failed to function properly in their accumulation, storage, and processing of mental sensory data, thereby producing delusion or odd behaviour?

The case of Margery Kempe, the comfortably off, early fifteenth-century merchant's wife from King's (or Bishop's) Lynn, Norfolk, has been a subject of much scholarly analysis and discussion, especially during the twentieth century. Most of what we know of Margery derives from *The Book of Margery Kempe*, said to be the first English autobiography, and dictated by her to a priest in or shortly before 1438, when she would have been in her late sixties. A very good age for any medieval person, even a solidly middle-class woman, to have reached – Margery was born around 1373 – especially considering her 14 or more pregnancies, along with the hazards she endured on pilgrimage to Jerusalem and elsewhere.

Margery Kempe's religious experiences began in her early twenties, following the birth of her first child, probably in the mid 1390s, when she underwent an eight-month period of intense psychotic or spiritual trauma. She had visions of demons and devils, and worst of all, of Christ himself telling her that, due to her partiality for lust and luxury, she had deserted him, though he would never desert her. Her visions and sense of demonic assault, counterbalanced by the ever-present support of the God whom, in spite of her traumas, she loved passionately, became part of an established syndrome. She was prone to loud and sustained weeping, and even disrupted church services by her intense emotional displays.

Margery's sense of the close, sustaining love of Christ also emboldened her to rebuke the slack and spiritually uninspired, especially among the clergy, and she was not afraid of undertaking a journey with the express intention of reading the spiritual riot act to a worldly priest or bishop.

Hardly surprisingly, these rebukes got her into trouble for heresy, and she was examined before ecclesiastical courts at Leicester, York, and elsewhere. One point of contention in these examinations was her orthodoxy: was she an orthodox Roman Catholic, or did her behaviour suggest that she was associated with the Lollards, that radical Christian movement coming out of the Low Countries which would form an ancestor of Protestantism? On every occasion, however, Margery's Catholicity was found secure.

Margery seems to have enjoyed a remarkable personal freedom, even for a well-off rich merchant's wife, going off on pilgrimages across England, Europe, and the Holy Land, and becoming one of the best-travelled women of the Middle Ages. She was also blessed with a husband, John, who seemed happy – or perhaps relieved – to be rid of her for a spell. At least, there would be no loud weeping and wailing to endure.

Was Margery mad or divinely inspired? In this respect, she was one of a rich late medieval tradition, including Mother Julian of Norwich (the anchoress who lived for years walled up in a cell in St Julian's Church, Norwich, to whom Margery made a visit), St Catherine of Siena, St Bridget of Sweden, and the Spanish mystics. These men and women lived lives of extreme asceticism and experienced profound spiritual insights focused on Christ. An established medical tradition of clinical insanity notwithstanding, the proof of the pudding was believed to be in the eating. Did the spiritual conversations with Christ, the weeping for sin, and the outspoken admonitions to repent bear fruit? In the case of Margery and the other great Christian mystics, it was unequivocally considered that they did. As time went by, Margery's power to make people take stock of their lives and become more devout and more charitable human beings certainly came to be recognized. Even lax-living monks mended their ways after encountering her spiritual power, and, like Mother Julian, Margery came to be reckoned a great Christian teacher. Shrieking apart, she was seen as a force for good and an agent of Christ in the world.

Attitudes to mental illness and unusual social behaviour are in many ways influenced by prevailing cultural norms. After the great innovations that took place in science and medicine in the seventeenth century, as well as those of the Protestant Reformation, the idea that the visionary was truly touched by God appeared less and less plausible. The full-scale Protestant assault upon the whole Catholic apparatus of redemption, an increasing "medicalization of madness" in the wake of the scientific Renaissance, undermined the very notion of God's fools. This would lead, in the centuries that followed 1600, to an essentially "physical" understanding of mental illness, even if the nature of the soul remained as elusive as ever.

But what could be done to treat the lunatic?

"BEDLAM": A PLACE OF ASYLUM FOR THE DISTRESSED?

Erudite academic discussions about epilepsy, brain cells, and being touched by God were all well and good; but how did the Middle Ages – and later centuries – deal with lunatics in their various manifestations, on a practical, daily basis? How did one cope with those who were incoherent, uncommunicative, living in a world of their own, or dangerously violent? A shared culture could be enormously important; and if a mentally ill individual were sufficiently coherent to be able to converse with the person trying to treat them, then shared beliefs could be invoked. Treatment could include taking a disturbed person to Mass to receive the Blessed Sacrament, convincing a self-harming patient that such behaviour would distress Jesus, as did the eighteenth-century clerical "mad-doctors", or explaining to a rationalist patient that his sexual obsessions were the re-emergence of long-buried subconscious infantile traumas, as an early twentieth-century Viennese atheist psychiatrist might have done.

But what about those so acutely disturbed as to be for the time being beyond talking cures? This is where the place of asylum, or refuge, came in. Europe's first mental hospital was the House of St Mary of Bethlehem – contracted to Bedlam – founded in London in 1247. For its first century of life, Bedlam would not be an asylum as such, but a fund-raising institution, under the Bishop, and then the City, of London for the English Crusader Church in the Holy Land. By 1370, however, it was serving, like most religious houses, as a *hospitium*, where travellers, the homeless, the sick, and the insane were given refuge. It is in the Report of a Commission of 1403 that Bedlam was first specified as a place for the insane, accommodating six men described as *mente capti*, or "deprived of reason".

The 1403 Visitation does mention the presence of chains, manacles, and locks, but whether these instruments of restraint were intended to be used on the inmates is not specified. During the fourteenth and fifteenth centuries, other refuges for the mad were established in Europe, such as in Paris, in Elbing, Germany, and in Granada, Seville, Valencia, and elsewhere in Muslim and post-Reconquista Christian Spain.

What asylum could a medieval or later madhouse offer? Little more, alas, than a roof over a deranged person's head, physical restraint to prevent self-harm or harm being done to others, and the use of purgatives and bloodletting to attempt to balance their agitated humours. Purging and bleeding the patient to a point of exhaustion was a way of inducing a brief semblance of calm.

These problems were to remain for several centuries to come as far as psychiatric medicine was concerned, although, as we shall see in Chapter 18, new ideas of moral management were beginning to emerge in the eighteenth century, which sometimes appeared to bear therapeutic fruit.

But now let us examine the diseases that truly terrified medieval people: sudden waves of epidemic disease, striking apparently from nowhere, and carrying away thousands.

CHAPTER 7

In Time of Plague

$\mathscr{P}$eople living in the modern Western world do not usually expect to die of something they catch from someone else. In our highly individualistic age, we expect, rather, when our time comes and hopefully after a comfortable old age, to finally succumb to something that goes wrong inside our own bodies: organic failure, such as heart or cardiovascular disease, a stroke, or when our immune system finally becomes incapable of destroying our own rogue cancer cells. Death by infection had been thought of as a thing of the past; and we are brought up sharp, panic, and become collectively terrified when we realize that we are not immune.

When the HIV virus first hit the West in a big way in 1981, many people went into something resembling shock. Kisses were refused, suspicious eyes were cast on eating utensils in pubs and restaurants, and even the Diocese of Oxford – among others – printed reassuring advice about "AIDs and the Chalice". Especially disconcerting was the fact that most people contracting, and tragically dying from, AIDS in the early days were *young* people. By the prevailing statistics, they should have had another half-century or more of life to look forward to. Since then, other scares, often originating in economically deprived countries, have hit the news headlines, such as the so-called "mad cow disease", bird and swine 'flu, and, most recently, the Ebola virus in West Africa.

Yet this freedom from death by infection which most people now take for granted is a very recent circumstance. It would not have been the case in 1840, when cholera, typhus, typhoid, scarlet fever, and a good few more infections were still scything down the populations of Europe and America, or even in 1940, when the dreaded "TB" (tuberculosis) bacterium was still reaping its grim harvest. How these diseases bit the dust, one after another, was brought about by the staggering progress made in microbiology, biochemistry, immunology, and industrialized pharmaceutical chemistry after *c.* 1900, meaning that a disease such as pneumonia, which in 1880 frequently had a fatal termination, can nowadays be reliably treated with a course of penicillin tablets taken at home.

How that transformation came about will be explored in later chapters. What concerns us at present is what happened "in time of Plague", as the Elizabethan poet Thomas Nashe put it.

EPIDEMICS: SIN, NATURE, AND THE PLAGUE OF THE PHILISTINES

Epidemic diseases, like mental illness, often seemed to strike out of the blue, but unlike mental illness they afflicted not just individuals, but whole cities, nations, and even continents. Why did the warm winds that heralded spring and brought about new life often usher in summer epidemics and fill the graveyards by October? Was it due to natural causes, ill luck, or individual or communal sinfulness? A variety of writers in the ancient, medieval, and early modern worlds attributed epidemics to all of these. The Hippocratic and Galenic tradition saw disease as a natural phenomenon; the devotees of the Roman goddess Fortuna, like the Philistine diviners and magicians of Ekron, in *c.* 1000 BC, might ascribe it to a "chance"; while numerous European writers after 1348 saw it as punishment for sin.

The unfortunate people of biblical Ekron, Ashdod, and elsewhere in Philistia around 1100 BC *may* have been the first recorded victims of a bubonic plague epidemic. What happened, as recorded in the Old Testament book of 1 Samuel, is that the Jewish children of Israel rebelled against their Philistine neighbour overlords and carried into battle the sacred ark of the covenant, in the hope that Jehovah would ensure them a victory. But the Philistines won the battle and carried off the sacred ark as a trophy,

114

putting it in their pagan temple of Dagon in Ashdod.

It was then that Jehovah acted: not only causing the statue of Dagon to mysteriously crash down in pieces, but also unleashing a terrible epidemic among the people of Ashdod and then Ekron. Piecing together the scanty medical details in the biblical narrative, this epidemic seemed to involve the victims developing lumps, swellings, "emerods", or tumours, and what were once called "buboes"; and, perhaps, the presence of a larger than usual number of rats, mice, or both.[1]

At this point, the Philistines, having had enough, decided not only to give the ark back to the Jews, but to accompany it with offerings symbolic of their affliction: five golden mice (or rats) and five golden "emerods" (tumours). Then having received back the ark, with its symbolic golden rats and tumours, the Jews themselves sickened with the same disease and began to die.

The medical hints contained in the passage in 1 Samuel are vague, but what they do tell us is that the disease that struck the Philistines was quite sudden, deadly, involved swellings, was somehow connected with vermin, and was unintentionally passed on to the Jews. And thence on to who else in the Middle East? We are not told.

Yet this divine punishment for sin explanation for sudden epidemics in particular would continue to roll down the succeeding millennia.

THE BLACK DEATH OF 1347 AND BEYOND

Although the term "Black Death" only dates as a historical term from 1823, the symptomology of that succession of related epidemics that hit Europe and the Arab world in 1347 and beyond (for which we have clear record) over the following centuries is well recorded. No one nowadays is claiming that it was a *new* disease, for as we saw above, it was probably not unknown in the Middle East of 1100 BC, though it does not seem to have spread across northern Europe before that date. There is a strong body of scholarly opinion that states bubonic plague may have been the horrible malady which

115

ravaged the Byzantine Empire in AD 541–42 during the reign of the Emperor Justinian, and killed tens of thousands of people. And while other scholars have suggested smallpox as the culprit in the "plagues of Justinian", from surviving details of symptomology, bubonic plague seems a likely candidate.

Regarding the fourteenth-century epidemic, however, good contemporary evidence suggests that deaths began in 1346 at Caffa on the Black Sea, where Mongols were besieging the city, which was a trading base for Italian merchants. As was not uncommon during the horrors of a protracted siege, the Muslim besiegers began to die of a nasty infection, and in the spirit of share and share alike it was soon spread to the entrapped Christians, it was said, by the Mongols hurling infected body parts over the city walls with their siege engines.

The siege ground to a halt, and the Christians sailed for Italy, stopping off at Constantinople on the way and inadvertently spreading the disease into the Byzantine capital city. The epidemic probably first broke out in Sicily, Venice, and Genoa late in 1347. It then spread like wildfire. The poet Giovanni Boccaccio in his *Decameron* (1353) described its impact upon his beloved Florence, where so many people either died or fled that this Italian centre of trade and culture soon had grass growing between the flagstones of its once-thronged streets and piazzas. And knowing what we know today, we can see that the pestilence travelled down fourteenth-century Europe's bustling trade routes. Across the Alps, into France, down the Rhine, and other arterial waterways, before finally entering England through the once-thriving port of Melcombe near Weymouth, Dorset. Once in the British Isles, it then went three ways: eastwards to London and East Anglia, westwards to Bristol, Exeter, and finally Cornwall, and north to Lichfield, Peterborough, York, and Durham. By 1349, Friar John Clyn was chronicling its devastating impact upon the people of Scotland.

No one was safe, be one a Spaniard, Pole, Swede, Russian, or Greek, as the plague travelled by land and by water, via the bodies of its future victims. It was also wholly democratic in its victims, for bishops, judges, and great ladies were no more immune to the pestilence than was the simple serf, and even great monasteries were decimated. It has been estimated by modern historians examining contemporary evidence that between one-third and one-half of Europe's population was wiped out over a few months. And not only that, but the plague, after its original devastation, tended to return every 20–25 years, to wreak yet fresh mayhem. It is hardly surprising, therefore, that it heralded what the historian Jan Huizinga styled "the waning of the Middle Ages" and inspired a quasi-comic *danse macabre* depicting skeletons and decomposing bodies dancing in the churchyard, and grisly *memento mori* ("be mindful of death") carvings on tombs in our cathedrals and great churches, showing the deceased as a shrunken cadaver, complete with rictus grin and worms feasting on the rotting flesh.

Various theories were proposed by the scholars of the time to explain the devastation. Divine punishment for sin was, needless to say, one of them; others were to do with malice, such as the Jews supposedly poisoning the wells and rivers – a suggestion rendered absurd by the fact that the plague devastated Europe's Jewish ghettos just as severely as it did everywhere else. But there were naturalistic or scientific theories as well. These included foul and noxious airs, especially carried on south winds, which was in the best tradition of Hippocratic medicine as presented in such works as the Hippocratic *Airs, Waters, and Places* of *c.* 430 BC. Foul miasmas were seen as disrupting the humours and leading to pandemic.

Proposed rational, as opposed to magically based, treatments were broadly in line with Hippocratic and Galenic teaching, in so far as they were essentially purgative. Bleeding was recommended for feverish patients, rose water and aromatics were used to counter the supposedly miasmic airs, while burning sulphur was also thought to cleanse the air. The eminent French physician and surgeon Guy de Chauliac in his *Chirurgia Magna*, or "Great Surgery" of 1363, ascribed the epidemic to a conjunction of Jupiter, Saturn, and Mars in Aquarius in 1345. The medical faculty of Paris issued its own therapeutic guidelines, while the pestilence generated a body of medical literature generally known as the "Plague Tractates", or tracts.

Among the victims of the 1347–50 plague were many notable persons, and Anna Campbell's *The Black Death and Men of Learning* discusses the epidemic's impact upon culture. Even the Archbishop of Canterbury, former Oxford mathematician and philosopher, Thomas Bradwardine, died in Rochester in August 1349, almost certainly a victim of the pestilence.

Then, by 1350 or so, it had died away. It is hard to be sure why, although, as is the situation with most epidemics, not everyone got the disease, and some may even have had a natural immunity. Unlike some epidemic diseases, however, such as smallpox, the plague did not confer immunity on its survivors, and in both the 1347 and subsequent bubonic plague epidemics there were well-documented cases of people who were carried off in their second infection, or even survived more than one attack.

So how lethal was this terrifying disease, and what do we know about it from a modern bacteriological and epidemiological point of view? Although bubonic plague seemed to leave Europe after the Marseilles epidemic of 1720, it did not by any means die away. When Napoleon Bonaparte invaded Egypt in 1797, the disease seems to have been active in that country and many other parts of the world. But it was not until an epidemic raged in China in the 1880s to 1890s, at the same time as Louis Pasteur and Robert Koch were discovering the bacterial aetiology of some other epidemics, that serious scientific progress became possible.

It was in the late nineteenth-century Chinese bubonic plague epidemic that the biological mechanism behind the disease came to be explicated. And while lumps, rats, and fleas were accompanying features of an epidemic, it came to be realized that the rats – and the fleas – far from being the villains of the piece were poor, hapless victims of the malady no less than were the humans. The Swiss-French bacteriologist Alexander Yersin and the Japanese bacteriologist Kitasato Shibasaburo discovered in 1894 that the real villain was the microbe *Pasteurella* (now *Yersinia*) *pestis* which, like most strains of bacteria, thrived and multiplied in warm body fluids such as blood.

This is what we now know. Plague tends to be a rodent disease, with a particular affinity for the black rat species, *Rattus rattus*. Rats tend to be timid creatures and prefer to live with their own kind in the wild. Their fur, however, combined with rodent body heat, is the home of choice for the flea *Xenopsylla cheopis*, which is partial to rat blood as a food. *Xenopsylla cheopis* feeds by inserting its hypodermic-like proboscis into the rat's skin and sucking up some blood.

If, however, an infected but still fairly active *Xenopsylla cheopis* has migrated to a healthy rat it will attempt to draw blood. But the sickening flea will reach a point where, instead of being able to draw blood, it will do something equivalent to coughing and choking, with the result that it spits *Yersinia pestis* bacteria into the body of the rat, condemning it to death. When the dead rat's body begins to cool, the still healthy yet infected fleas in its fur will attempt to migrate into the bodies of warmer, healthier rats. As fleas can jump 12 to 13 inches, this means that new hosts are well within migrating range. Several short-lived "generations" of healthy, infected, and terminally choking fleas can therefore disseminate themselves through a pack of rats.[2]

The crunch comes, from the human point of view, when a scarcity of food compels the rats to come out of their holes in the wild, to seek edibles lying around in a human community. Being good jumpers, the still frisky fleas could desert the cooling body of a dying rat and land on a warm human host. If those humans were already hosts to livestock, and were none too clean, the disease could pass to them via the same bloodsucking route that it passed between the rats, and a human plague epidemic could break out.

The plague could then be passed by the fleas from human to human, through different human groups and down trade routes, travelling – we think – across the Asiatic Steppe, through Syria and the Arab world by the Black Sea, into Greek Byzantium, and across Europe as far as Iberia, Ireland, and Scandinavia. As this could take place over no more than a few months, one can fully understand the panic and horror that its advance engendered, as wholly vulnerable and defenceless communities heard from pedlars and travellers that the "pestilence" was coming forever nearer!

The advancing terror drove some to grovelling repentance, others to foolhardy bravado or crime, and yet others to wild drunken

119

revelry, taking the advice tendered in the Old Testament Ecclesiastes "a man hath no better thing... than to eat, and to drink, and to be merry".[3] Geoffrey Chaucer's "The Pardoner's Tale" (*c.* 1391–92) in his *Canterbury Tales* mentions such a group of young men, drinking, wenching, and revelling as yet another funeral passes their tavern door. Asking who is in the coffin, they are told by a tavern boy that the dead man was an old drinking companion of theirs:

> *And suddenly, last night, the man was slain,*
> *Upon his [tavern] bench, face up, dead drunk again.*
> *There came a privy thief, they call him Death,*
> *Who kills us all round here, and in a breath*
> *He speared him through the heart, he never stirred.*
> *And then Death went his way without a word.*
> *He's killed a thousand in the present plague.*[4]

Though Chaucer lacked our modern knowledge of the aetiology of plague, this well-educated and observant civil servant possessed a sharp eye for detail – as his *Tales* show us time and time again – and in "The Pardoner's Tale" he supplies us with medical data whose significance he could never have realized. People did die suddenly during plague epidemics, like the drunkard in his tale, but the great majority took around a week from initial infection to perish. The bacterium would incubate for several days, then would come the fever and the buboes or tumours as the infection ran riot through the patient's lymphatic system, producing the painful blackened swellings in the neck, armpits, and groin: the "secret parts" of the biblical plague of the Philistines, or the lymph nodes of modern medicine.

Yet the bubonic form of the disease – infecting the lymph system – was by no means universally lethal, and if a patient were fortunate enough to have a courageous person to care for them, to help cool their fevered bodies and give them plenty of fluid to drink, then one had a 30–40 per cent chance of recovery. The liquid would have been especially important, for with a raging fever dehydration could be a major contributory factor to death.

Yet what about the *sudden* deaths, not only firmly established in the popular pathology of the disease by the time of Chaucer's "Pardoner's Tale", but also recognized by Guy de Chauliac in his brilliant clinical descriptions of the first outbreaks after 1347?[5]

The microbe *Yersinia pestis*, it was discovered by the early European and Japanese bacteriologists, could attack two other major bodily systems in addition to the lymph nodes. These were the lungs and respiratory system, and the blood. *Septicaemic plague* was also insect-borne, yet instead of the lymph tissue, the blood took the brunt of the microbial assault, and as warm blood is a perfect culture fluid for bacteria, the germs "bred" at an alarming rate. Thus, a victim might be hale and hearty at breakfast, and dead by dinner – and in earlier centuries, that was around midday! This was probably the plague type contracted by Chaucer's "Pardoner's Tale" drunkard and could lead to the victim simply dropping dead.

Pneumonic plague, however, attacked the lungs and respiratory system, leading to coughing fits, with clouds of bacterially charged moisture being disseminated into the surrounding environment. And in any human-occupied enclosed space – be it a cottage, a church, or a castle hall – the bacteria could have spread at an alarming rate by the victims doing nothing other than breathe. Pneumonic plague was also fast acting, and both it and the septicaemic variety would have been over 95 per cent lethal, judging from post-1894 studies of plague.

Yet why did plague suddenly strike Europe in 1347, then keep regularly scything down the population in return epidemics, only to go away, seemingly of its own accord, after the last major outbreaks in London in 1665 and Marseilles in 1720, never, mercifully, to return? All sorts of theories have been put forward, usually focused around the story that the "Hanoverian" brown rat ousted the black rat. The chronology does not fit, for if brown rats supposedly came into England with King George I of Hanover in 1714, then why had the black ones caused no more epidemics in the half-century after 1665?

I would conjecture, although we have no histological specimens to go on, that the bacterium mutated so as to be less likely to affect humans. As we shall see in the next section, infectious diseases do come and go, as in the case of the Tudor "sweating sickness". Bacteria, after all, are the fastest-breeding things in nature, and as they are no less susceptible to evolutionary change than men or monkeys, it is possible for a microbe "key" to mutate so as to become incapable of entering the human "lock" and thereby restart the disease.

Plague is, however, very far from extinct in the modern world. As it appeared in 1100 BC Palestine, sixth-century AD Byzantium, and fourteenth-century Europe, so it resurfaced in China in 1855, killing un-reckoned numbers of people over the following decades (100,000 in Hong Kong alone in the late 1890s). Then it moved on to India, Hawaii, and elsewhere, as fast steamships were, by that time, capable of carrying *Yersinia pestis* along with its flea and rat hosts around the globe. There was even a *contained* epidemic in Glasgow in 1900.

Luckily for everyone, however, the nineteenth-century medical revolution and Yersin's and Shibasaburo's pioneering epidemiological work were themselves part of that growing advance in technology, science, and global trade which also brought the steamships and railways. So as one branch of advancing technology – trains and steamships – inadvertently played a part in spreading the Victorian Asiatic plague epidemic, so another branch, rooted in microscopes, laboratories, and developing biochemistry and bacteriology, played a vital part in finally elucidating the mechanism by which plague spread. And while in 1900 no drug or procedure was capable of curing the plague once a person had caught it, the new science opened up lifesaving possibilities. For one thing, by understanding the vectors through which the disease spread – germs, flea, rat, human – it became possible to curtail the extent of the disease by instituting public health measures. While this was miserable consolation for the millions of poor souls who died across the Far East, at least the Victorian plague never became a global pandemic.

Once it was known how the disease acted once in a human body, it became possible to manage an outbreak in a given environment. Strict quarantine, rodent control, and other hygienic measures could be instituted, if circumstances allowed. Then with the antibiotics revolution of the 1940s, the scales could at last be tipped in favour of survival for individual sufferers, with streptomycin, doxycycline, and other antibiotics being found to have a lifesaving effect. Nowadays bubonic plague sufferers have well over a 90 per cent chance of survival when properly treated, whereas without modern treatments, well over one-half died. The septicaemic and pneumonic varieties are still much harder to treat today, largely because they are so fast acting, and a patient is already likely to be far gone before a doctor can even get to them.

In the late 1970s and early 1980s, plague struck the heartland of the United States of America. I have a large newspaper cutting sent to me by an American medical friend entitled "The Land of the Flea and the Home of the Plague". The early victims of this "epidemic" were veterinary doctors who looked after people's sick pets.

It seems that through some bio-mechanism or other the *Yersinia pestis* bacillus got into small furry animal populations in America's wild places, such as state parks. People would let their dogs off the leash to chase and catch chipmunks and other little creatures, only to find that some days later the animal would become ill. The pet would then be taken to a veterinary doctor, who in turn would sicken, and, in the early cases, die from a mystery infection. And when biopsy checks were made on a feverish vet, bubonic plague was not a disease that, at first, would naturally be looked for. When it was recognized, however, proper tests could be done and antibiotics given.

The real tragedy, however, lies in those parts of the world where *Yersinia pestis* is still active, but where economic and other circumstances put twenty-first century people at risk from a disease which is now generally curable – if only access to doctors and antibiotics is possible.

A MISCELLANY OF MEDIEVAL MALADIES

As we have seen above, the main killer diseases of medieval Europe and Arabia were not the organ-failure-related ones which we fear today, but things one might catch in the market-place. But while plague was the killer most popularly associated with the medieval centuries, it in no way had the field to itself.

So what other maladies were doing the rounds of Europe in the "High" Middle Ages, *c.* 1100 to 1400, and how did they compete with each other in the *danse macabre*? We saw in Chapter 1 that a collection of skin diseases which the early Bible writers termed "leprosy" became the first disease to be the subject of quarantine regulations, but what was *real* leprosy, and how common was it in medieval Europe?

Leprosy proper, or Hansen's disease, known in the Middle East in classical times, first seems to have hit Europe, and especially northern Europe, in a serious way in the eleventh century. How it made its appearance in Europe has been much debated by scholars, but it is likely that it came through increasing trade with the Middle East and the growing number of pilgrims visiting the Holy Places: a profitable trade tolerated by the more sophisticated Arab rulers of Jerusalem. But after the fanatical Seljuk Turks displaced Jerusalem's Arab rulers in the eleventh century, thus effectively closing the pilgrim routes after 1071, and leading to the First Crusade in 1095 as an attempt to reopen them, very large numbers of Europeans began to flow into the Holy Land, and often, to return home. All these factors contributed to the burgeoning European leprosy problem, as disease movement goes along with that of populations. And leprosy seems to have spread relatively quickly, given the slow rate at which the disease normally progresses in individuals. One way, in that non-statistical age, in which the impact of the disease can be reckoned is in terms of the opening and endowing of hospitals and refuges in which to isolate lepers. (When, for instance, the Old Testament king Azariah (also known as Uzziah) of Judah was stricken with the disease, he spent the rest of his days in a "several house", or leper community.[6])

It has been estimated that medieval Europe may have had as many as 20,000 leper houses, large and small, by *c.* 1300. And as far as we can tell from recorded sources, they seem to have been humanely run institutions, usually located outside major cities, their unfortunate inmates being cared for by religious or civic charities. Indeed, I live within a short walk of one such twelfth-century leper community in Oxford, its honey-coloured stone buildings long since converted into private houses, and its chapel now a daughter church of an Oxford parish: Bartlemas (St Bartholomew's) Chapel.

Pictures and descriptions of medieval lepers of *c.* 1200 indicate that this is real leprosy, Hansen's disease, and not a medley of skin infections. Now we get depictions of the leper with his hideously blotched face and clapper bell, living out a purgatory on earth. Indeed, in some respects lepers were regarded as privileged persons, who, in paying for their original and specific sins in this life, were more likely to have a faster passage to salvation after death.

Yet because European leprosy, at its height around 1100–1300, generated so many buildings and endowments, one can trace the decline of the disease simply by the rate at which leper houses, or leprosaria, were converted to other purposes, such as ordinary small hospitals or almshouses for the infirm, or were simply allowed to fall into decay, like Oxford's Bartlemas Chapel, which was taken over by a local farmer and used as a cowshed before being reconsecrated as a church. By 1450–1500, it is clear that Europe no longer had enough lepers to fill them. So what had happened to the disease?

It is not only the lives of plant and animal species that are controlled by competitive evolutionary forces, but also those of diseases. And in the long-term "survival of the fittest" struggle in the disease world of medieval Europe, leprosy was a dinosaur. It was a very "slow-burn" disease, taking, perhaps, a couple of decades to kill a person and hence likely to become extinct in the face of new, deadly "fast burn" diseases, which killed within days or even hours. Bubonic, pneumonic, and septicaemic plague were such "fast burn" diseases, devastating heartland Europe for the first time in 1347–50, and under the onslaught of *Yersinia pestis*, the plodding *Microbacterium leprae*, one might suggest, did not stand a chance.

125

Nor was bubonic plague the only vicious beast on the post-1300 medical block. Two others in particular were scything their way into the population: the viral infections measles and smallpox. Both diseases caused a sharp rise in temperature, then, very conspicuously, on outbreak of pustules and spots – the "pox" – across the head, eyes, neck, and chest, and sometimes blindness. Both were very contagious. You could only catch measles and smallpox once: one exposure conferred lifelong immunity, if you survived the attack.

Few people in the West would now think of measles as a major killer, yet evidence suggests that when the disease began to attract comment in the late Middle Ages it may have been much more virulent than it is today. It had been the great tenth-century Arabic physician Rhazes who wrote the first major treatise on smallpox and measles. There is also evidence to suggest that typhus and typhoid too were known by the late Middle Ages: typhus being an insect-communicated disease, likely to break out in prisons, in armies, and on board ships, and other places where unwashed people were herded in together and personal livestock could circulate. Typhoid, was passed on through bacterially contaminated drinking water.

The above were all fast burn killers, yet a more medium burn disease was tuberculosis. Sometimes this "White Death", as it later came to be called from the emaciated faces of its victims, could even outpace leprosy as a slow-burn killer. A patient might take a couple of decades to slowly decline and become thin and deathly pale as they coughed themselves to death. The consumption could gallop and bring about death in a few months or a year or two, as would later happen to the poet John Keats.

More insidious, however, were those tubercular complications which produced crippling skeletal damage. It was not until many intermediary clinical advances had been made that these complications came to be connected with tuberculosis. The landmark discovery was made by the St Bartholomew's Hospital surgeon Sir Percivall Pott between 1779 and 1782, when he recognized that the deforming, paralysing appearance of pustules and abscesses in the spinal vertebrae was connected with an underlying tubercular condition, a condition since immortalized

126

in his name: Pott's disease. Yet alas, this discovery lay centuries ahead for medieval patients, and another 170 years would have to elapse after Pott's clinical diagnosis before it could bear fruit, when antibiotics such as streptomycin were found to be capable of destroying the tuberculosis bacterium.

By the late fourteenth century, large parts of Europe had become very dangerous, from a disease perspective. It is true that, meanwhile, improved agricultural machines and techniques, such as wind- and watermills, iron ploughs, and the introduction of new breeds of agricultural horses that were stronger than oxen, were all allowing more land to be cultivated and used to produce food. However, by wiping out up to one-half of Europe's population, plus return epidemics of bubonic and other "pestilences", disease had a stalling effect on the burgeoning technological and agricultural advances already taking place before 1347. The drastically reduced labour force meant that harvests went ungathered, and as time went on, land fell out of cultivation. And by devastating the prosperous monasteries, which owned up to one-third of the best, and best-managed, land in Britain and elsewhere, epidemic disease sent economies into sharp contraction.

Yet the demographic devastation would come to have a beneficial side for survivors, as enterprising serfs, peasants, and small tenant farmers, seeing the financial problems suddenly faced by many of their manorial lords, struck good business deals with them. The shrewd Norfolk peasant Clement Paston senior was one of them. Clement appears to have come to an agreement with his lord that gave him greater independence as a tenant farmer and enabled him to grow rich. His sons and grandsons entered the legal profession, went to Cambridge University, and joined the landed gentry, and then the aristocracy. There is no suggestion that the Pastons were unique in their rapid upward mobility. Their claim to fame lies in the large body of letters and family papers which have fortunately survived and are now easily available as *The Paston Letters*.

It was only slowly, and after the plague had abated somewhat, that the population began to rise above the pre-1347 levels. And in addition to the epidemics mentioned above, the Tudor age of the sixteenth century was hit by a seemingly new epidemic: the "Sweating Sickness". Perhaps because of its impact upon the aristocracy and upper social echelons – King Henry VIII was terrified of catching "the Sweat" – it was attributed to gluttony, excess, and luxurious living, though in reality it was probably a bacterium or a virus that mutated into human susceptibility, and then, after 120 years or so, mutated out of it again. As its name suggests, the disease was characterized by fever, delirium, and copious sweating.

Yet no matter how precise or how confused the recorded symptomology of a given outbreak might be, what cannot be denied is that epidemic diseases were as commonplace as they were dreaded. And as Thomas Nashe so poignantly expressed it in his "In Time of Pestilence", *c.* 1592, verse 1:

> *Adieu, farewell, earth's blisse;*
> *This world uncertain is;*
> *Fond of life's lustful joys;*
> *Death proves them all but toyes;*
> *None from his darts can flye;*
> *I am sick, I must dye;*
>
> *Lord, have mercy on us.*[7]

CHAPTER 8

Medicine and Surgery in High Medieval Europe, 1200–1500, Part 1

*I*n medieval Europe, medicine and surgery were seen primarily as conservative, rather than as exploratory or experimental disciplines. But it is misleading to confuse this essential conservatism and respect for classical discovery with being in any way unenterprising. For if, as most medieval academics believed, the essential groundwork of anatomy, surgery, and medical practice had been firmly established by one's wise professional ancestors, an essentially didactic approach by the present generation made good sense.

Consequently, in the medical schools of Paris, Montpellier, Toulouse, Bologna, and elsewhere that were already thriving by the fourteenth century, a teaching tradition was in place that was analogous in its approach to the learned commentary traditions which operated in the theological, literary, and scientific divisions of the curriculum. Just as astronomy produced works that were essentially amplifications of Ptolemy, and literary studies were often glosses on the works of the classical grammarians such as Priscian and Donatus, so anatomists, physiologists, and medical practitioners regarded Hippocrates, Aristotle, and Galen as their founders, with subsequent amplifications by Arabic writers and commentators such as Albucasis or Avicenna.

What, therefore, were the forces that were changing medieval European civilization and advancing its capabilities from the twelfth century onwards?

POPULATION GROWTH, PROSPERITY, AND
INNOVATION

While Western Europe had yet to learn the significance of statistics and exact demographic recording (that really started in the seventeenth century), we know from a variety of records that population was on the increase by AD 1100. Previously neglected marginal land was coming into cultivation to increase food supply, and surviving legal documents tell us that small towns were becoming bigger towns or cities, and villages were expanding into urban centres of trade. The names of traders, shopkeepers, lawyers, money-lenders – and doctors – begin to appear in the records of York, London, Antwerp, Paris, Hamburg, Venice, and numerous other "municipalities". By 1340 the Europe-wide luxury goods trade was in full swing, as Chinese silk, ivory artworks, expensive jewellery, silver and glass tableware, Indian spices, and gloriously illuminated private prayer books were appearing in well-built merchant houses where glass windows and chimney-flue heating were coming to be taken as standard. By 1450 even clocks were to be found on pictures of domestic interiors, and by 1480 one could buy a spring-powered watch, or a pair of spectacles, in somewhere like Nuremberg or Florence.

In addition to the mercantile cities, there was that rapid expansion of new and vigorous monastic institutions, especially in the wake of St Bernard of Clairvaux's Cistercian Order, an Order characterized by its combination of spiritual commitment, prayer, simplicity of living, humane duty, and commercial and managerial good sense. This movement was carried across Europe by predominantly young men, inspired by the message of the Christian gospel.

Perhaps a "climatic warm" after around AD 900 played a part in facilitating food production, enabling Europe to sustain a larger population. Incidental references in literature and in legal records suggest that growing seasons may have been getting longer and that grapes were even being grown on English monastic walls. Climate does move through natural cycles: perhaps a thinly populated Dark Ages cold was relaxing into a High Medieval warm, only to be followed, around 1350, by a gradual descent into a cooling which would hit rock bottom in the mini ice-age of the seventeenth century. This climatic "cool" could have driven hungry rats out of their natural wild habitats into human communities, bringing bubonic plague with them.

Irrespective of the climatic and other factors that may have occasioned the High Medieval warm, however, what is beyond doubt is that it coincided with the beginnings of a distinctly European as opposed to a purely post-classical culture. Among its features were Christianity, corporate banking and finance, stable currencies, monasteries and universities, polyphonic music, self-governing civic institutions, shops, religious and secular career structures, rags to riches careers via the burgeoning professions, chivalry, vernacular literature, and, in England, the English common law and Parliament. And then there was an array of "modern", technological inventions such as clockwork, large musical organs, firearms, naturalistic oil painting, and printing. These brilliant cultural innovations not only survived, but even continued to thrive and develop during the plagues, the harvest failures, and the fifteenth-century Islamic destruction of the Byzantine Empire and all the subsequent horrors of the "Waning of the Middle Ages".

All of the above factors are important in putting into context that blossoming of scientific medicine during the High Middle Ages. The approach to medicine was "scientific" in so far as it drew upon the growing number of rediscovered classical and Arabic texts flooding into Europe and being translated into Latin by 1250 – Hippocrates, Aristotle, Galen, Avicenna, and others – which, as we saw in previous chapters, defined medicine as based upon rational principles, involving the careful observation of the patient, case histories, and dietary and manipulative cures, all founded upon an established knowledge of human and animal anatomy.

This did not in any way exclude the use of prayer, spiritual discipline, saintly invocations, and even charms from the wider therapeutic package, but it firmly wedded medicine to a public rational tradition with roots extending back into antiquity, and with shared intellectual standards from Salerno in southern Italy to St Andrews in Scotland. These rational standards were shared with

131

other classically derived disciplines, such as astronomy, geometry, music, grammar, and even architecture. Yet while some of these disciplines, such as grammar and geometry, were already mature and functionally precise by 1300, others were not. The earth-centred cosmology of Ptolemy still led to problems in calendrical and other astronomical predictions, while medicine was bedevilled by a confusing tangle of imponderables: for how could a doctor successfully treat a fever, a plague epidemic, or madness when so many maladies seemed to strike out of the blue?

TEACHING ANATOMY, CHALLENGING MYTH, AND THE STATUS OF EXPERIMENTAL KNOWLEDGE

The received wisdom intones that anatomical and physiological research was banned by the church in the Middle Ages on the ground that it was wicked to dissect dead bodies. This legendary prohibition was supposed to exist for two reasons. Firstly, dissecting human beings ruined their chance of being resurrected at the Day of Judgment. And secondly, intellectual curiosity per se was banned by the ecclesiastical authorities. Both are patently *false*, as the well-documented historical record makes clear. It is true that dissection dismembered a body; yet so did being blown to smithereens by a cannonball or being eaten, either by a shark at sea or by cannibals. Indeed, in the intellectually daring world of medieval philosophical theology, such matters were openly discussed; and if a dismembered soldier, sailor, or cannibal victim could be resurrected, then why could not a dissected person?

As for a ban on intellectual curiosity, how then does one explain the very existence of Europe's burgeoning schools and universities? Intellectual hothouses have always pushed the barriers, in medieval Europe no less than today; and outrageous new ideas came in as more and more classical texts were translated into Latin, only to be devoured by eager young minds. Certain churchmen may have issued their prohibitions, as did Stephen Tempier, Archbishop of Paris in 1277, when he banned the reading of Aristotle by the students of the Sorbonne. But the archbishop might as well have saved his breath, for the students continued to gobble up the newly translated Greek philosopher who knew everything. Only a few decades previously the great Italian scholar St Thomas Aquinas had already absorbed much Aristotelian teaching into mainstream Christian theology, where it is still to be found today.

The reason why medicine did not seem to "progress" very much in the medieval centuries was because there was a lack of fundamentally *new* factual data. All the sciences, medicine included, need fresh physical *facts* to wrestle with before advances can be made – not just intellectual speculations divorced from demonstrable reality. A cataract, a fever, or a cancerous lump in Paris in 1350 was not much different from a similar malady in Greece in 400 BC, so what could a medical man say that was especially new, or likely to lead to a fundamental rethink of pathology, physiology, or therapeutics? It was much easier to come out with something new and outrageous in a purely intellectual literary or philosophical discipline than it was in a hard-fact discipline such as medicine. Medicine, like the other sciences, advanced to pastures new on the strength of fresh data which in some way challenged and demanded a rethink of the accepted norms. And over the last 500 years or so, this has tended to come about as a result of some kind of experiment, instrument, or technique that had not been available previously.

In the fourteenth century, however, experimentally derived knowledge had a distinctly dubious status, especially if – unlike optical physics – it was obtained by non-mathematical means. Optics was based on straight lines and precise refraction and reflection angles measured against a 360° scale. In 1350, no such verifiable mathematical criteria could be applied in medicine or physiology, where everything depended on the doctor's very individual previous experiences which enabled him to compare symptoms that were generally *visual*.

A major reason why experiments tended to be regarded with scepticism in medieval times had nothing to do with superstition or suppression, but was because of the often unpredictable nature of experimental results. Working with the pharmacologically impure drugs and chemicals of the day, a given quantity of opium – often of unknown biochemical strength – might simply make one patient drowsy, then kill another. Likewise, how could an experiment performed upon an animal have any credible parallel with one performed upon a human?

Not only did medicine lack easily verifiable or measurable criteria, but many medieval – and classical – scholars even doubted the fundamental value of sensory or physical knowledge in the first place. Was not *truth* only attainable through disciplined intellectual procedures, such as logical deduction and axioms based on first principles, rather than through variable and messy physical objects such as human and animal organs, or chemicals, objects that could display baffling variations even between specimens of the same type? Logic, reason, and deduction, to the contrary, being attributes of humankind's divine soul, displayed no such confusions; and the correct thinking procedure could always be relied upon to give the correct answer.

What science needed to advance, therefore, was the shock of unparalleled, new physical evidences that challenged accepted ideas and demanded new systematic tests to verify the truth status of such evidences. Only in this way could an experimental method be built up. This process started with the great geographical discoveries of the late fifteenth and sixteenth centuries, followed by the breath-taking finds made with the telescope after 1609, all of which conspired to change old ways of thinking to incorporate the hitherto unimagined continents and oceans, the new stars and worlds, and the microscopic images of flies and fleas that made them appear as big and anatomically detailed as would a lion or an elephant to the naked eye. It demanded an influx of new, quite unfamiliar, data of which Hippocrates, Aristotle, and Galen had been wholly ignorant, yet of which the men of the fifteenth, sixteenth, and seventeenth centuries would be compelled to make sense. And this had to be done across the entire gamut of natural knowledge, from astronomy and geography to cardiovascular physiology.

This need to wrestle with strange new physical facts was not yet a serious problem for medieval people. Consequently, the ancients

134

could still be revered as the eternal touchstones of their various disciplines, non-mathematically demonstrable experimental knowledge could be regarded as no more than conjectural, while *truth* continued to be associated with the disciplined intellect rather than with the eyes or the fingers. Yet in a subject so practically and pragmatically driven as medicine, and in which so many variables might come into play, ancestral truisms could be challenged by the appearance of new cases and by the experiences of observant doctors, as thirteenth- and fourteenth-century academic discussions about the "laudable" or poisonous nature of wound pus make clear.[1]

PUS: LAUDABLE OR A LIABILITY?

As we shall see in the next chapter, the illustrious Guy de Chauliac hedged his bets regarding the "laudable" nature or otherwise of pus in wounds in his deeply influential *Chirurgia Magna* or "Great Surgery" of 1363. But there was no shortage of experienced surgeons, even of the earlier generation, whose writings make it clear that they believed pus to be a bad thing and best avoided. Well over a century before Guy rose to medical eminence, the early thirteenth-century surgeon Hugo (or Hugh) Borgognoni of Lucca was arguing against it. Around 1220, clearly on the basis of extensive practical experience as a physician and surgeon, Hugo was arguing for a simple, clean treatment of surgical wounds. Instead of inserting various salves (of which Guy also disapproved) into a fresh wound, invariably inadvertently causing infection and hence pus formation, Hugo advocated simply cleaning out the wound with wine and then wrapping it up in a clean bandage. And the man whom some sources suggest may have been Hugo's son, and was certainly his pupil, Theodoric of Lucca, further advocated the "first intention" or pus-free approach to wound management in Bologna. We will discuss Theodoric in more detail in the following section.

Then there was the illustrious Bologna surgeon William of Saliceto, who died in 1277. In his own *Chirurgia* of *c.* 1275, William was an unequivocal advocate of what would be styled "hygienic" methods. William argued against the Arabic practice of cauterizing wounds, for while we now know that heat kills germs, the uninformed use of a hot iron on a wound could cause horrible burns, which could easily become susceptible to infection from unsterile fingers, instruments, and dressings. Practical experience indicated that minimal interference in a wound was the best policy. Simply cleanse a wound with wine, or a mixture of rose water and beaten egg-white, and let it heal naturally.

The man who is often seen as the founder of French surgery, the Norman Henri de Mondeville, won a reputation for his skill as a manager of wounds. This graduate of Paris and Montpellier universities, then student of Theodoric in Bologna, went on to become surgeon to King Philippe the Fair of France. Henri's ground-breaking *Cyrurgia* (1312) likewise proposed a clean and simple wound regimen, with no nasty ointments and wound salves aimed at provoking suppurating pus, but just letting nature take its course as well as nature could, in an age still innocent of the existence of germs.

The fourteenth-century English surgeon and older contemporary of the poet Geoffrey Chaucer, John of Arderne, also seems to have favoured a clean and simple treatment of wounds, and we will say more of John below.

THEODORIC BORGOGNONI OF LUCCA: SURGEON,
HYGIENIST, FRIAR, AND BISHOP

At Bologna, Hugo of Lucca had two students who themselves rose to eminence in the history of medicine and surgery. One was the above-mentioned Theodoric Borgognoni; the other was Bruno da Longoburgo.

Bruno's own mid thirteenth-century treatise on surgery contained several ideas later taken up in Theodoric's own and more famous *Cyrurgia* ("Surgery"), written some years later. This does not necessarily imply, however, that Theodoric was deliberately plagiarizing Bruno, for the two men, after all, had shared the same teacher and both epitomized the already established Bolognese tradition of rigorous scholarship, clinical curiosity, and thinking for oneself if what the ancients said seemed, from observed fact, to be wrong. And both men agreed that pus was neither "laudable" nor praiseworthy, but was what we today would call a "surgical complication", or an impediment to the natural healing of a wound.

Theodoric's career displayed several parallels to that of Guy de Chauliac in the following century. Both Theodoric and Guy were not only eminent surgical and medical men, with impeccable academic credentials, but both were also churchmen, and Theodoric may have joined the Dominican Order as a friar while he was still a student at Bologna. This would have been at a very early stage in the history of that Order, for Dominic had only just begun to assemble his original band of followers and had obtained papal sanction in 1216. The Dominicans were to share the joint ideals of the pursuit of truth through learning and academic rigour and a life of poverty, simplicity, prayer, and service to the poor, following in the footsteps of Christ. The new universities of Europe, from Oxford to Bologna, were often energized by the radical and controversial presence of Dominican friars on their faculties. Their rigour and radicalism was shared by their brother mendicant friars, the Franciscans. This radicalism did *not* aim to challenge the basic truths of medicine, learning, or the faith, but rather to confront academic ossification and smugness, and the corruption and wealth of the church.

Theodoric, born around 1205, became physician to Pope Innocent IV in the 1240s, probably while still in his late thirties. He rose high in the church, becoming a friar-bishop, serving the cathedral church of Cervia, near Ravenna, from 1266 until his death aged about ninety in 1296. Theodoric's dual career path puts paid to the myth that the medieval church somehow anathematized surgery, the cutting of the living with the inevitable spilling of blood, and the dissection of the dead.

We saw above that Theodoric advocated the clean, pus-free treatment of wounds, but his hefty mid thirteenth-century *Cyrurgia* covered many other topics as well. Like many of the "new breed" of university-trained surgeons, Theodoric saw the advantage of wine

137

as a cleansing agent in surgery; for while wine is not an "antiseptic" in the modern sense, it was safer to use on wounds (and to drink) than river water. Theodoric, like several contemporaries, saw the benefits of soaking bandages in wine before applying them to a wound.

Cyrurgia also deals with injuries to the abdomen, emphasizing once more the importance of cleanliness and attempting to avoid what a post-bacterial age would define as "cross-contamination" when dealing with faecal matter in abdominal injuries. Similarly, Theodoric discusses chest and head wounds, and how best to close them, along with possible treatments for some external cancers.

Theodoric is not saying anything especially new here. Celsus had discussed how best to deal with abdominal and thoracic injuries in *c.* AD 30, as a string of other pre-antiseptics surgical writers would do, extending into the nineteenth century. Even the Edwin Smith Papyrus of 1500 BC had discussed treating a serious head wound in which the brain coatings had been exposed. Yet what was happening in thirteenth- and fourteenth-century Europe was that surgical discourse was coming to be wedded to the academic life of the great universities and beginning to lay the foundations for a "medical profession" possessing formal training and qualifying procedures and internationally recognized ethical codes.

THE FIRST ACADEMIC MEDICAL SCHOOLS: A EUROPEAN INNOVATION

Though the ancient world had illustrious places of medical and scientific learning, such as in Alexandria and then Baghdad in medieval Arabia, the self-conscious, academic medical school was really an invention of post-twelfth-century Christian Europe. Medicine and surgery soon came to be embedded within the wider arts curriculum of Paris, Bologna, Montpellier, Oxford, and other universities to take their place alongside philosophy, theology, law, astronomy, rhetoric, and grammar, as truly *learned* disciplines. Away from this intellectually elite world of scholars, many perfectly

138

respectable local healers learned their medicine and surgery through a combination of apprenticeship and practical experience, but the universities nonetheless set a new standard towards which to aspire.

This High Medieval marriage between medicine and surgery, especially in the great Italian and French universities, sadly was not sustained. As the centuries passed, while medical training remained firmly within the universities, surgery – in spite of its continuing technical advancement – migrated away from academia, to be taught through apprenticeship. This led to an odd division in the medical profession, for while the leading surgeons were Latin-literate and cultured men, often learned in practical medicine as well as being masterly operators, they were no longer awarded degrees – a situation which continued, especially in the British Isles, well into the nineteenth century.

Yet our contemporary use of academic and professional titles dates from this, especially thirteenth-century, period. After a set period of training – irrespective of his later specialism – the young (always male) student would proceed to the degrees of Bachelor, then Master, of Arts. Then, if he had the time and the resources, he could proceed to a Doctorate: in Divinity, Civil Laws, or Medicine. The MA was rapidly recognized as the qualification required to teach in Bologna, Paris, or Oxford, and once this *gradus*, or degree, had been obtained, a man was deemed qualified to teach, in Latin, in any university in Europe, be he monk, friar, priest, or layman.

The title "doctor", while more correctly the designation of a high-level specialist teacher (but *not*, unlike today, a researcher), would come to replace the Latin *physicus*, or healer, leading to the modern colloquialism of "doctor" for any medical professional. It was also in the medieval European universities that our modern titles of academic rank were first used. The "Professor" would profess his subject with magisterial authority. His junior, the "Reader", would read out the appropriate piece of text – medical, legal, theological – upon which the professor would lecture. The "Demonstrator" (sometimes *Ostensor*, or "one who shows"), in the medical schools, would show the relevant dissected organs to the audience. Nowadays, the title "demonstrator" usually signifies a junior scientific or medical researcher at the beginning of his or her career.

This new academic professionalism even came to acquire its own dress code, so that one could spot at a glance exactly who was what. Bachelors and Masters would both wear black gowns of different sartorial cut, along with hoods of designated colours. Doctors, however, would be resplendent in gowns of scarlet, with different-coloured sleeves and other trimmings to indicate whether the wearer was a Doctor of Divinity, Medicine, or Law. Many of Europe's older universities still cherish their ancient regalia, worn on special or ceremonial occasions. They are still very much in evidence in Oxford. Indeed, I recall, a few years ago, at an academic-dress "Encaenia" midsummer garden party, seeing a college friend who was already a distinguished senior consultant physician in the university's John Radcliffe Hospital and Medical School rejoicing in the scarlet robes of his newly conferred "DM" or Doctor of Medicine degree, which is the highest medical qualification conferred by Oxford University.

MONDINO DE LIUZZI OF BOLOGNA AND HIS *ANATHOMIA*

We saw above how, during the thirteenth century, Europe was developing a vigorous medical and surgical tradition that was growing up as a natural part of the great new universities of the High Middle Ages. Mondino de Liuzzi (Mundinus in Latin), whose death in his mid 50s in 1326 made his life a relatively short one compared with the 90-odd years of Theodoric Borgognoni of Lucca, was nonetheless destined to become the most influential anatomist of late medieval and early Renaissance Europe.

9. Anatomical dissection, *c.* 1310. Note that the professor speaks from his great "chair", with a text, perhaps Galen's *De Usu Partium*, before him. His "Ostensor" or Demonstrator opens the cadaver and displays the required organs. (Traced from Mondino's *Anathomia* (first printed edition, Venice, 1493) by A. Chapman.)

141

This came about through his *Anathomia Corporis Humani* ("Anatomy of the Human Body") of *c*. 1316, which became the definitive atlas of the human body down to that of Andreas Vesalius in 1543.

Unlike the surgical treatise of Theodoric, or even that of Guy de Chauliac in 1363, Mondino's *Anathomia* was not primarily intended to be a book of therapies or surgical procedures. Rather, it discoursed on how best to dissect a human body, demonstrated where all the organs lay, and indicated how they might interrelate in life. Like Galen and Celsus in antiquity, Mondino regarded a coherent understanding of the human body not simply as a curious piece of academic learning, but as the foundation of rational medical practice. His dissections were intended for *human* cadavers, rather than those of animals such as pigs and monkeys from which human parallels might then be inferred. None the less, he retained many of the old errors deriving from veterinary anatomy, such as the liver possessing five lobes and the heart three ventricles. Yet Mondino was the first, or at least the first major, public anatomist to teach directly from human cadavers, a procedure authorized by the Pope at the beginning of the fourteenth century.

Very much within the prevailing taxonomic and commentary approach to learning that one finds in the other disciplines of the medieval universities, Mondino classified the parts of the body into *three*. Taxonomic "trinities" abounded in medieval thinking: think of the animal, vegetable, and mineral divisions of nature and in science; the Bile, Phlegm, and Blood trinity of vital bodily juices; body, mind, and spirit; and that trinity of trinities, the nine spheres of the planets. In Mondino's case, the three body parts were the abdomen, thorax, and skull, which related to the three "spirits" of Galenic physiology: the natural, digestive spirit in the stomach; the warm, life-imparting vital spirit in the heart; and the intellective animal (*anima* = "soul") spirit in the brain.

This hierarchy of body parts would have been seen as possessing a logical parallel to the hierarchies within human society, within the animal, vegetable, and mineral kingdoms (we might recall the *Bestiaries*, *Herbaria*, and *Lapidaria* mentioned in Chapter 5), and within the angelic realm. Aligned with the medieval ranks of medical status mentioned above, the anatomical lecture would become a piece of threefold theatre: with the professor, the reader, and the demonstrator (or *ostensor*).

When commencing a dissection, Mondino tells us, one should begin with the abdomen, making a vertical incision extending from the pectoral muscle above down through the abdomen below, then from side to side across the navel, allowing the four quarters of flesh to be folded back or cut off, to reveal the muscular wrapping that enfolded the guts. He advocated starting here because the contents of the abdomen putrefy before all the other body parts, and Mondino's corpses were not likely to be fresh. Many were those of criminals who had probably been required by law to hang on a gibbet for several hours, perhaps in the warm Italian sunshine, before being handed over to the medical school. So public academic dissections came increasingly to be held in winter, although in Italy that was no guarantee of freshness. On the whole, the corpses handed over to the doctors, in Bologna and other medical schools, tended not to be those of criminals executed for some run-of-the-mill crime. Rather, the perceived humiliation of being dissected before an audience was often reserved for those who had committed especially loathsome crimes, such as killing their parents or masters, or being traitors.

Within the thorax, the next part of the body to be dissected, one found the lungs and the heart. As in other parts of the *Anathomia*, one encounters beautifully detailed descriptions, in this case of the vena cava and the pulmonary artery and vein and their relation to the lungs, and of some of the heart valves. But, in this pre-blood-circulation stage of medical understanding, Mondino misunderstood the function of many of these structures in life. He followed Galen in the belief that blood could pass from the right to the left ventricle of the heart, via a supposed middle "ventricle" within the thick muscular wall – the septum – that divides the heart. Mondino also had only an imperfect grasp of the heart–lung functional relationship: of how respiration and air-supply were related to cardiac function.

This is not to condemn Mondino – or Galen, or any other early anatomist or physiologist – for failing to grasp how the heart, lungs, and blood system actually operated. Before the advent of William Harvey's blood circulation model of 1628, the inside of the thorax of a dead human or animal looked baffling: all tubes, bends, chambers, and soft bags. Trying to work out exactly how the whole system worked in life, and which bag or valve sent blood, air, or humoral juices down which tube, was by no means straightforward. To untangle all of that, one needed the insights of an experimental approach to science, which would not develop for over two centuries after the *Anathomia* was published, as we shall see in later chapters.

Mondino's third division of the body, to which the dissector must come last, was the head and the brain. He followed what we saw in Chapter 6 as the acknowledged medieval model of brain function, namely, the three "cells" or chambers. The frontal or anterior cell dealt with incoming data from the eyes, ears, and other sensory organs, conveyed via the cranial and other nerves. Then the incoming sensory data were believed to pass to the middle "cell", where they somehow interacted to produce *imaginativa*: imagination, or creative thinking. Finally, sensory and "imaginative" experiences were passed to the posterior or rear chamber, to form memory. The neural structure called the choroid plexus, sometimes referred to in medieval brain anatomy as the *vermis* or "worm", was believed to be a pathway facilitating movement between the cells, resulting in a sort of "feedback" process, whereby senses, imaginative cogitations, and memories interacted with each other to constitute mental life.

In addition to the three main body cavities, Mondino discussed the muscles, nerves, and blood vessels. The *Anathomia*, indeed, was a fascinating mixture of acute personal observation and Hippocratic, Aristotelian, Galenic, and Arabic "truisms". These truisms included, the location of the lens in the middle of the eye, and the placing of a *rete mirabile* network of blood vessels at the base of the brain. They are factually incorrect, many of them deriving – such as the *rete mirabile* – from the earlier tradition of reading veterinary into human anatomy.

Yet the *Anathomia* abounds with practical tips which give us a glimpse of the working methods of fourteenth-century anatomists. When studying the muscular and skeletal structure of the body, for example, don't use a fresh corpse, which will be already decomposing, and probably getting a bit sloppy – and stinking! Rather, leave an eviscerated corpse out in the warm sunshine for a while and let it dry out. You will then find that the desiccated cadaver, having lost much of its subcutaneous moisture, is much easier to work with, enabling you to trace the musculature to its attachment to the bones, until you finally work your way down to the skeleton itself.

The impact of Mondino's *Anathomia* upon the development of European medicine in the late Middle Ages and Renaissance was huge. It not only presented a coherent system and method for anatomy; it also emphasized the importance of using *human* cadavers for the study of *human* anatomy. While the well-known woodcut from the 1493 printed edition shows the professor on his *sedile*, or great seat, with a classical text open before him, and a young man – the demonstrator or *ostensor* – at the table, the eviscerated corpse upon it is unmistakably that of a human being; a tall bearded man with long hair, so it appears. The young *ostensor* has his hands in the cadaver's guts, while his knife lies alongside. Magisterial as the formal anatomy lecture was, it is clear, from the detail in his book, that Mondino was fully familiar with all the procedures of dissection at first hand. Mondino's pervasive influence is evident from the way in which more experimentally driven anatomists took a delight in pointing out his perpetuation of ancient errors.

In this chapter we have seen the vigorous growth of formal academic medicine and surgery in thirteenth- and fourteenth-century Europe, and how that tradition grew up within the intellectual hothouse world of Europe's great new universities, most notably in Bologna, Montpellier, and Paris, and soon thereafter in that powerhouse of medicine and anatomy during the Renaissance, the Venetian Republic's University of Padua. In that world where bright young medical students lived and worked cheek by jowl with academic brethren studying law, theology, mathematics, astronomy,

145

classical literature, and philosophy, it may not seem surprising that they saw their anatomy and physiology as part of a wider cultural tradition. That tradition was founded in the world and wisdom of Greece and Rome, with supplementary insights drawn from Arabic and contemporary Western writers. It would be from this tradition that the modern global medical profession would be born. And one of its most influential teachers was a fourteenth-century Frenchman: Guy de Chauliac.

CHAPTER 9

Medicine and Surgery in High Medieval Europe, 1200–1500, Part 2

*M*óndino de Liuzzi's *Anathomia*, as we saw above, had an enormous impact upon late medieval anatomy. Half a century later, it would be complemented by a treatise that would become the most influential medical text of the Middle Ages: Guy (or Guido) de Chauliac's *Inventarium, sive Chirurgia Magna* ("The Inventory/Catalogue/Encyclopedia, or the Great Surgery") of 1363.

A SCIENTIFIC PHYSICIAN AT THE PAPAL COURT IN AVIGNON

In many ways, Guy was a characteristic product of that culture and society outlined above: a truly "rags to riches" young man, from Chaulhac in the lush and beautiful lands of the Auvergne, in south-central France. He is said to have been born around 1298–1300 into a humble family, probably of peasant farmers, and was fortunate because the local lord of the manor spotted his unusual potential and sent him to school. From there, Guy went on to study in several of those unique medieval institutions – universities – in Toulouse, Bologna, Montpellier, and Paris. By 1320, he held the degree of Master of Surgery from the illustrious University of Paris; but the truly formative influence upon his future career was

Nicola Bertuccio, from whom he learned anatomy and surgery at the illustrious, subsequently papal, University of Bologna.

Guy then seems to have made a career choice which, while perhaps seeming odd to us today, would have appeared eminently natural to a fourteenth-century scholarship boy: he entered holy orders in the church and also became a medical doctor, thereby combining the two noble Christian callings of cure of body and cure of soul. The exact status of his orders is not entirely clear, there being several ranks in the medieval church, but it is likely that (like the Polish physician–astronomer Nicholas Copernicus two centuries later) Guy held the orders of a deacon. These orders would later enable him to hold canonries at St Just and Rheims. Guy then seems to have practised medicine at Lyon, before being invited by Pope Clement VI to become papal doctor at Avignon. And when Clement died in 1352, he would continue in that office under his successors, Innocent VI and Urban V, until his own death in 1368. From such a career, under three pontiffs covering a span of around 25 years, one can only assume that Guy was not only thoroughly competent as a physician, but also had a good general and bedside manner, and was considered thoroughly learned and theologically orthodox.

Why were Guy and his pontifical employers and patrons in Avignon in south-east France, near the Italian border, and not in Rome itself? This circumstance derives from one of the less edifying aspects of the medieval papacy: one that would add fuel to the simmering fires of discontent which would feed the followers of John Wycliffe in later fourteenth-century England, Jan Huss in Bohemia and Holland, and finally Martin Luther and his fellow reformers after 1517. This was the so-called "Babylonish Captivity" of the church by the French.

Following a dispute between King Philip IV of France and popes Boniface VIII and XI, a Frenchman was elected pope, taking the papal title of Clement VI in 1305. Relations between the powerful kings of France and the Italian cardinals were rarely good, as the church was a powerful international political as well as a spiritual corporation. Consequently, Italian, French, and Spanish dynastic politics often decided the outcome of papal elections and the patronage prospects of the respective national cardinal factions in the Vatican.

Withdrawing from Rome altogether in 1309, the French Clement VI established his court at Avignon. Safely over the French border, yet near enough to Italian territory to be influential in Rome, the new French papacy was deemed to have been taken into "captivity", occasioning the Italian poet Petrarch and others to draw a parallel with the Old Testament Jews being taken captive both by Sennacherib and then Nebuchadnezzar, kings of Babylon. And, as one might expect, a French pope wanted a French doctor, leading, via various convolutions of circumstance, to Clement's wise patronage of the man who would later become medieval Europe's most illustrious physician and surgeon. (The "Babylonish Captivity", incidentally, would drag on until 1370, although further complications set in, to be finally healed by the Council of Constance, between 1414 and 1418.)

Yet what was Guy's significance when it came to systematically describing and healing the human body?

CHIRURGIA MAGNA, OR THE "GREAT SURGERY": A
MEDICAL ENCYCLOPEDIA FOR FUTURE AGES

Guy's great masterpiece, extending across seven "Books", "Tractates", or chapters, became the enduring medical *vade mecum* for late medieval and Renaissance Europe, the Latin text being destined to be translated into most of the languages of Europe – including English. It was also one of the first great medical books to be printed: in a French edition, in Lyon, in 1478. A Latin edition came from Venice in 1498. It is likely that its *ex cathedra* status was enhanced by the fact that its author did not claim it to be a book of new discoveries, so much as a meticulously presented compilation of the best and wisest medical thought and experience of the centuries. A wise, learned, meticulous display of scholarship in the noblest tradition of that quintessential medical scholarly genre: the learned commentary. *Chirurgia Magna* was a logical work of scientific intellectual integration: a sort of medical equivalent to the *Summa Theologica* of St Thomas Aquinas. It also encapsulated a lifetime of learning and medical practice for Guy himself when it was finally completed in 1363, and respectfully dedicated to his long-deceased teachers at the great French and Italian universities where he had once studied and whose qualifications he had received.

We might be puzzled that a work of such conspicuous conservatism should have been so praised and treasured across the following centuries, and still be cited in the seventeenth century. We must remember that before the great discoveries of the seventeenth century, a truly wise and learned doctor was deemed to be a respectful and a conservative one. Illness was seen as an innately mysterious process, as eternal as the seasons; and while some maladies seemed to come and go, such as the arrival of the terrifying new bubonic plague, and, by 1400, the waning of leprosy, illness, like death itself, was an enduring fact of life.

It was the doctor's job to give ease and comfort to the sick more often than to effect actual cures, for most serious maladies were simply incurable. Drugs and diets might re-balance the humours, and the planets might move into more auspicious "aspects" in the patient's horoscope, but ultimately life and death were in the hands of God. The physician aspired to cure or ease the body, and the priest gave comfort and a safe passage to the departing soul. If, as in Guy's case, both offices were held by the same man, so much the better.

If you were looking for *innovation*, however, you went *not* to an MD of one of Europe's great universities, but to a *quack*. He would sell you a bottle containing a secret preparation that would drive out all your aches and pains, "sweat away" cancerous lumps, dissolve your bladder stones, cure your blindness, deafness, or madness, and, as an added bonus, enable you to live to be at least as old as the biblical Methuselah! (As we shall see in Chapter 19.)

But given that illness was an eternal bane, how could a doctor in 1363 presume to be wiser than Hippocrates or Galen? Consequently, if the patient had any sense or education, and could afford the fee, he or she would summon a conservative doctor to the bedside. And if that doctor happened to be of the stamp of Guy de Chauliac, and even have studied the *Chirurgia Magna*, then all to the good.

Everything discussed so far has related to *medicine*. So how does surgery itself fit into Guy's masterpiece? The received "wisdom" tells us that in the Middle Ages, and for some time thereafter, surgery was a barbarous business, performed by ignorant men of no education or intellect, who were only one step above pig-gelders and butchers. Like so many popularly held myths about our scientific past, however, it is untrue. While in many a village Henri the butcher might oblige a sufferer by calling around with his sharp knife to make war upon a nasty tumour or abscess after a day in the slaughterhouse, this was not the case in the great urban centres. Guy is quite explicit in *Chirurgia Magna* Book I that surgery at its best was considered – even in 1363 – a *learned* discipline, demanding a man of education for its correct performance. Guy himself held a Mastership in Surgery from Paris – a formal academic qualification.

For Guy, surgery ranked as a liberal art. Yes, it involved hands-on work, and not just the oral, visual, and gentle touch examination of the pure physician prior to writing his prescription. Medieval Europe acknowledged other hands-on liberal arts that occupied a quite different status from that of simple handicraft. Think of music, for example, where through viols, flutes, and organs the soul was lifted aloft through the skilled hands of an accomplished performer; or a great architect or visual artist, whose works engaged with the profoundest truths. So why should surgery be any different? In making this point, Guy says that any surgeon who takes a knife to the human body without a prior training in anatomy is like a blind man attacking a piece of timber with sharp tools and expecting to produce a beautiful carving![1]

Guy begins *Chirurgia Magna* with a course of descriptive anatomy, clearly based on that of Galen. He is quite open about his lack of originality, and it has been reckoned that he made no fewer than 3,299 references to sundry classical and Arab medical writers, including 890 to Galen alone.[2] This may not be the way to write a benchmark medical treatise today, but it proclaimed to the world of 1363 that Guy knew thoroughly what he was talking about.

151

Essentially, however, Guy was a book-trained doctor and surgeon rather than a hands-on dissector, as he makes clear. For some 1,200 years previously, the very hands-on Galen had on numerous occasions been up to his elbows in blood: from the Pergamum gladiators, from apes and pigs, and maybe from human dissections in Smyrna, Alexandria, and Rome. Galen, quite simply, had seen it all, done it all, and written it down meticulously for future generations of pagans, Jews, Christians, and Arabs to read. As no "modern" could reasonably expect to improve upon antiquity's greatest anatomist and surgeon, why not simply follow behind at a respectful distance? This was a very different attitude from that which Vesalius, Harvey, and Willis would advocate in the sixteenth and seventeenth centuries, but undoubtedly wise in 1363.

It is highly likely, however, that Guy would have witnessed at first hand dissections and post-mortems in the great medical schools which he had attended, but what he may have done hands-on in the line of dissection, as opposed to having witnessed, is a moot point. Occasional public dissections, usually of executed criminals, were part of a medical education by Guy's time, although they were more in the tradition of formal set-piece medical "theatre", or demonstrative anatomies, than of the *research* dissection of 1600 or 1650. In the Mondino tradition, before *c.* 1550, the relevant sections of Galen's *De Usu Partium* ("On the Function of the Parts") or other works would be read out to the listening students, the cadaver opened by a demonstrator, and the organs removed and displayed to the audience.

Guy does describe several of the "standard" operations of the period, however, such as attempts to close anal fistulae and reduce hernias, operations as old as surgery itself. But how much actual operating Guy did across his career is hard to ascertain, though one suspects that it was not a great deal. For Guy's real talent, as comes over clearly in his *Chirurgia Magna,* is that of the teacher, not the working surgeon. And like all great teachers, Guy displays a love of his subject, a wide-ranging power of synthesis and organization, and an ability to impart his knowledge in a clear and precise fashion. The first word of the title, *Inventarium,* which one might translate as an inventory, catalogue, or encyclopedia, says it all. For *Inventarium, sive Chirurgia Magna* is both a showcase of medico-surgical scholarship and a valuable book of reference dealing with diet, bloodletting, "regimen" or healthy lifestyle, and wider patient management.

When discussing surgical and other wounds, however, Guy does devote space to the perennial problem of infection, and the natural tendency of wounds to suppurate and generate pus – a concern which he shared with the thirteenth- and fourteenth-century medical writers of Bologna and elsewhere, as we saw in Chapter 8. Was this pus a good or a bad thing? In spite of the earlier medieval debate, opinions still differed between surgical and medical writers. Did a wound healing naturally close up by "first intention", or did the healing process really begin with the oozing of pus, which Galenic wisdom said cleansed away the poisons? The problem would exercise the minds of medical men down the centuries to come, until first John Hunter in the late eighteenth century and then, decisively, Joseph Lister in 1865–67 showed pus to be an avoidable surgical complication rather than a benign natural process. (We shall return to Hunter and Lister in Chapters 17 and 24).

Many medical men still abided by the Latinized Galenic tag *pus bonum et laudabile* ("pus is good and praiseworthy"). Pus was not merely good; it should be encouraged. Were not all wounds infected by their very nature, and was not pus the natural vehicle by which they were cleansed? (This "cleansing" would later be encouraged by inserting "drains" into the wound, such as stitches or bits of cloth, which would form a pathway for the pus. The wound was often left partially open, and only closed up gradually, so as to prevent the oozing pus from getting trapped deep inside. It was the unsterile hands, instruments, and pieces of fabric used by the surgeon that caused the infection which encouraged the pus!)

Wounds apart, however, one aspect of Guy's *Chirurgia Magna* gives us an insight into his originality as an astute clinician.

GUY DE CHAULIAC: VICTIM, SURVIVOR, AND
STUDENT OF THE BUBONIC PLAGUE

In 1348 the bubonic plague epidemic reached Avignon, and Guy, the papal physician, caught it. Perhaps because there were people willing to look after him in his raging fever, he passed the five-day crisis point, lived to tell the tale, and in *Chirurgia Magna* left an account of the disease from the perspective of both a doctor and a patient.

He tells us that when the epidemic first struck, victims displayed the characteristic symptoms of painful neck, armpit, and groin tumours, along those bodily pathways which seventeenth-century anatomists would discover and identify as the *lymphatic* system (from the Latin *lympha*, "water"), which is now known to be an adjunct of the wider blood circulatory system.

After about five days, the patient either died – the more likely outcome – or came through the crisis and recovered, with the lymph buboes subsiding. As we have seen, these were classic bubonic plague symptoms. Yet as the disease continued in the population, new sufferers were noted to be displaying a quite different set of symptoms. In this case, it was the lungs and respiratory pathways that appeared to be affected, for they began to cough violently and spit blood. This was the pneumonic or pulmonary strain of the disease, which was much more deadly, invariably carrying the patient off within three days, and it had a much lower survival rate. It would seem likely, therefore, that Guy's own attack of plague was of the milder bubonic variety.[3]

SO WAS MEDIEVAL SURGERY BARBARIC?

As we have seen, then, medieval Europe had a rich and noble tradition of surgery, with men such as Guy and his colleagues setting the highest standards of academic learning, meticulousness, and skill, allied with a sense of duty to do the best they could for their patients. This approach to surgery saw itself as residing within a clear academic context, intimately related to medicine, based upon dissection and organized teaching, and grounded in wider learned culture.

Inevitably, in an age without antiseptics or effective anaesthesia, surgery was a ghastly affair, causing distress to both patient and surgeon. Hence, it was only resorted to when gentler medical techniques had failed. Guy describes a wide variety of procedures in *Chirurgia Magna*. These include the treatment of a diverse assortment of wounds, for which he recommended suturing, or stitching, and supportive bandaging. Also, contrary in some respects to William of Saliceto whom we saw in Chapter 9, Guy advised the *cautious* use of the cautery, or hot iron or knife, on some wounds, following certain Arabic writers.

Tracheotomy is also discussed for the treatment of throat constrictions which could otherwise lead to death, along with the rules for venesection, or bloodletting. Nowadays we may regard deliberately bleeding a patient as foolish, but within the physiology of Aristotle and Galen it made eminently good sense scientifically. If a raging fever were deemed to have been caused by overheated blood, or a badly inflamed hand laceration by a build-up of poisoned blood in the arm, then it made good sense to open a vein and let the offending blood out. Besides (as indicated in Chapter 3), phlebotomy could even make a patient feel better and more relaxed.

In addition to the procedures mentioned above, there was that time-honoured repertoire of operations that went back to antiquity: operations mentioned by Albucasis and other medieval Arabs and Europeans, which *could* work, if one were lucky enough to survive. Think of the lithotomy procedure, relegated to specialist operators in the Hippocratic Oath, where the patient's bladder was opened to remove calculi, or urinary tract stones. Similarly, couching for cataracts to treat blindness had a good 2,000-year history behind it by the time of Guy de Chauliac, as had treatments for fistulae, hernias, running sores, broken noses, broken bones, nasal polyps, and head injuries, and the amputation of gangrenous limbs – especially risky before the discovery of blood circulation in 1628 – and even of cancerous breasts. Then in addition to these "elective" or non-injury-related operations, military surgery had its own repertoire of techniques and therapies, as we shall see in subsequent chapters.

Out of bounds for *any* surgeon before the late nineteenth century, however, were non-emergency operations on the abdominal, thoracic, or cranial cavities – other, that is, than the time-honoured "trephination" procedure, or cutting a hole in the skull to release pressure, usually following a head injury. But go poking around in the guts, liver, lungs, or brain when there was no prior injury, before *c.* 1880, and you knew the outcome would be infection-related complications, almost certainly to be followed by a funeral.

Yet surgeons did their best to be as gentle and humane as the ghastly necessities of their liberal art allowed, and Guy was not alone when, in *Chirurgia Magna*, he discussed soporifics and sleeping agents. Various "soporifics" were used, usually based upon opium: a drug indigenous to the Far East, but known in the West from the earliest times. Others too were recommended, such as henbane and mandrake (mandragora) extracts. These botanical juices would sometimes be dissolved in water, or, better still, wine (alcohol being a good solvent for organic substances), and soaked up into a sponge, which could be dried afterwards. The sponge was then reactivated with warm water or steam when needed for surgery, then placed over the patient's nose for him to inhale the contents.

I must admit to being very sceptical about accounts of "soporific sponges" and sleeping draughts for use in medieval surgery. While they might make a patient drowsy, they could not have been able to produce the deep anaesthetic sleep that was needed for serious surgery. I suspect their value lay in calming a frightened patient prior to surgery, or aiding relaxation and sleep in a traumatized post-operative patient.

The anaesthetic ineffectiveness of such soporifics is obvious from records in subsequent medical literature, especially by the surprise and delight expressed by historically literate surgeons when effective ether and chloroform anaesthetics first became available after 1846. Terror, pain, and trauma would remain the inescapable accompaniments of surgery for a full five centuries after Guy.

So while there would have been no shortage of showy, over-confident self-styled surgeons in the Middle Ages who were ignorant of the underlying anatomy of the limbs they were cutting

– Hippocrates castigated such fellows in *c.* 430 BC,[4] and humanity does not change – there was also no shortage of operators of a higher calibre and with a deeper sense of responsibility. Guy de Chauliac was not only one of them, but he drew upon the already established tradition of case histories and recorded techniques, and wrote *Chirurgia Magna* for men of a similar outlook who would enter the profession in the centuries to come. Yet while *Chirurgia Magna* became a benchmark of medieval academic medicine and surgery, and he himself enjoyed the kudos of being papal physician, Guy was by no means unique, either in his medical skill or in his sense of a doctor's duty to his patient.

Yet how could a doctor so openly traditional and conservative as Guy de Chauliac be seen as a "medical scientist"? In the context of the curative possibilities and social expectations of fourteenth-century Europe, the range of his learning and techniques constitutes a stunning display of that *scientia*, or organized knowledge, which was so prized in medieval Europe: a display further enhanced by his brilliance as a medical teacher. It is Guy the lucid, learned, systematizing medical teacher, *not* Guy the as-yet-to-be-born hospital or laboratory researcher, that had an enduring impact upon the future development of medicine. Living and practising in an age before the concept of experimentally derived new knowledge had really come to be understood, Guy represented an approach to medicine that could truly be regarded as "scientific".

In the next chapter, however, we shall see surgeons in action in a more hands-on context than in a university lecture hall.

CHAPTER 10

Prince Hal and the Surgeons: The Rise of Medical Professionalism

*A*s we saw in the previous chapters, medicine and surgery came, between *c.* 1200 and 1400, to be embedded not only within the academic curricula of the great new universities of Europe, but also within wider cultural understanding. Chaucer, for instance, includes a "physician" in his *Canterbury Tales* of *c.* 1392, and from what Chaucer says, his Canterbury physician was a classically educated gentleman and not simply one of those individuals which the Company of Barber Surgeons would castigate in 1376 as one of those "countryfolk who came to London and without training intermeddle with barbery, surgery, and the cure of maladies".[1]

Yet while both Oxford and Cambridge Universities had been thriving for a couple of centuries by Chaucer's time, neither developed medical schools after the manner of Paris, Montpellier, Bologna, and then Padua. They could, however, award medical licences and even degrees, although these would have been based upon an intimate knowledge of the classical medical writers available in Latin, such as Hippocrates, Aristotle, and Galen, along with the encyclopedic works of Arabs such as Avicenna and Albucasis. Oxford certainly possessed book-learned physicians by the time of Chaucer, an early one being the late thirteenth to early fourteenth century Nicholas of Tingewick, who combined the callings of priest, physician to King Edward I, and early teacher of medicine in Oxford.

Yet neither Oxford nor Cambridge – the only English

universities before 1826 – nor Scotland's St Andrews (1413) aspired to formal clinical training before the early nineteenth century. This did not mean, however, that none of these universities produced individuals who would dissect and experiment in a *private* capacity, as we shall see with William Harvey and Thomas Willis in subsequent chapters. What it meant was that if a person from the British Isles wanted to do systematic clinical work before *c.* 1720, it was wise for him to do "postgraduate" work on the Continent – or, after *c.* 1760, in Edinburgh, as vast numbers did, and still do.

JOHN OF ARDERNE: MASTER SURGEON OF THE AGE OF CHAUCER

Though there is no significant corpus of English learned surgical writing before the fourteenth century, it is unlikely that John of Arderne came out of nowhere. Born, it was said, in Newark-on-Trent in 1307, he appears at first to have lived and practised in Nottingham before moving to London.

Where he first became a surgeon of note, however, was when he served as a military doctor in the Hundred Years War, which raged between England and France in the fourteenth century, and stemmed from the claims of the English kings to the French throne. We know that John served in the Algeciras campaign of 1342–44, which was part of the strategy to win back Spain from the Muslims. As an experienced military surgeon at that time, he would have been among the first men in his profession to have to deal with the new horrors of gunshot wounds: wounds far more varied, nasty, and dangerous than the more straightforward puncture and slash wounds inflicted by arrows and swords. But we will return to gunpowder injuries and their impact upon surgery below.

John of Arderne was significant in a number of respects. For one thing, in that age in which battlefields were still dominated by heavily armoured men on horseback, there was a particular "occupational" injury to which such men were prone: *fistula in ano*, or a painful condition of the bottom, a condition brought about by spending much of one's life in the saddle.

Fistula is a clinical term used to describe an unnatural orifice which forms alongside a natural one and which interferes with the latter's efficient operation. The accumulated strain and jolting resulting from years in the saddle could lead to a rupture of the anal coatings near the exit point of the intestine. Consequently, an unnatural hole, or fistula, could form through the skin of the lower buttock. This could lead to bowel leakage, abscesses, and persistent pain, especially in circumstances of poor hygiene, such as on military campaigns. Fistulae could also occur as a result of sustained, chronic constipation.

This was an ancient problem and is mentioned by earlier writers; yet one suspects that it was particularly prevalent on the horse-dominated battlefields of medieval times (putting an entirely new complexion on the "Age of Chivalry"). John is noteworthy for his development of a simple, efficient, and relatively quick operation for *fistula in ano*, excising with a sharp knife the painful abscess lump and sealing the aperture with a ligature. His written account of this procedure still survives.[2]

In common with his Continental brethren, John was an advocate of hygiene and what were, by the standards of the day, clean (yet by no means sterile) procedures. As a military surgeon, he would not just have treated wounds or performed operations, he would, like his counterparts today, have been the regimental general physician. On a daily basis, he would have dealt with a wide variety of ailments, from bad teeth to alcoholic poisoning to dysentery, to men going mad under the pressures of what we now recognize as battle trauma.

John of Arderne had a deep sense of the ethical responsibility of his calling, as was enshrined in the Hippocratic Oath. He suggests, for example, that a doctor should not charge a patient a fixed sum, but set the charge in accordance with his ability to pay. Furthermore, he should treat the poor free of charge. He also spoke of opiate and henbane-based soporifics, some to be applied locally, though as was mentioned in Chapter 9, one must harbour reservations about surgically effective "pain-killers" before the development of sulphuric ether general anaesthesia after 1846. He is a fascinating figure, when seen in context. His technical skill, his learned authorship of *Practica Chirurgiae* (*c.* 1370), combined with a deep sense of the Christian duty of care, all mark him out as a truly *professional* medical man. And while John was some 33 years older than Chaucer, he predeceased the poet by only eight years, John dying in 1392 at the grand old age of 85. As the two men moved in some of the same circles, such as working for members of the nobility and perhaps frequenting the itinerant royal court, in London, Windsor, and elsewhere, one wonders if they ever met?

AN UNFORTUNATE INCIDENT OF AN ARROW IN THE FACE

In 1403 Henry IV, the Lancastrian king of England, was locked in a cycle of wars with the Scots, the Welsh, and the powerful Percy clan of English barons led by the charismatic Henry Percy, or "Harry Hotspur", who took it as read that *they* ruled the north. And on 21 July 1403, some three miles to the north of Shrewsbury, bordering on the Welsh Marches, the armies of King Henry and Harry Hotspur met in battle.

The Percys were defeated and Harry Hotspur killed, and the knightly cavalry charge which carried the day was led by the 15- or 16-year-old "Prince Hal" (born in either 1386 or 1387), later to be Shakespeare's hero of Agincourt, 1415, after becoming King Henry V. In the Shrewsbury battle, however, Prince Hal was hit in the (probably) left side of his face, in the upper jaw, or maxilla, by an arrow. One can only assume that it was a "spent" arrow shot from 400 or more yards away, for had it come from a powerful English or Welsh longbow at a mere 100 yards, it would have gone straight through Hal's head and probably skewered the man behind him.

In the adrenalin rush of battle, into which he was about to lead his own victorious charge, the fearless 15-year-old pulled the arrow out, declaring it to be but a "shallow scratch". Medieval longbow arrows, however, were not designed for easy extraction, the barbed iron points being joined to a short cylindrical section that was only

161

attached to the shaft by finger-pressure. Consequently, when Prince Hal pulled out the wooden arrow shaft, the barbed point, probably a couple of inches long and three-quarters of an inch across at the barbs, remained firmly embedded in his left maxillary bone.

Then the battle was won. The Percys were defeated, and Prince Hal was declared a hero. Yet the 15-year-old still had a nasty lump of iron wedged into his skull. Eventually, by which time he had ridden over 50 miles south to Kenilworth Castle, Warwickshire, treatment was finally obtained. An operation was performed by the eminent London surgeon John Bradmore, who seems to have been in the royal employ. We know exactly what Mr Bradmore did, because both he and his pupil and professional successor, Thomas Morsted or Morestede, left a detailed account of the operation, complete with a sketch of the extractor instrument that Bradmore employed. It appears in Bradmore's *Philomena*, written at some stage before his death in 1412, and is also preserved in Morsted's note in the British Library manuscript.[3] Bradmore's *Philomena* was written in Latin, although an English translation was later made.

By the time that Bradmore saw his royal patient, several days after the battle, considering the distance between Shrewsbury and Kenilworth, the wound was beginning to close, as such puncture wounds do, if the patient survives; yet the arrowhead was still embedded inside. Bradmore, however, used an instrument of the "manner of tongs": a type of extractor that was of known design and perhaps a standard component in instrument chests of the day.

The extractor seems to have consisted of a metal cylinder, the same diameter as an arrow, divided longitudinally into two semi-cylindrical halves. What appears in the drawing to have been a screw mechanism could then be operated to open the semi-cylinders to make them expand like jaws, or "tongs".[4] If the instrument were inserted into the hollow tubular end of an arrowhead, it could be eased into place, to occupy the position of the original arrow shaft. Then, by turns of the screw, the instrument could be expanded so as to fit tightly inside the shank of the arrowhead. Next, and with a gush of blood and a grinding of Hal's teeth, the adroit Bradmore would have pulled out the arrowhead! An audience of courtiers, clergy, knights, and servants were present, and they heartily thanked God when the arrowhead came out. Ointments, most notably the *unguentum fuscum chirurgicum* ("the dark ointment of the surgeons": no formula given), were then applied to facilitate healing. One suspects that Bradmore, like Galen and many other surgeons before and after, knew how to perform to best effect in such a nail-biting piece of surgical theatre.

Before the operation could even be attempted, however, the surgeon had to reopen the wound. We are told that Bradmore employed a set of "tents" (ancestors of "stents" or dilators), being smooth rods of elder wood, dipped in honey, each rod being wider than its predecessor, until a rod was inserted that was the diameter of an arrow shaft – thus producing a neat aperture of the right size down which the extractor could be inserted.[5] And while no one in 1403 had any idea of the true causes of infection, Bradmore knew on purely empirical grounds that honey is good for wounds.

Bradmore's operation on Prince Hal tells us much about the sheer professionalism, skill, and ingenuity of a top-class medieval surgeon. It also tells us what kind of instruments such men employed. Bradmore, Morsted, John of Arderne, and their English and Continental European colleagues were a million miles away from the grunting medieval butchers of popular legend. They were also well paid, and Morsted in particular died in 1450 a very rich man, with landed property and his own coat of arms: the social equal of any physician and a formally registered gentleman.

TOWTON MAN: SOPHISTICATED FACIAL REPAIR
SURGERY IN EARLY FIFTEENTH-CENTURY ENGLAND

In 1996, builders digging at Towton, near Tadcaster, Yorkshire, accidentally uncovered a mass grave. It turned out to be the burial place of men killed in the Battle of Towton Moor, on 29 March 1461, generally accepted as one of the bloodiest battles fought on English soil. Like Shrewsbury half a century before, Towton was part of the Wars of the Roses, that ongoing saga of baronial and dynastic wars which would finally end in 1485 with the death of

Richard III, of Shakespeare fame, and the accession of the first Tudor king, Henry VII.

Of the many hacked and damaged skeletons unearthed on Towton Moor, one was especially fascinating. Not from the wound that killed the anonymous soldier in question, but from the massive prior trauma inflicted upon his skull in some previous battle, which by 1461 was completely healed. The previous wound had been so severe that it is a miracle that it did not kill him, either on the battlefield or from haemorrhage, infection, and other complications soon after. Yet the anonymous soldier not only survived, but had recovered sufficiently to fight and then die in a subsequent battle.

One assumes that his return to fighting indicates that Towton Man was a professional soldier, and considering the remarkably sophisticated surgical treatment and sustained post-operative aftercare which he received, he was probably a soldier of high rank, with the resources to benefit from the best care that the mid fifteenth century could offer. For in a previous battle, Towton Man had received a massive broadsword blow across the left side of his face. The weapon damaged his left jawbone, in the upper part of that bone, probably destroying teeth and gums, and leaving a six-inch or more lateral slash which, in addition to osteological damage, would have severed the left cheek muscles, which connect the lower jawbone to the upper skull, and which, among other things, enable us to bite and chew. Several major blood vessels supplying the face would also almost certainly have been cut in two. The wound would have been ghastly and should have been lethal. But it was not. For this tough soldier, whom osteological evidence suggests was about 40 years old in 1461, lived on to die at Towton Moor from quite different wounds.

So what kind of surgical attention could he have received? It is impossible to give a definitive answer because no record of his case survives. Yet we can make a number of educated guesses, based upon what we know had been attempted elsewhere in medieval battle surgery. Once the bleeding had been staunched, Towton Man's lower face was probably encased in a tight-fitting yet pliable splint, made, perhaps, of soft leather and cut, stitched, and contoured to fit his face tightly. The inside of this splint may have been thickly coated with something soft like fresh beeswax, and then the whole thing stitched around his head, like a helmet. While the front of his mouth and front teeth were not damaged, it would have been very difficult to eat and drink normally when wearing such an encasing artefact, especially as it would most likely have been left *in situ* for several weeks to ensure correct healing particularly of the severed muscles. During convalescence, nourishing meat broths and wine were probably sucked through straws.

Irrespective of what was done to Towton Man, however, the treatment was resoundingly successful, for what was left of his mandible bone healed beautifully, as did his left facial muscles and blood vessels. How normally he could have eaten and spoken thereafter, when his possible face-splint had been removed, is a moot point, but apart from a nasty scar, which he probably covered with a beard, he would probably have lived a normal life. He must have he recovered his full strength and vigour, otherwise he would never have fallen at Towton.

So while the great academic surgical treatises of Theodoric and Guy de Chauliac discuss a wide variety of operative procedures, it is in John Bradmore's treatment of Prince Hal and his later anonymous counterpart's treatment of Towton Man that one sees the real-life medieval master surgeon in action. The obvious skill and care which they lavished upon their patients speak volumes in terms of their sheer professionalism.

THE ANONYMOUS SURGEON OF HMS *MARY ROSE* IN 1545

When King Henry VIII's great battleship, HMS *Mary Rose*, suddenly heeled over in a gust of wind and sank outside Portsmouth Harbour on 19 July 1545, in the act of repelling a French attack, she dragged some 500 men to their watery graves, with only 35 escaping. But after Alexander McKee, Percy Ackland, and other divers located and began to excavate the wreck after 1971, and she was brought up to the surface and conserved after 1982, this "Tudor time capsule" revealed a myriad of fascinating facts about Tudor life. Surviving animal bones found in the remains of storage barrels enabled modern butchers to identify the very cuts of meat which would have been fed to the crew. Most fascinating of all, from a medical historical point of view, was the discovery of the great surgeon's chest and its contents.[6]

This beautifully made great oak chest contained drug jars (some with traces of their original contents still present), medical and surgical instruments, and even the surgeon's close-fitting leather cap. Very clearly, the owner of this chest – and military and naval surgeons at this time usually supplied their own kit – was a man of education and standing. He had his own private cabin on the ship; and, while all iron and steel blades have been eaten away by the sea, their wooden and bone handles and fittings survive.

Two instruments in particular, made respectively of pewter and brass, are in wonderful condition. These were two syringes. Not hypodermics intended for venous injection (the hypodermic injection of drugs only dates from 1855, in Edinburgh), but for the injection of mercury-based compounds via the urethra into the bladders of sailors suffering from the then new syphilitic infections. (We do know exactly, from extant medical textbooks of the sixteenth century, how these syringes were used to treat sexual complaints.)

Quite incidentally, it had been the appearance of new sexual diseases – most notably syphilis – in Europe at the end of the fifteenth century that had caused one of these "shocks of the new" which had led doctors to search for and experiment with new cures, for diseases unknown to Hippocrates and Avicenna, against which the ancestral pharmacopoeias were impotent. Mercury-based compounds, fluids, and salves were among these new remedies, and doctors were soon finding, by experiment, that metals such as mercury, antimony, and arsenic, given in small doses, provoked all kinds of interesting reactions in the human body in a way that traditional herbal treatments did not. Mercury, for instance, led to the generation of copious saliva, and as syphilis was believed to be a "moist" disease, it was – incorrectly – believed that mercury-induced salivation would purge and cleanse away the syphilis, in accordance with the principles of humoral pathology.

This was a curious logic, and wrong, but it came about by a "let's give it a try" experimental approach, rather than simply following the ancients. It was a new technique for a hitherto unknown disease, although mercury had previously been mixed into pastes to treat tumours and skin blemishes. It is interesting that around the time that *Mary Rose* sank in 1545, carrying her surgeon and his beautifully fashioned urethral syringes to the sea bed, the Italian physician Girolamo Fracastoro was publishing his landmark treatises *Syphilis sive Morbus Gallicus* ("Syphilis, or the French Disease"), 1530, and *De Contagione et Contagiosis Morbis* ("On Contagion and Contagious Diseases"), Venice, 1546, works which could clarify European thinking about treating syphilis.

GUNPOWDER, GOD, AND EUROPE'S SURGICAL RENAISSANCE

The introduction of firearms into the battlefields of Europe in the early fourteenth century, and their early employment at Crécy (1346), Agincourt (1415), and other landmark actions heralded not just a revolution in warfare technology, but also in surgery. For while in these Hundred Years War battles the English longbow still played a far more decisive role than did guns, the writing was now on the wall.

Gunshot wounds constituted another "shock of the new" which demanded that surgeons *experiment* with possible new techniques and therapies by which to treat injuries quite unlike those faced by a Roman or a medieval military surgeon. For the incredible force that a charge of black powder could put behind a projectile might shatter a limb at 300 yards, leading to lacerated bones and muscles, and sudden appalling blood loss. A soldier might be partially disembowelled yet still be alive several hours later. Pieces of skull might disappear, exposing cranial membranes, yet once again the patient continued to live, and accidental explosions produced the most frightful burns, which men somehow survived.

167

With the development, in the fifteenth century, of the early arquebus, or musket – handguns firing a soft lead ball about three-quarters of an inch in diameter – a veritable myriad of complicated injuries started to appear. A ball, for example, might deform or flatten on passing through a leather-padded garment, then do all manner of nasty things once inside the body. For a spherical musket shot rarely made a neat hole in the patient. A ball might, for example, get inside the musculature of the chest, thighs, or buttocks, follow the muscular wrappings around the body, and come to rest a foot or more away from the entry hole.

Ambrose Paré, the great early sixteenth-century French surgeon, was one of the great pioneers of gunshot wound surgery, and his writings include several examples of his skill in tracing musket balls from their point of entry to their point of lodgement: an essential preliminary if the patient were not to be cut unnecessarily. Not only did bullets enter wounds, but they also dragged in with them pieces of long-unwashed clothing, to provide a perfect breeding-ground for the bacteria that cause septicaemia and gangrene.

The problem was still present in the Victorian age 300 years later, when soft lead bullets were still in use. Robert Druitt's *The Surgeon's Vade Mecum* (1878) mentions an old case related to him by a colleague, in which a soldier "was struck about the *pomum Adami* [Adam's apple] by a bullet, which passed completely round the neck and was found lying in the very orifice at which it entered".[7]

All of these myriad complications were a far cry from the days when it was a military surgeon's primary job to stitch slashes and extract arrowheads. Apart from anything else, the sheer complexity of gunshot wounds demanded a much more thorough knowledge of internal muscular and skeletal anatomy. Exactly where was it safest to cut? What muscles lay where, and how did they interact? And where, for any part of the body, were the main blood vessels, and how deep were they?

Amputations, among other things, became much more common after the advent of gunpowder. How exactly *do* you take off a leg above the knee without the patient bleeding to death? How can useful artificial limbs be developed to make the patient's life a little bit easier, if he survives the operation? It was gunpowder warfare which, I suggest, perhaps more than any other single factor, helped stimulate the creation of orthopaedic and prosthetic surgery. And firearms-related surgery gave rise to the first "controlled experiment" on the nature of infection. It was widely believed from the first use of firearms that a gunshot wound was *ipso facto* poisoned. This was thought to derive from the very nature of the chemicals, flames, and choking sulphur stinks required to shoot the bullet. The accepted theory was put to the test, however, by an ingenious experimentally minded surgeon named William Clowes around 1580. He asked the master gunner serving with the Portsmouth garrison to load his musket with a gunpowder charge. But instead of loading it with a ball, Clowes asked him to drop an archer's heavy war arrow down the barrel and fire it at the town gate. On inspection, the arrow was found to be completely unburnt or even singed by the explosion, for both it and the gun wadding had left the barrel *ahead* of the flames. So if the flames and gases of the explosion did not singe an arrow's flight feathers, how could they poison a bullet? Ergo, gunshot wounds were not, of necessity, poisoned.[8]

Gunpowder and gunshot wounds occasioned a new surgical literature, especially in the wake of Hieronymus Brunschwig's *Buch der Wund-Artzney* ("Book on Wound Surgery"), 1497, and Giovanni da Vigo's *Practica in Arte Chirurgica Copiosa* ("Complete Practical Art of Surgery"), 1514. Much of it was in Europe's newly burgeoning vernacular languages, rather than in Latin. A whole new world of surgical and anatomical therapeutics began to unfold, for in addition to the old disputes about the "laudability" or otherwise of pus, new questions were being asked about wound toxicity, hygiene, and infection, as surgeons began to report their cases, discuss new points of technique, and describe new instruments. Surgery was always the most pragmatic and experimentally driven branch of medicine, and was becoming even more so as the medieval age blended into what the nineteenth-century historians came to style the "Renaissance".

And how does God (or perhaps, the devil) enter into all this? I would suggest this was because the chivalric knightly classes saw gunpowder as the work of the devil. The stench, noise, and flame of firearms certainly appeared satanic, while firearms clearly upset the natural social order, by making it possible for a common peasant armed with a gun to kill a chivalrous knight or even a prince. The terrible wounds inflicted by guns brought all the horrors of hell into the midst of the living.

THE ROYAL COLLEGE OF PHYSICIANS AND THE WORSHIPFUL COMPANY OF BARBERS AND SURGEONS

As we saw with Chaucer's *Tales* and with Oxford and Cambridge Universities, there were trained academic physicians – learned "doctors" – established in England by the fourteenth century, while similar numbers of trained, as opposed to country folk, surgeons were probably to be found in London, Bristol, Norwich, and other cathedral cities as well. Then in 1390 the Lord Mayor of London sanctioned a City Livery Company for surgeons, and in 1423 Thomas Morsted, who had assisted in John Bradmore's operation upon Prince Hal in 1403, established a college for both physicians and surgeons.[9]

By the early sixteenth century, medical professionalism was beginning to take off, and in 1518 King Henry VIII gave a Royal Charter of Incorporation to the Physicians, to become "The Royal College of Physicians". Perhaps more than any other single individual, the driving force behind the establishment of the Physicians' College was Thomas Linacre, fellow of All Souls College, Oxford, a friend of the new House of Tudor, and, after 1509, physician to the young King Henry VIII.

Linacre was fully within the classical tradition of European medicine, not only being at one with those intellectual and spiritual forces which characterized the great Continental universities, but also being both a product of and a leading participant in a new movement: that of Christian "humanism". This was not humanism in the modern-day sense, which is generally secular and humanity-centred, but rather an enriched Christian approach to all scholarship.

It was a movement inspired and enlivened by classical texts in the original Greek language, many of which were pouring into Europe as Greek Orthodox scholars fled their homeland in the wake of the Muslim destruction of the Christian Byzantine Empire after 1453 – one of the key triggering forces behind the Renaissance.

Europe's young scholars were electrified by the prospect of being able to obtain the classical writers in their own, original Greek words, rather than filtered through Syriac, Arabic, and medieval Latin intermediaries, often replete with errors and confusions. In their quest to go back to the fountainhead of Western culture – *ad fontes*, as they put it – Latinate young men from northern Europe began to flood into Italy, where refugee Greek scholars could teach them the language of Plato, Aristotle, and the Christian Gospel and Epistle writers. William Grocyn, John Colet, and Desiderius Erasmus were among those who joined this new humanist movement.

Just as "humanism" stimulated new literary, philosophical, and theological studies, so it stimulated medicine. It even captivated royalty, and the highly intelligent and cultured 25-year-old King Henry VIII was keen to learn and enjoyed conversations with these brilliant young scholars. Discussions ranged widely over disciplines such as astronomy, poetry, medicine, and theology. It was this way of thinking that enabled Dr Thomas Linacre – just like Theodoric of Borgognoni and Guy de Chauliac in previous centuries – to combine the careers of physician and Roman Catholic churchman. Linacre was in priest's orders, holding several ecclesiastical posts including the precentorship of York Minster, in addition to being a physician of distinction: yet another example of an individual combining the cure of body with that of soul.

These were the circumstances that led to Henry's chartering the Royal College of Physicians in London in 1518. While, like an Oxford or Cambridge college, it had some teaching functions, including the right to the bodies of executed criminals for dissection purposes, it also possessed wider regulatory powers. Henry VIII's Royal College was authorized to regulate medical practice in London, decide who was, and who was not, a proper doctor; prosecute quacks, and, soon after, issue an official *Pharmacopoeia* of permitted drugs and procedures and destroy the wares of any apothecary found selling adulterated or substandard drugs.

Things soon became contentious, especially when non-fellowship doctors objected to the college's draconian powers, and surgeons and apothecaries could be hounded. It was the apothecaries and druggists, as well as the "non-elite" surgeons, who were the "cheaper" healers, to whom ordinary folk resorted when sick. Undeniably, the Royal College set vital new professional standards and curbed the abuses of quacks, but its intense academic conservatism, reverence for the classical Hippocratic tradition, and aura of "jobs for the boys" often set it at odds with more experimental or intellectually adventurous approaches, and the practices of popular and more accessible healers.

In the same way that the physicians sought to confirm their own corporate professional identity with a college, so, in 1545, the London surgeons, largely at the instigation of the London master surgeon Thomas Vicary and his senior colleagues, obtained their own Royal Charter: not for a Royal College, but for a City Livery Company of Barbers and Surgeons. In many ways, it was an odd marriage, for by 1545 learned surgeons did *not* cut hair, nor did barbers amputate limbs, though they might have shared some common ground when it came to the ancient art of phlebotomy, or bloodletting. When one looks at the great multiple portrait of the named, black-robed gentlemen, kneeling respectfully before the now obese King Henry VIII as they receive their new dignity, it is quite clear that these men did not see themselves as beard-trimmers so much as medical gentlemen. They, like their physician brethren, received powers to train and supervise surgeons; to examine, to confer qualifications, and also to dissect. Inevitably, the surgeons were always going to be a numerical minority within the Worshipful Company, although they were the biggest earners, and by 1745 the groups officially parted, to form separate and distinct companies of surgeons and barbers. Then in 1800, the surgeons received their own academic endorsement upon being chartered as the Royal College of Surgeons.

So while England never had a major medical university of the standing of Paris or Bologna, it is nonetheless clear that, from the early fourteenth century onwards, a very self-conscious professional community of both physicians and surgeons existed, especially in London. This was brought into being by a whole body of circumstances, extending from rapid advances in military surgery to classical humanist scholarship, and it was from this community that the British medical profession would evolve. Yet let us end with a speculation.

We do not know the name or place of abode of the surgeon who was serving in HMS *Mary Rose* in 1545. Yet he was clearly a man of some education, who possessed a fine chest of medicaments, instruments, and probably books – the paper of which has sadly not survived the sea. Could he, perhaps, have been involved in any way with those colleagues in London who, in that same year, were in the process of negotiating the formation of a professional Company of Surgeons with King Henry VIII: the king in whose great warship he was serving, and in which he died?

CHAPTER 11

Antiquity Found Wanting in
Renaissance Italy

*I*n the early and middle sixteenth century, attitudes towards medical knowledge, and especially towards anatomy and physiology, were changing. This did not happen, as some historians in the past claimed, because the Dark Ages somehow gave way to the Renaissance, but because of other powerful movements that were under way in European society. Medicine and science do not inhabit sealed boxes, and were inevitably open and responsive to culturally formative processes going on in Europe.

On the one hand, while bubonic plague may have scythed through the population since 1347, causing a waning of the Middle Ages, there was abundant evidence of a remarkable new dynamism abroad in Europe by 1500. That essential burgeoning of medieval prosperity which we mentioned in Chapter 8 continued on its course, as Europe began to develop those features which remain its global hallmarks today and are still present wherever Western-style societies have taken root: in the USA, Australasia, Russia, China, Japan, and elsewhere. These include progressive technologies; corporate and institutional, as opposed to tribal, social structures; increasingly complex free mercantile and international economies; a growing sense of national identity; a love affair with pictures and naturalistic art; and an approach to human dignity and morality which ultimately springs from the Judeo-Christian religion.

There were other factors: the love of argument, criticism, and debate which became the very life-blood of the medieval universities

174

– rooted in the "sciences" of dialectic (argument) and rhetoric – had created a state of mind which was far from the supine obedience to authority which many still think of as medieval. This passion for debate, contradiction, argued response, and critical reappraisal was rooted in the very structures of medieval education. An ability to argue one's case lay at the very heart of the degree-granting process, as it still does today: in the *viva voce* ("living voice") cross-examinations based upon a *thesis* (proposition), through which doctoral students – myself included – have passed in Europe's great universities from the thirteenth century onwards.

Printing further excited the medieval passion for critical thinking, for it enabled newly found Greek texts of Aristotle, Hippocrates, Galen, and others to be collated with existing translations into Latin or Arabic. Printing made old and new texts available in identical copies, from St Andrews in the north to Salerno in the south. It was this passion to explore and collate Greek, Latin, Hebrew, and other biblical texts that acted as a major spur to the Protestant Reformation and then the Catholic Counter-Reformation. The Dutchman Desiderius Erasmus, who produced the first critical edition of the Greek New Testament in 1516, was a good friend of Dr Thomas Linacre, a fellow Greek scholar and the driving force behind the founding of the English Royal College of Physicians in 1518.

But a love of argument, criticism, and textual interpretation were only part of those changes going on by 1500 which would lead to fundamental revisions in medical thinking. Crucial in altering wider cultural perspectives were new and hitherto unimagined discoveries in the *physical* world. Think of the discovery of that unimagined continent, America, by Christopher Columbus in 1492; Vasco da Gama's discovery that the Atlantic and Indian oceans were connected, giving a sea route to India and the Far East by 1499; followed by Ferdinand Magellan's epic voyage around the world in 1519–22, which showed that there was more ocean on the earth's surface than there was land.

Then in addition to these very "hands-on" geographical discoveries, which showed that Ptolemy and the classical geographers had got it badly wrong, there was that myriad of discoveries coming in their wake: rattlesnakes, tomatoes, potatoes, tobacco, quinine bark, ocean currents, compass needles not pointing north, strange wind systems, and high civilizations, such as those of the Peruvian Incas, of which scholars in the "old world" had been entirely ignorant in 1450.

All of these factors had their role in that wholesale burgeoning of natural, or scientific, curiosity, of which changes in medical thinking formed a part. If the ancient geographers, botanists, and zoological writers had got it wrong when it came to the establishment of universal principles by which we might understand nature, then where did this leave the *ex cathedra* status of Aristotle, Galen, Dioscorides, or Avicenna in the field of medicine? Could it be that when it came to fathoming the workings of human and animal bodies, and understanding those maladies to which humans are prone, there were, perhaps, equivalents of Columbus and Magellan, who might overthrow the wisdom of centuries, and set us upon quite new paths of understanding and acting?

This was the challenge of the "new knowledge", and it took place in that heartland of medieval European medicine, and in that country where naturalistic oil painting, neoclassical architecture, vernacular poetry, credit banking, and the outrageous political ideas of Niccolò Machiavelli were all causing a sensation: Italy.

RENAISSANCE ITALY AND THE "LESSER CIRCULATION" OF THE BLOOD: ANDREAS VESALIUS, PADUA, AND THE NEW ANATOMY OF THE RENAISSANCE

In the same way that "new knowledge" in the form of geographical discovery, book availability, textual analysis, artistic imagination, and technological innovation entered late fifteenth-century European civilization, so did new approaches to medicine. For by the 1530s, approaches to anatomy and physiology were beginning to change, and nowhere more than in the Italian University of Padua.

176

The city of Padua stands on the "terra firma", or the Italian mainland, some 24 miles up the Brenta Canal from the great lagoon and islands upon which the Serene Republic of Venice – *Serenissima Repubblica di Venezia* – stands. Encouraged and protected by the great merchant state of Venice, with all its wealth and political and cultural clout, the University of Padua dates back to 1222, its institutional symbol being the Ox, leading to its nickname of the "Bo": *bös* being Latin for "ox". The University of Padua was renowned, and even notorious, for its independent and often unconventional thinking. It was, in many ways, a student-driven university, where the academic professoriate had to keep on the good side of the student "Nations", or regional factions, if they wanted to be sure of their salaries and continuing employment.

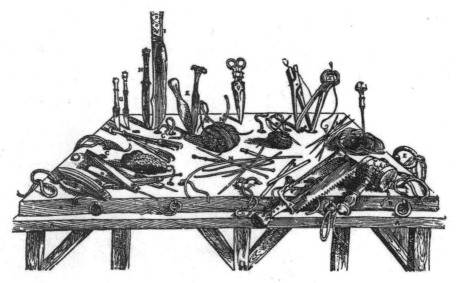

10.   A table of surgical and dissection instruments. (Vesalius, *De Fabrica* (1543), Book II. Christ Church Library, © Governing Body of Christ Church, Oxford.)

Perhaps it was these circumstances that led to Padua becoming the first university in Europe where the classical medical writers came to be challenged not only on specific points, but on a sustained basis – by professors, by their pupils, then by generations of succeeding students – so that men of 1600, such as William Harvey, would trace their intellectual ancestry as doctors back to the great Paduan teachers of 60 or more years before. This new thinking was very much bound up with fresh approaches to anatomy and physiology which originated in Padua in the 1530s, as Andreas Vesalius, Realdo Colombo, and others began to dissect in a different way from Mondino and Theodoric.

Instead of "an Anatomy" being based upon the professor's reading of a text by Galen and illustrated by parts of the body being demonstrated to the audience by the "ostensor", Vesalius came to adopt a more "hands-on" approach. How Vesalius first came upon this method is not entirely clear, but it seems to have developed through practical experience, which led him to discover errors in the classical anatomists and physiologists such as Aristotle and Galen, just as Columbus discovered new lands and Erasmus found discrepancies between the Greek New Testament and the Latin Vulgate. Vesalius adopted a more *inquisitive* approach, based on physical evidences.

This new approach had not yet developed in 1538, when the 24-year-old Professor Andries van Wesel published his *Tabulae Anatomicae Sex*, or "Six Anatomical Tables". Better known under his Latinized name, Andreas Vesalius, van Wesel was a Roman Catholic of Dutch descent born in Brussels, who had studied at Leuven and Paris before being awarded his doctorate in Padua. Young Vesalius's brilliance led to his being offered an anatomical chair in Padua upon graduation, and it would be in Padua over the following years that he would play a leading role in the transformation of anatomy from a *demonstrative* to an *investigative* discipline. He would also abandon the traditional schematic representation of the human body in favour of a more lifelike one, in collaboration with the young Jan van Calcar, a former pupil of the great Renaissance artist Titian, thus making the *Tabulae* of 1538 a landmark in the accurate scientific depiction of the human body. Gone were the wooden-looking cadavers accompanying the medieval anatomy texts, and in their place were the beautifully delineated, if somewhat grizzly looking, animated

178

bone and muscle men of the *Tabulae* – exquisite works of art in their own right.

Yet beautiful as the *Tabulae* may have been, scientifically Vesalius still followed Galen as far as basic anatomical structures and functions were concerned, and the artist van Calcar dutifully represented accordingly.

THE ART OF THE ANATOMICAL ILLUSTRATOR

Vesalius owed an enormous debt to both van Calcar and to other artists who illustrated his *Tabulae* and subsequent works, for they played a major role in the advancement of medicine. By the early sixteenth century, science and art had come together in the "new" anatomy of Renaissance Europe. We are all familiar today with photographs of Leonardo da Vinci's exquisitely detailed anatomical drawings – based on actual cadaver dissections – of the human heart, foetus, muscles, and so on. In their own time, however, these masterpieces were largely unknown, being Leonardo's private unpublished studies, stored away in private collections. They remained so until their discovery by art collectors in the late eighteenth century.

Such drawings, both "photographically accurate" as scientific studies and also works of art in their own right, were of immense importance for the wider advancement of anatomy. And here van Calcar was truly a pioneer. Yet a drawing inevitably remains of limited use if it cannot be easily reproduced, and so another new Renaissance technology, the copper plate engraving, made its impact, and would continue to do so until displaced over four centuries later by photography. (The art of anatomical draughtsmanship is very much alive today, as many twenty-first-century anatomy textbooks still include beautiful coloured drawings. An artist can give a visual emphasis to structure and detail which is, even now, not easy to catch with a camera.)

Early copies of Mondino had been illustrated, to a limited extent, with woodcuts, where a paper drawing was transferred to a flat block of wood, and a craftsman cut through the paper and

11. Andreas Vesalius, *De Fabrica* (1543), frontispiece. Unlike Mondino's anatomy depicted in Plate 9, the professor, Vesalius himself, is no longer aloof in his 'chair', but doing his own dissecting at the cadaver. This new 'hands-on' approach would elevate anatomy from a theatrical demonstration of classical wisdom to a medium for original research. One suspects that the fashionably-dressed yet bloody-fingered Vesalius enjoyed performing before a packed audience, complete with dogs and monkeys. (Christ Church Library, © Governing Body of Christ Church, Oxford.)

into the wood with a sharp tool. By *c.* 1470 the new printing press could be used to duplicate copies from the block, but these early woodcuts were relatively crude as far as detail was concerned. By the early sixteenth century, however, rapidly improving engraving and printing techniques were making it possible to duplicate drawings showing exquisite details of line, form, and delicate cross-hatch shading. These new drawings were engraved into polished sheets of copper – "copperplate printing" – as well as into very fine-grained wood blocks. Such plates were capable of expressing breath-taking detail. One of the first great artists to use this new medium (albeit not for anatomical illustration) was the celebrated Albrecht Dürer, and it was in his tradition that the Titian-trained van Calcar worked.

In his recognition, then, of the power of high art in the service of both medical research and teaching, Andreas Vesalius invented the anatomical atlas and the scientific medical plate.

VESALIUS AND HIS *DE FABRICA* OF 1543

While the anatomy so elegantly depicted in the *Tabulae* was still Galenic and conservative, Vesalius was coming to adopt a significantly new approach by the time of his monumental *De Humani Corporis Fabrica* ("On the Construction of the Human Body"), published five years later, a new approach that went even further by its second edition of 1555 and in its illustrated abridgement, the *Epitome* of 1543.[1]

Vesalius's doubts about Galen seem to have begun soon after 1539, when he came to realize that Galen's anatomy was based not on human cadavers, but on those of Barbary apes. For example, the human mandible, or lower jawbone, is significantly different in apes from in humans. Even more problematic for Vesalius was the anatomy of the heart. Try as he might, he could find no evidence of any tiny holes, or pores, in the thick central muscular wall – or septum – that divided the left from the right ventricles of the heart.

If the septum was solid, however, how could the blood entering the right ventricle of the heart from the vena cava get into the *left* ventricle? Dissection revealed that blood was actually present in the left ventricle. Could it somehow distil through the septum to transit the heart? Vesalius never came to a firm conclusion about the heart in *De Fabrica*, at least in the first, 1543, edition. Real progress towards a post-Galenic understanding of the heart would take a little longer, in the hands of Vesalius's contemporaries and immediate successors, as we shall see below. Vesalius did recognize that the mitral valve (which takes its name from its flaps which somewhat resemble a bishop's mitre) plays a significant role. In fact, the mitral valve – which separates the left auricle and ventricle chambers – regulates the blood flow through the heart itself, as the blood passes from the lungs, up the pulmonary vein, through the auricle and ventricle heart chambers, and out through the aorta and into the arteries.

Vesalius's *De Fabrica*, however, takes its significance not only from his specific discoveries, but also from his new attitude to anatomy. For while post-mortem dissections, for legal purposes, went back centuries in 1543, Vesalius came to epitomize what we might call pathological anatomy: not just trying to establish the cause of death in a cadaver, but attempting to associate changes in the body with specific diseases in life, in order to understand how health, disease, and body changes interrelate.

At the heart of Vesalius's whole medical enterprise was a concern with how living structures held together and how they worked: an approach that Vesalius shared with Aristotle, who was also fascinated by how bodily mechanisms worked and interconnected. Seventy-odd years later, this quest for an understanding of structure and function would inspire Vesalius's "grand student", William Harvey, and would lie at the heart of his discovery of the full circulation of the blood around the body, as we shall see in the following chapter.

In some respects, one might say that Vesalius's approach to anatomy drew upon architecture. Just as the medieval Gothic cathedral builders, and then the Renaissance Italian architects such as Michelangelo, Bramante, and Serlio, balanced and juxtaposed compression and lateral forces in the building of their sweeping arches, soaring spires, and "floating" domes, so, Vesalius realized, did the human body in the manner of its construction. How did the skeleton, with its complex system of vertical (spinal), lateral (rib), lever (arm and leg), and dome (skull) bone systems give both strength and incredible flexibility to the body? How were the muscles anchored on to the bones to form lever points, rather like the jibs of builders' cranes? And how did this amazing structure of rigid and flexible parts hold the internal organs secure?

In *De Fabrica* Vesalius divides his material across seven "books" or divisions, and of these Books 1 and 2 deal with the skeleton and its muscular appendages. Vesalius saw the skeleton as having a parallel with the poles of a tent, which hold the rest of the structure together. This section is illustrated with three exquisitely drawn and detailed skeleton plates, showing the bone structure in a variety of positions. Vesalius's plates are not stiff, rigid, and schematic, as had been the custom in earlier medical illustration, but attempt to capture the postures that a skeleton would occupy when clothed in flesh, inside a living human being.In fact, it was probably Vesalius's comparisons of preserved and articulated human skeletons with those described in Galen that brought him to his first realization that the skeletons described by Galen were those of apes. He also drew some cogent conclusions about how the muscles worked. If, for example, one made a longitudinal cut down a muscle, it could still exert a mechanical force, yet if one severed a muscle cross-wise, it became useless. Muscles, it seemed, pulled length-wise against a bone, while bones hinged against each other in a variety of different types of joints, depending upon the work they had to do. Vesalius's "muscle men" plates, showing the muscles peeled and hanging from the body – to illustrate their points of skeletal attachment – are, it has to be said, some of the most gruesome in *De Fabrica*, especially the engraving of a flayed cadaver held upright by a rope passed through its eye sockets, to show the muscles in a backwards-writhing posture!

It was in *De Fabrica* Book 7 that Vesalius gave the best illustrated description of the human brain, and his work on the brain would not really be superseded until the researches of Thomas Willis in the 1660s, as we shall see in Chapters 13 and 14. In his treatment of the brain and nerves, Vesalius was to diverge yet further from

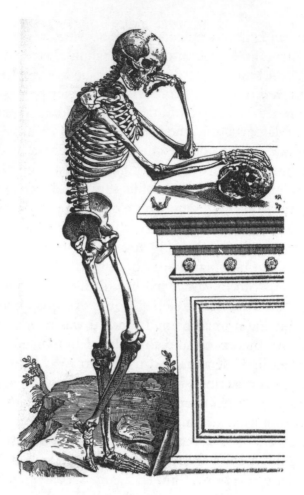

12. A human skeleton contemplating mortality upon a tombstone. Vesalius emphasized scientific accuracy and attention to detail in his depiction of the human body, as opposed to the schematized cadavers of medieval anatomies. His skeletons and cadavers are set in naturalistic poses, emphasizing the way bones and muscles interconnect when in different positions. Vesalius's artist – perhaps the Titian-trained Fleming Jan van Calcar – breaks new ground in the art of scientific depiction by bringing the Renaissance stress upon naturalism into medical illustration. (Vesalius, *De Fabrica* (1543), Book I. Christ Church Library, © Governing Body of Christ Church, Oxford.)

Galenic orthodoxy. He came, for example, to abandon the classical idea that the nerves were hollow tubes down which animal spirits from the brain moved both to facilitate sensation and to activate muscular action. He also departed from Aristotle in suggesting that it was the brain, rather than the heart, that was the seat of the sentient soul.

It would probably not be much of an exaggeration to suggest that Vesalius was the first to think of living bodies, human and animal, as machines: as wonderful examples of bio-engineering. While Aristotle had seen living organisms as integrated structures, each part fulfilling a specific function, Vesalius's approach is more mechanical. Indeed, this approach to anatomy and physiology would become a hallmark of the "Paduan School", with its emphasis upon hands-on dissection and its concern with fathoming out from a cadaver how the body actually functioned in life. It was a way of thinking that would be shared by Vesalius's successors and would underlie the work of William Harvey, who took his MD at Padua in 1602. This "engineering" approach was held in tandem with the belief that living things, and most obviously humans, were divine creations, for God had built the fabric of the human body no less than, in the view of the astronomers of the day, he had created the mathematical geometry that governed the motions of the planets. Both were amenable to rational enquiry.

None of this radical anatomical investigation, however, seems to have acted as an impediment to Vesalius's appointment as physician to two of the most actively Catholic, anti-Protestant monarchs of the sixteenth century: Charles V, Holy Roman Emperor, and then his son, the young Philip II of Spain. It was Philip, who in 1588 would send his legendary Armada to subjugate Protestant England under the Catholic faith, and fail disastrously.

By the time that Vesalius published the second, 1555, edition of *De Fabrica* – a work which even by that date had assumed iconic status – he had lost all of his prior reservations about criticizing Galen, Aristotle, and other classical anatomical writers. He was now fully convinced that Galen's dissections had been on animals and not humans, that there were major differences of structure, and that one could not use one as a guide to the other. What Vesalius did advocate were comparative human and animal dissections where, organ by organ, the differences became clear. He detected,

185

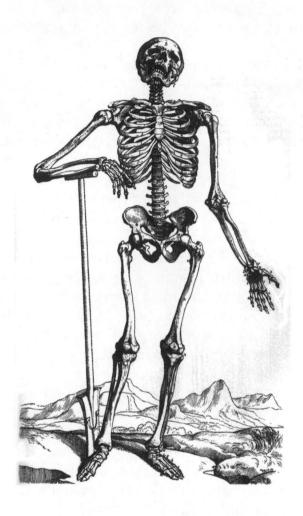

13. A free-standing skeleton, now posed as a grave-digger, looking as though
he is about to burst into song! (One wonders whether Vesalius and Calcar
shared a macabre sense of humour from the way they arranged their
cadavers.) (Vesalius, *De Fabrica*, (1543), Book I. Christ Church Library, ©
Governing Body of Christ Church, Oxford.)

among other things, fundamental differences in the structure of
the sternum or breastbone, the liver, and the uterus, not to mention
many divergences in skeleton structure. He even experimented
with artificial respiration, demonstrating how, if the chest wall were

14. In Book II of *De Fabrica* (1543) Vesalius clothes his skeleton in muscle, showing how the muscular wrappings are attached to the bones and interconnect to produce flexible movements, such as expanding the chest in breathing, walking, or gesturing with the arms. To Vesalius, the skeletal-muscular system was a living machine, within which an immortal soul resided. (Christ Church Library, © Governing Body of Christ Church, Oxford.)

penetrated, an animal could still be kept alive, and its heart-rate stabilized, by inserting a pair of bellows. This experiment would be taken further, under different circumstances, by Robert Hooke in 1667, as we will see in Chapter 15.

187

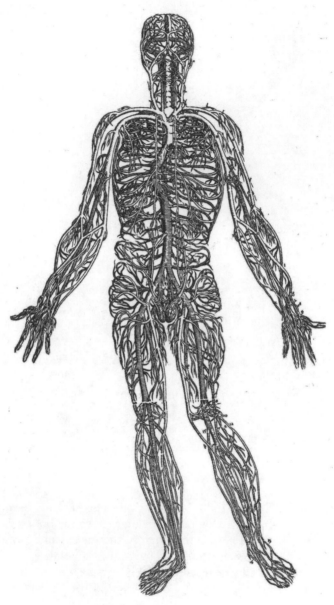

15. Vesalius's (or Calcar's) depiction of the dissected arterial and venous systems. The vein and artery structures are traced with great accuracy, but not until William Harvey in 1628 would their function in life be resolved. (Vesalius, *De Fabrica* (1543), Book III. Christ Church Library, © Governing Body of Christ Church, Oxford.)

Not everyone was in agreement with Vesalius's increasingly radical approach to anatomy, one being the orthodox Galenist Jacobus Sylvius Dubois of Paris (not to be confused with the seventeenth-century anatomist Franciscus Sylvius, or Franz de la Boë). Indeed, Jacobus Sylvius, who was a brilliant teacher, was primarily literary in his approach to medicine, placing Galen on a pedestal and even arguing that if structures in the body now appeared different from what had been seen in antiquity, then the fault lay in the cadaver, rather than in Galen!

Knowing what we do today, we may be inclined to laugh at Sylvius as a reactionary pedant, but by no means all Renaissance scholars were convinced by the "progressive discoveries" argument, and Vesalius himself had been reluctant to challenge Galenic and Aristotelian ideas as a young man. His new, investigative, questioning approach only really began in his late twenties, on the basis of explicit physical evidence. The compelling, all-embracing intellectual, moral, and cultural power of antiquity could not be shed like a coat; in its very comprehensiveness and integration, classical learning, from medicine to astronomy to poetry to philosophy to mathematics, and down to Christian theology, was a truly awesome edifice. A very great deal of that legacy still forms the warp and weft of present-day Western civilization.

Yet what Vesalius began to do at Padua resulted in the evolution of a new tradition in the medical sciences, starting most notably with anatomy and physiology. As far as we can tell, no one had the opportunity to dissect Vesalius's own cadaver. For sailing with the Venetian fleet, on his way to the Holy Land on pilgrimage, imperial physician Andreas Vesalius was shipwrecked off the island of Zakynthos (now Zante). He was buried in an unknown, unmarked grave on the nearby island of Corfu, sometime in October 1564. He was a couple of months short of his fiftieth birthday.

REALDO COLOMBO, THE VESALIAN TRADITION, AND THE SECRETS OF THE HEART

Colombo was only one year younger than Vesalius, being born in Cremona, Italy, in 1515, and both men were the sons of parents with medical connections. Both had fathers who had practised as apothecaries and druggists, in Flanders and Germany and in Italy respectively. Vesalius's career, however, had been very largely academic, but while Colombo had studied at Milan before going to Padua, he had served a practical seven-year apprenticeship as a surgeon. Perhaps as a result of the sheer pragmatic nature of his surgical training, Colombo was not especially inclined to textbooks, being not only a keen human dissector but also a vivisector of living creatures. His historical role model was not Galen, but Erasistratus and the third-century BC Alexandrian anatomists.

Colombo enrolled at Padua in 1538 and rapidly showed himself a masterly anatomist. So much so that, when the youthful Professor Vesalius was absent in Basel to see his *De Fabrica* through the presses, Colombo sometimes deputized for him. It appears to have been at this early time that Colombo began to criticize Vesalius, arguing that he was still too much guided by Galenic anatomy in his thinking and did not systematically vivisect. How warm the relationship between the two men was is unclear, although Vesalius, who seems to have had the instincts of a diplomat, claimed that they were friends.

Realdo Colombo's own *magnum opus*, *De Re Anatomica* ("On Anatomy"), which came out in 1559, shortly after its author's death, appears to have been researched and written over several years. While having notable parallels to *De Fabrica*, especially the 1555 second edition, it contained several significant original statements of Colombo's own. In Book 15, for instance, he stressed the primacy not only of human dissection, but also of animal vivisection, in the understanding of the functional relationships between organs. The dynamic nature of Colombo's thinking is seen in his fascination with fluid flows within living as opposed to dead things. This flow, or river-like action, lay at the basis of his "trinity" of primary organs: the liver generated natural spirit blood, the heart vital spirit blood, and the brain animal spirits, which somehow distilled from

190

heart blood. Like Vesalius, Colombo came to see the brain as the primary organ of sense and cognition.

It was in his "re-plumbing" of the heart that Colombo was at his most original, and once again, this was another fruit of dissection. Starting, like Vesalius, from the realization that the septal wall which divides the two sets of heart chambers (auricles and ventricles) is thick, solid, and without holes, he had to explain how blood could cross the heart from right to left: which dissection demonstrates that it does, in both humans and animals. Galen had taught in *De Usu Partium* ("On the Function of the Parts") that the pulmonary vein (styled "arterial vein" by Galen) linking the lungs to the left-hand side of the heart conveyed only air (and spirit, or life-giving *pneuma*) and not blood. Yet when Colombo opened up the thoracic cavity of a living dog, he found the pulmonary vein to be delivering blood into the heart. But how could it get there?

Colombo came thereby to a radically different interpretation of cardiac and lung function. In his new vivisection-based model, the lungs were not bags that blew inhaled air into the heart, but places where blood and air physically mingled before entering the heart. In Colombo's model, blood entered the heart in the usual way, from the "Great Vein" or vena cava on the right-hand side. Some of it then passed *downwards* (the direction of flow, as would later be discovered, being controlled by the tricuspid and semi-lunar valves) into the lungs via the pulmonary artery; then through the lungs, up the pulmonary vein, and back into the heart, and on to invigorate the brain.

Colombo's model was limited. It still had an imperfect understanding of the arteries and only explained the "thoracic transit", or how the blood crossed over from right to left within the chest. But it would come to play a vital part in William Harvey's complete model of whole-body circulation, as we shall see. Whether Colombo had any acquaintance with the work of the thirteenth-century Damascus physician Ibn al-Nafis (whom we first met in Chapter 3), who also had developed a model for the thoracic transit, is not known, although al-Nafis's work would not really be known in the West until 1924.

In the new dissection- and vivisection-based world of Renaissance anatomy, others had followed, sometimes

191

independently, in Colombo's wake. The Spanish physician Miguel de Serveto (Servetius) had also developed a thoracic transit or "lesser circulation" model after the discovery of the solid septum, while the Aristotelian Andrea Cesalpino would coin the term *circulatio*. However, Cesalpino's model was based upon a cyclical flux and reflux action rather than upon a circulation proper.

While the role of the heart and lungs in the blood movement process would become of pre-eminent importance to future anatomy, there were also major dissection-based researches into other regions of the human body. These included Bartolomeo Eustachio's researches into the oral and auditory functions within the human head – leading to the discovery of the ear's tympanic nerve and those hearing-related tubes which immortalized his name – the eustachia – and Colombo's and Gabriele Falloppio's researches into the female reproductive system. Colombo's work on the blood supply to the foetus led to his coining the term *placenta*, while Falloppio's name would be permanently associated with the tubes in a woman's reproductive system. How the human body actually worked as a *machine*, based on the complex interrelationship of working parts and systems, would soon become one of the pre-eminent driving forces in medical research. All this was going on in a society that was becoming increasingly bedazzled by clocks, great lifting machines, guns, windmills, watermills, and a whole world of complex machines. One branch of medicine in which this approach particularly manifested itself was surgery.

AMBROSE PARÉ: RENAISSANCE MASTER SURGEON

While Renaissance anatomy arose, in many ways, from the surprising discovery of errors of fact in Galen (due largely to Galen's not dissecting humans), Renaissance surgery was much more of a sequential development of the surgery of classical and medieval times. But as we saw in the previous chapter, one great spur to surgical innovation came from having to cope with gunshot and gunpowder burn wounds, for by the time that Paré was in practice, in the late 1530s, guns of all sizes had become the predominant instruments of death on the battlefields of Europe.

192

Ambrose Paré was born in north-western France in 1510 and was apprenticed to his elder brother, who practised as a barber-surgeon in Paris. His education, therefore, was not driven by classical languages and precedents, but by practical training. As the years went by, French would become his language of speech and writing, for unlike his Italian surgical contemporary Realdo Colombo, who combined his surgical training with a university education, Ambrose Paré was never a "university man". He was, rather, in the tradition of those Englishmen we saw above: John of Arderne, Thomas Bradmore, John Morsted, and Thomas Vicary, who enjoyed great professional success and reputation *outside* academia.

Paré, however, was certainly a cultured man who not only wrote French with seeming ease and left behind some three major treatises dealing with various branches of surgery, but also wrote autobiography and travel literature, and served four French kings in a professional capacity. The fast-moving necessities of battlefield surgery were not for the pedant, and the men who practised it were likely to be instinctive experimentalists, as is evident in Paré's great surgical works. His *La méthode de traicter leys playes faictes par hacquebutes, et autres bastons à feu* ("The method of treating wounds made by arquebuses/hackbuts and other firearms"), 1545, for example, tells us how, after a battle in 1537, there were so many wounded that he ran out of the boiling oil traditionally used to "cleanse" bullet wounds. He improvised with a mixture of egg-white, rose oil, and turpentine, and found to his surprise and delight that it was more effective and caused much less discomfort than did the hot oil.

Gunshot wounds, as we have seen, also necessitated the development of new and often radical techniques of limb amputation, and to hope for any success at all, a surgeon needed not only a good knowledge of anatomy, but also a well-trained team of assistants. In his *Dix livres de chirurgie* ("Ten Books of Surgery"), 1564, Paré described the technique for amputating a leg at the thigh: a horrendous operation for the pre-anaesthetic era, requiring the stitched application of no fewer than 53 ligatures. Another of Paré's innovations was the use of needle stitches to close the bleeding vessels following an amputation, rather than cauterizing them with a hot iron. To plan such an operation and prepare his assistants, Paré must have needed no less strategic skill than the general commanding the army.[2]

In his *Cinq livres de chirurgie* ("Five Books of Surgery"), 1572, Paré ventured into what was then the almost wholly female world of obstetrics, studying childbirth complications and reviving and perfecting the classical technique of podalic version, whereby the operator might turn a complicated "breech birth" baby in its mother's womb and bring it safely into the world. The great, 1641, folio edition of Paré's works brings over the sheer range and erudition of his medical and surgical understanding.[3]

Paré's younger French contemporary Guillaume Chamberlen would take obstetrics even further, devising for use in complicated births the earliest version of those special large padded forceps to take a gentle hold of a baby's head and ease it from its mother's womb. Owing to his Huguenot or Protestant religious sympathies, however, Chamberlen was obliged to flee to England, where his male descendants – several named Peter – followed him into surgery and medicine, to become pioneers of obstetrics and consultants to several English queens and other royalty.

Ambrose Paré's religious beliefs too were not without ambiguity. Although living, marrying, and having his children baptized as Roman Catholics, some events in his life – such as the danger some of his friends believed him to be in in August 1572, when large numbers of French Protestants were massacred on St Bartholomew's night – have suggested that Paré had secret Protestant Huguenot sympathies. Irrespective of denomination, however, he was a devout man, for when speaking of his own humble role in the business of curing, he simply stated "Je le pansai, Dieu le guérit", or "I dressed his wound, but God healed him."[4] By the time of his death in 1590 the anatomical and physiological branches of medicine in particular were changing fast under the impact of new discoveries. This tide of new thinking was influencing surgical techniques in particular.

In 1593, a 15-year-old boy from the King's School, Canterbury, went up to Gonville and Caius College, Cambridge, on the Archbishop Matthew Parker Scholarship. After graduating from

Cambridge, he would take his MD degree at Padua, in 1602, prior to returning to begin a distinguished medical career in London. Over the next 20-odd years, he would develop the first coherent account of the action of the heart and blood in the body. His name was William Harvey.

CHAPTER 12

William Harvey and the Circulation of the Blood

*I*n the early and mid twentieth century, it was often the fashion to depict the great Renaissance scientific discoverers, such as Copernicus, Galileo, Vesalius, and Harvey, as somehow standing up to and challenging ossified tradition: as self-conscious revolutionaries. But this radical scenario simply does not fit the documented historical evidence.

In addition to the obvious respect that astronomers such as Copernicus paid to the ancients, one finds a similar attitude of mind in most of the medical researchers. As we saw with Vesalius, Colombo, Eustachio, and others, all of these men took a great deal of convincing that Aristotle, Galen, and the wisdom of many centuries might be wrong on key points of anatomy and physiology. Medicine, no less than astronomy or geography, was part of a broader scheme of interpretation, with its own long-accepted truisms about how natural phenomena operated, and how the human intellect – be the writer a Greek pagan, a Judeo-Christian, or a Muslim – related to such phenomena.

One man who also drank deeply from this stream of classical wisdom was the instinctively conservative admirer of Aristotle, William Harvey. Upon leaving Padua with his MD degree in 1602, he would have been shocked had he known that, 30 years hence, his own discoveries would set in motion a whole train of ideas and researches which would fundamentally undermine classical medicine. For while first in Cambridge, then in Italy, Harvey would

196

have learned about the newly discovered "pulmonary circulation", and the tricuspid, mitral, and venous valves (and Vesalius's, Colombo's, Falloppio's, and Eustachio's newly discovered "tubes"), he, like these discoverers, would have seen such findings as *amplifications* and *clarifications* of classical anatomy rather than as fundamental contradictions of it. Classical medicine was such an elegant intellectual structure, with its four humours, three "spirits", brain cells, body cavities, and interrelating parts, that the idea of overthrowing it must have seemed ridiculous. The entire therapeutic rationale of classical medicine, such as bloodletting, purgation, diet, hygiene, and exercise, rested upon this structure, for in classical, just as in modern medicine, physiological explanation and the practical treatment of the sick were bound up together.

And in much the same way that Jacobus Sylvius had, from his own classical anatomical viewpoint, taken issue with Vesalius's discoveries, so the classical physicians Jean Riolan, Caspar Hofmann, and James Primrose after 1628 would take issue with William Harvey's. A significant number of senior medical men remained sceptical about both the reality and the significance of Harvey's 1628 discovery because it had no foreseeable application to *practical* medicine – nor would it do so for over 200 years!

Looking back with our gift of hindsight, it is hard to deny the monumental significance of Harvey's theory of total blood circulation. Without a knowledge of blood circulation "modern" medicine as we know it could not exist: nothing from the simple hypodermic injection of drugs into the arm, to post-1950 chemotherapy, to the blood-filtering techniques used to treat certain types of encephalitis, would have been possible.

Who was the youth who went up to Cambridge in Michaelmas Term 1593 on an Archbishop Matthew Parker Scholarship of £3. 0s. 8d. per annum, and would go on to elucidate how the cardiovascular system actually functioned?

ORIGINS AND EDUCATION

In the reign of Queen Elizabeth I, in the late sixteenth century, the Kentish Harveys seem to have been an upwardly mobile family. William's father, Thomas, had two businesses on the go, combining farming with operating a transport business between Folkestone and London. Thomas and his second wife Joan had nine children, and William, the future doctor, born in 1578, was the eldest of their seven sons, there also being two daughters. It is not clear where his brothers were educated, although they clearly did well for themselves, making money in business and even developing important City of London and Royal Court connections. We do know, however, that by the time William was 10, he was at the King's School, Canterbury. There, within the precincts of the great cathedral, he would have acquired a fluency in Latin, a knowledge of Greek, basic Anglican theology, geometry, natural philosophy (classical science), and a training in rhetoric and the arts of speaking – the usual curriculum in a school whose graduates would have been expected to go on to the universities, the professions, and into public life. It appears, however, that Harvey acquired what might be termed a practical or functional rather than a literary knowledge of the classical tongues, for as his admirer and biographer John Aubrey would record many decades later, he was "no Critique, and he wrote very bad Latin". Indeed, Aubrey tells us that his 1628 treatise on the blood circulation was "donne into Latin by Sir George Ent", Harvey's friend, to give it a more polished literary appearance.[1]

The Archbishop Parker Scholarship that took Harvey to Cambridge apparently had the explicit intent of enabling a Kentish lad from the King's School, Canterbury, to go on and study medicine. It is also significant that Harvey's Cambridge college had been re-founded and amply re-endowed by Dr John Kays (Latin "Caius"), to become Gonville and Caius. Kays, or Caius, had been an eminent physician, president of the Royal College, and a royal doctor of 50 years before, and his college would have a tradition of educating medical men. Almost certainly in Cambridge – for Dr Kays left provision for the dissection of two executed felons each year – then in London, Harvey would have witnessed the

198

dissection of human cadavers, in addition to receiving a systematic grounding in the medicine of Hippocrates, Aristotle, Galen, and the Arabic and medieval European writers. Harvey graduated with the usual Cambridge Arts or first degree in 1597. The Gonville and Caius Registers suggest that, while still on the college books, he had several quite long absences before formally moving out of residence in October 1599.

In the early months of 1600, Harvey left England to study in Padua: Europe's pre-eminent medical school, the heir of a great tradition extending back through Vesalius to Mondino and beyond. According to a story the elderly Harvey told John Aubrey, his first biographer, his life was saved by a curious circumstance when he arrived at Dover *en route* for the Continent. The governor of Dover, it seems, had had a strange dream telling him to detain a young traveller matching Harvey's appearance, thus causing him to miss the Calais boat. Tragically, this boat was lost in a sudden storm and all the passengers drowned. The following day, the commander saw fit to release him![2]

Upon his safe arrival in Padua, Harvey studied under the great Hieronymus Fabricius ab Aquapendente, a man whose approach to anatomy in this post-Vesalian world would be fundamental and far-reaching. Already 63 years old when Harvey arrived in Padua, Fabricius had held the anatomy and surgery chair in the university since 1565. His stress upon meticulous dissection had helped form generations of European medical men. Fabricius, indeed, had even designed and built the famous anatomy theatre at Padua (which still stands), where the students stood in an elliptical arrangement of steeply tiered viewing positions around the central dissection table. Leaning on the rail, and looking down, everyone was assured of a good view of the cadaver. And in addition to classroom teaching, the students were encouraged to do "hands-on" dissections of their own.

One of Fabricius's great discoveries were the *ostioli* or "little doors" of the veins. These were flaps of tissue, rather like a pair of double doors that opened in one direction yet shut together in the opposite and were especially noticeable in the large veins of the legs and arms. What could their function be? Despite his emphasis upon practical anatomy, Fabricius was a traditional classical physician and had wrestled with these doors. They seemed to shut against blood coming down the vein *from* the heart, yet open to anything moving upwards *towards* the heart. Yet this was most definitely not how the body worked, according to Galen. As you will recall from Chapter 2, in Galenic physiology the blood came *down* the vein from the heart, to nourish and be consumed in the extremities, as the venous blood distributed natural and vital spirits from the liver and heart, along with heat.

Fabricius had an explanation. Surely, the volume of blood descending down the arms, legs, and other organs, first from the liver and then from the heart, would be so great that it might flood out the extremities. So did the "little door" valves in the veins act as descending stepladders, or weirs on a river, to steady the stream of blood? This was Fabricius's suggestion, in a work that was to influence William Harvey's future thinking, *De Venarum Ostiolis* ("On the Little Doors of the Veins"), 1603. Fabricius was an Aristotelian in his broader philosophical thinking, seeing teleology or directional purpose as lying at the very core of nature. Was not the entire human (and animal) body an elaborate system of mutually interactive parts, where the *structure* or design of an organ was postulated by its physiological *function* and *purpose*? This manner of thinking was perfectly displayed in Fabricius's understanding of the vein valves, blood, and heart, and would help to mould the physiological thinking of William Harvey. Yet Harvey would, as we shall see, come to a very different teleological explanation for the "little doors" in the veins.

HARVEY ESTABLISHES HIS PROFESSIONAL CAREER IN LONDON

Having taken his Padua MD on 25 April 1602, Harvey returned to England, was incorporated at Cambridge, took a house in the parish of St Martin, Ludgate, London, applied for membership of the Royal College of Physicians, and began his practice as a young doctor. He married Elizabeth Browne, the daughter of a physician, towards the end of 1604. William and Elizabeth had no children, and sadly, we know little about Elizabeth, except that she seems to have kept a pet parrot. Elizabeth was well connected, however; her father, Dr Lancelot Browne, was physician to King James I, and a leading fellow of the Royal College of Physicians.

Harvey was admitted to the licentiate of the college in 1604, then, as the years passed by, doors were opened to him through his younger, very upwardly mobile brothers, John and later Eliab, both of whom became rich City of London merchants, with royal court connections. John, four years William's junior but already doing extremely well in the City of London, may have used his growing influence to help secure William's appointment to the important post of physician to St Bartholomew's Hospital in 1609. By that date, Rahere's twelfth-century foundation was already a long-established city charity, being taken over by the city following the monastic dissolution 70 years before, to become a lay charity, as it remains, now in conjunction with the National Health Service.

The Bart's appointment, however, only required William to attend the hospital one day a week and to treat the sick poor without charge. The appointment was not well paid in itself, but carried considerable prestige, especially for a young doctor, and gave him the rest of the week to make a good living treating the well-to-do. Yet in the great tradition of Christian charity, it meant that on at least one day a week even the poorest of the poor could be examined and treated by a top-flight doctor. And for the rest of the week, the poor were tended by a surgeon, an apothecary, and nurses, as the hospital foundation provided a bed, food, care, and shelter.

The hospital also provided an opportunity to study all manner of maladies, perform post-mortems on dead inmates, and, by definition, dissect. For in 1609, just like today, a major hospital was a fascinating museum of human infirmities, where doctors and students "walking the wards" could greatly expand their clinical knowledge and see in reality what they might otherwise only have read in a textbook. Such ward-walking could be a humbling experience – "there but for the grace of God go I" – and could help teach a vigorous, confident young gentleman the importance of compassion. And it still can.

OF HEARTS, PARADOXES AND PURPOSES: HARVEY'S
ROAD TO THE BLOOD CIRCULATION

Now that he was firmly settled within the medical establishment and the Royal College of Physicians, Harvey's professional career truly began to blossom. In 1618 he became physician extraordinary, or consultant, to King James I, his professional superior at court being the eminent Huguenot Sir Theodore Turquet de Mayerne, who, from his travels and high-profile consultancies, was probably the most experienced physician in Europe, and an authority on the occasional and hereditary diseases of the crowned heads of Christendom. Harvey's royal appointment continued under King James's successor, Charles I, and Harvey and Sir Theodore became not just colleagues but also friends. Following Sir Theodore's retirement, Harvey would succeed him as King Charles I's physician and would come to be regarded as a friend both by His Majesty and the royal family.

As a court physician he would not only meet, treat, and get to know great figures of state such as the Lord Chancellor and scientific philosopher, Sir Francis Bacon, but also join the stately summer court progresses. He was invited by the king, along with other medical men, for example, to examine some of the poor women arrested in the Lancashire witch trials of 1633, where his advice led to the "witches" being pardoned. (Their "sister" witches in the previous Pendle, Lancashire trials of 1612 had been hanged.) Then in 1635, at the behest of the king, Harvey dissected the body of "Old Parr", who claimed to have been born in *1483*, and to have performed penalties for adultery when 100, or some 52 years before! Harvey found no obvious pathological irregularities within Old Parr's body.[3]

William Harvey was also with the king at Edgehill, Oxfordshire, in 1642, the first battle of the English Civil War, where he was given charge of the young Prince of Wales (Charles II) and Duke of York (James II). Apparently, the three of them were reading under a hedge when a cannonball shot dangerously close, obliging them to move.

Back in 1615, however, the then young Dr Harvey was appointed Lumleian Lecturer at the Royal College of Physicians – a series of lectures intended for physicians and surgeons. Like Harvey's own education and practice, these lectures were essentially Galenic in their anatomy and physiology and Aristotelian in their "structure and function" approach to organ action, yet sufficiently flexible to incorporate new ideas and interpretations from Vesalius, Cesalpino, and others, such as skeletal differences and the "pulmonary circulation".

Harvey delivered the Lumleian Lectures in 1616. Central to them was the Aristotelian idea of structure and function. This implied that functional duties should be evenly distributed between organs, and yet, with the heart and lungs, there were problems. For example, in Galenic physiology the great vena cava appeared to be "overworked" in comparison with the aorta, for the Galenic vena cava both supplied blood into the *right* auricle (or atrium) and ventricle chambers of the heart, and then sucked most of it back out again (heated and enriched, with vital spirits) and sent it into the rest of the venous system. The aorta, by contrast, simply discharged a smaller volume of spirit-enriched blood into the arteries. The Galenic pulmonary artery took some blood from the right ventricle to enrich the lungs, whereas the left pulmonary vein simply blew *pneuma* (life force) enriched air into the right ventricle to make arterial blood. Was the "mitral valve" between the left ventricle and left atrium less efficient than its corresponding "tricuspid" or three-flap valve on the right-hand side? For the mitral valve allowed fumes to get back into the lungs, to be exhaled as bad breath. These were just some of the functional asymmetries in Galen-based cardiovascular function that were being pondered by doctors by 1618.

Stemming from what he had learned in Padua, and from his skill as a human dissector and animal vivisector, Harvey was considering the problem of the pulse. It was a crucial diagnostic technique, supposedly indicating how the humours and spirits were acting together within the body.

According to Galenic orthodoxy, the essential role of the heart was to initiate and sustain the pulse. In this Galenic view, the *active* phase of the pulse was when the heart swelled up and expanded in diastole. This regular swelling-up of the heart was believed to be caused by a sort of spontaneous tidal action from the liver, portal vein, and vena cava, to inflate the heart with a great rush of blood. The systolic phase, by contrast, was viewed as *passive*, rather like the tide ebbing away as it lost its force. Very importantly, in Galenic thinking, all this was seen as part of a spontaneous surging and ebbing process rather than some kind of "mechanically" driven one. One might suggest an analogy: laying aside our modern knowledge of air pressure, think of the blowing-up of a party balloon as an *active* phase and its deflation as a *passive* one.

Experiments on fishes and animals, however, were leading Harvey to a diametrically opposed explanation: namely, that the systolic, or contractive, phase of the heartbeat was the dynamic one. The systole, it appeared, corresponded with the expulsion of blood from the heart into the arteries, causing them to *swell*, thus making it the dynamic phase and the diastole the passive phase. Harvey suggested that the swelling of the arteries at right ventricle systole was analogous to the inflation of the fingers of a glove when one blew air into the glove.[4] He demonstrated this by puncturing the artery of an animal and noting at which stage in the cardiac cycle it spurted blood. This was an awful experience for the animal and far from pleasant for Dr Harvey, but the only way forward at a time when CAT scanning machines lay 350 years into the future.

Another problem lay with the very status of the heart itself. The Galenic tradition taught that in spite of its tough and flexible structure, the heart could *not* be a true muscle (and therefore, could not be a driving agent), because its action was *involuntary*, and we do not consciously will it to beat. To the classical anatomists, a muscle was generally understood as a *voluntarily* acting structure, such as when we *choose* to activate the muscles of our arms and legs.

If the contractive, systolic phase was what really directed the blood into the aorta, then was not the heart was performing some kind of *mechanical* action, perhaps like that of a water pump? In his 1616 lectures Harvey had spoken of the "clacks", or valves, of a water pump as an analogy to cardiovascular function.[5] His careful study of the physical structure of the heart led him by 1628 to suggest "its actual movement seems to resemble that of a muscle contacting in the line of its tendinous and fibrous components".[6]

In this increasingly technological age, he would not have needed to look far to see great mechanically driven water pumps in action, their "clacks" moving in exact sequence with the gushes of water into a pipe or tank. Only just down the road from his house in the city there was the great river-driven pumping station below London Bridge, where the force of the Thames turned waterwheels which operated great pumps, to drive a piped water supply, uphill, to the City of London. And in northern Italy, and especially in the polder drainage system of Holland, water- and windmills drove yet more pumps, some of which Harvey would have seen on his Continental travels.

Just like Vesalius with his stress on skeletons, muscles, and lever actions, so Harvey's gradually clarifying concept of cardiac action drew upon analogies from contemporary technology. Were not clocks self-acting devices, complete with their own heartbeat, or "tick-tock", and a necessary source of power, deriving from a spring or falling weight? Even by his Lumleian Lectures, after 1616, the forceful, systolic, phase of the heart was becoming uppermost in Harvey's physiological thinking. But we must be cautious about reading too much mechanism into Harvey's descriptions, for unlike the weights of a clock, or the river turning the waterwheels, there seemed to be no obvious source of power driving the heart which, to all appearances in 1628, still seemed to act *spontaneously* rather than *mechanically*. Perhaps, therefore, we should see of the pump more as a metaphor, rather than a functioning machine, in Harvey's thinking.

In his Lumleian Lectures Harvey would lay the foundations upon which his full circulatory system of 1628 was based. As a trained scholar as well as a physician, fully versed in the academic

205

procedures of the day, he presented his experimental evidence as an architecturally conceived structure, sincerely respectful to the ancients, yet making his original points with elegance and clarity.

ANNOUNCING THE WHOLE-BODY CIRCULATION OF THE BLOOD IN 1628

Harvey's book, *Exercitatio anatomica de motu cordis et sanguinis in animalibus* ("An anatomical exercise concerning the motion of the heart and blood in living creatures"), known thereafter as *De Motu Cordis*, had parallels to Galileo's *Sidereus Nuncius* ("Starry Messenger"), 1610. Both books related what were destined to become world-changing discoveries in their fields of research: medicine and astronomy. Both hinged on newly discovered physical evidences rather than philosophical deductions, and both were quite short: indeed, little more than thick pamphlets. *De Motu Cordis* consisted of 72 pages of Latin text, divided into 17 chapters, with illustrations. While Galileo had dedicated his book to Cosimo II, Grand Duke of Tuscany, Harvey dedicated *De Motu Cordis* to his patron, patient, and friend King Charles I of England. Both Harvey and Galileo were Padua men, although poles apart temperamentally. Galileo had a taste for provocation and a love of mocking those who followed the ancients, whereas Harvey was an instinctive conservative and respectful to the wisdom of antiquity, even when his experiments fundamentally challenged classical norms.

In *De Motu Cordis*, Harvey presents an elegantly structured case for a radically original interpretation of cardiovascular function, based upon meticulous experiments, many performed on cold-blooded eels and reptiles, with their slower and hence more easily visible heart action. But it is in Chapter 8 that he breaks radically new ground and becomes the first researcher to apply mathematically based, carefully quantified experimental criteria to physiology and medicine. It all had to do with the apparent "missing mass" of blood in the human body, especially if the *active* heart phase was the systolic, and the left-hand chambers of the heart disgorged blood into the arteries in enormous quantities.

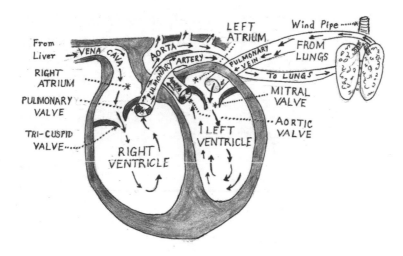

16. Heart function according to William Harvey, 1628, which is essentially how
it is understood today. The blood moves thus: vena cava, right atrium, right
ventricle, pulmonary artery, lungs, pulmonary vein, left atrium, left ventricle,
aorta. (Drawing by A. Chapman.)

He determined experimentally that each beat, or "pump", of the
heart discharged, at a *low* estimate, just over half a fluid ounce of
blood in the old apothecary's measure (14.2 ml). From this, Harvey
went on to calculate that the heart might, in different individuals,
beat between 1,000 and 4,000 times in 30 minutes, driving at least
"three and a half pounds of blood" into the arteries – a quantity
which, if it did not circulate, would simply cause the limbs to burst!
The arithmetic logic of his argument was impressive, even if his
figures were less precise than what we would reckon today. And
to show the range of his experimental investigation, he added that
heart action was "similar in the sheep or the dog".[7]

He also worked out the route of the blood circuit with astonishing
accuracy. Taking the sixteenth-century discovery of the pulmonary,
heart–lung–heart circulation as an established fact, William Harvey
proposed that, first, the blood enters the right-hand side of the
heart from the vena cava. It then passes through the atrium and
ventricle chambers and out via the pulmonary artery into the lungs.
The blood is somehow enriched in the lungs, making it lighter in

colour, before rising up through the pulmonary vein into the left ventricle and atrium, and leaving the heart by the great aorta artery. The blood's path through the heart is no mere sluicing action, but is beautifully orchestrated and volumetrically controlled by the regular movements of the tricuspid, pulmonary, mitral, aortic, and other flapped valves, or "clacks", as well as the sequential expansions and contractions of the four muscular heart chambers.

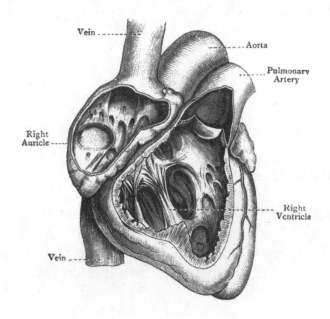

17. The human heart, showing the right auricle, right ventricle, and pulmonary artery (conveying the blood to the lungs). The left auricle and ventricle are not shown. (*The Family Physician* (*c.* 1905), Vol. V, p. 35. A. Chapman collection.)

Unlike the single-chamber "clack" and piston pumps, or water bellows, the heart was a truly amazing structure, with its four chambers, four principal valves, and lungs working in perfect sequence. It is not hard to understand why the anatomists came to see the hand of God at work in this glorious piece of bio-engineering.

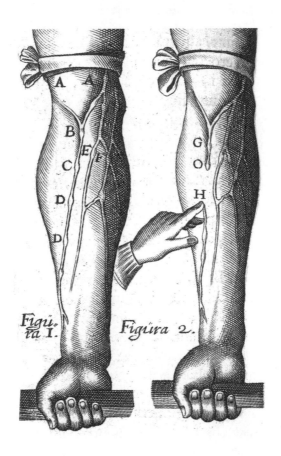

18. The valves in the veins. Re-interpreting a similar experiment depicted in
Hieronymus Fabricius's *De Venarum Ostiolis* (1603), Harvey demonstrated
that by placing gentle finger pressure upon the veins in a lightly-ligatured
arm, one could make the blood pass *upwards*, towards the heart, but not
downwards, or away from it, as the vein valves open to an up-flow of blood,
but close against a down-flow. (Harvey, *De Motu Cordis* (1628). Christ
Church Library, © Governing Body of Christ Church, Oxford.)

Then, having entered the aorta and its branching system, the
effervescing blood somehow *gushed* into the whole arterial system.
Among other things, this explained a phenomenon which military
surgeons in particular had known for millennia: why a severed
artery spouted huge quantities of blood in regular fountain-like

gushes, whereas a severed vein merely bled.

Ever since the days of Galen and before, it was known that the vein and artery systems paralleled each other throughout the body, almost like mirror images, as is clear in the plates of Vesalius's *De Fabrica* (1543). Yet while the peripheral branches of each system became more and more delicate, it was not believed that they interconnected, but that they remained quite distinct. And it was at this point that Harvey, whose whole case so far was built upon an edifice of elegant experiments and quantifications, made a leap of faith. With no optical device more powerful than a magnifying glass at his command, he proposed that the venous and arterial systems were linked by very tiny tubes, too small to see even with a lens, that permitted arterial blood to enter the veins. He fully understood, in 1628, this was the Achilles heel of his whole model. Yet once the blood had crossed safely into the veins, it became possible to explain the function of Fabricius's *ostioli*, "little doors", or valves, each of which opened upwards *towards* the heart. Instead of breaking the downward descent of the blood to the hands and feet, as Fabricius had proposed in *De Venarum Ostiolis* (1603), the "little doors" now provided an elegant stepladder by which the blood rose back to the vena cava – to resume the cycle all over again!

De Motu Cordis contained a simple, elegant, yet clinching experiment, borrowed from Fabricius but reinterpreted by Harvey, which anyone can try for themselves with the help of a friend (I have demonstrated this to many of my own students). An arm firmly grips an upright rod or staff. Then a light tourniquet is wrapped around the upper biceps. As this tourniquet temporarily prevents the blood from rising up those veins just below the skin and returning to the heart, the veins begin to swell up as the tourniquet dams the blood. The vein soon begins to resemble a string of sausages, with swollen chambers and depressions between. These depressions indicate the positions of the valves.

Now apply gentle finger-pressure, and try to push the blood back *down* the vein towards the hand. You simply cannot do so, because the valves are locking tight against any downward movement, in plain contradiction to Fabricius's explanation for the valves. Yet if you get a friend to apply finger-pressure on a "sausage" a bit lower down the arm and release the upper tourniquet, then the blood between the tourniquet and the spot where finger-pressure is being applied instantly shoots upwards towards the heart. But while pressure is kept on the vein lower down, the upper part subsides, quite empty, and it only refills when the finger is removed, allowing fresh blood to shoot up the vein once more from below.

The lesson to be learned from this experiment is this: blood cannot be made to go *down* a vein, for the valves close against it. Conversely, blood rises *up* the vein under the force of the heart's systolic pressure, and onwards back to the heart.

Many medical men, and philosophers, across Europe were fascinated by Harvey's elegant experimental results. The French philosopher René Descartes admired Harvey's circulatory system (as he did Copernicus's heliocentric solar system), largely on the ground that circularity as a phenomenon chimed in elegantly with his own philosophy of swirling "vortices" of physical forces. Yet many of Harvey's professional colleagues had their reservations and some were actively hostile, for a variety of what, in 1628, were perfectly sound reasons. Three of his most forceful critics were the German Caspar Hofmann, the Frenchman Jean Riolan, and the Englishman James Primrose, who between them articulated a series of perfectly reasonable objections.

In a learned culture dominated by Aristotelian teleological thinking, what on earth could be the *purpose* of the blood merely circulating around the body, rather than going to specific organs or body regions to accomplish specific tasks? What was the Aristotelian "final cause" of circulation? If the blood merely circulated, this must undermine the whole classical idea of the four humours and three spirits that lay at the very heart of physiology, health, and disease. For, as the ancients fully believed, the humour "blood" acted in partnership with the other humours and spirits to nourish and sustain life. The physician cured the sick by rectifying disturbed states of the humours and spirits by bleeding, purging, heating, or cooling the body on a regional basis: through the arms and legs, intestines, brain, or chest by influencing arm, chest, intestinal, and other *local* bloods. But if the same blood went to every organ of the body, and did so very rapidly, then the whole rationale of practical therapeutics was undermined. In an age lacking knowledge of brain, lymph, and glandular chemistry, these essentially purgative regimes of treatment could produce what we might call "placebo" effects, such as encouraging endorphin secretion.

When one adds to that the Achilles' heel of Harvey's inability to actually *see* the crucial arterial–venous bridge of the capillaries, one realizes that his critics were not so blind or reactionary as they are often depicted. (It would be the Bolognese papal physician, Professor Marcello Malpighi, who would finally use the microscope to see the capillaries – in the lungs of a frog – in 1661, some four years after Harvey's death, and thereby provide the key missing piece of the circulatory jigsaw puzzle.)

The Oxford scholar John Aubrey, whom we met above, felt privileged to get to know the elderly Harvey when he was young. He records details of the young Harvey's appearance and temperament, such as his being short in stature, with raven-black hair, and "very Cholerique", or of a "yellow bile", or hasty, humoral temperament. As an old man, Aubrey relates, Harvey found it hard to keep warm in bed and apparently followed the impeccable Old Testament precedent of the elderly King David, for Harvey "kept a pretty young wench to wayte on him, which I guesse he made use of for warmeth-sake as King David". In other words, as a human hot-water bottle. Harvey was generous to her in his will. Aubrey also tells us that the announcement of his circulatory theory damaged his professional standing, it being "beleeved by the vulgar that he was crack-brained; and all the Physitians were against his Opinion".[8]

In spite of the fact that Harvey's circulatory physiology fundamentally undermined the traditional rationale behind bloodletting, Aubrey tells us that nonetheless, when the dying Harvey suffered a paralysis in his tongue, "Sambrooke, his Apothecary" still bled him from the tongue, on the Galenic assumption that it would somehow release the locally pent-up humours causing the stroke. The bleeding did him no good, said Aubrey, although the palsy may have helped to facilitate Harvey's "easy Passe-port" to death.

THERAPEUTIC INNOVATIONS AROUND HARVEY'S TIME

While his professional colleagues acknowledged Harvey to be a consummate anatomist, and he enjoyed the prestige of being a royal doctor, he was *not*, recorded Aubrey, "admired [in] his Therapeutique way", nor were his prescriptions valued. He does not seem to have been outstanding as a working *healer*, as opposed to a medical scientist.

Sixteenth- and seventeenth-century Britain and Europe witnessed several significant improvements in practical therapeutics. One came from the growth of popular, English-language, "Poor Men's Physic" books, made possible by cheap printing and probably stimulated by the sudden disappearance of monastic health-care facilities after the Reformation. One early writer in this vein was the ex-monk doctor Andrew Boorde, whose *Dyetary* (1542) a *Breviary of Helthe* (1547) initiated a tradition in medical self-help books aimed at an "English literate tradesman" social bracket.[9] Their "cures" for a wide variety of maladies were generally based upon tried and trusted Galenic and Hippocratic medicine, now made available to anyone who could afford a modest sum and could just manage to read Tyndale's English Bible. Much more expensive was the "London Master in Chirvrgerie" John Gerard's Bible-sized *The Herball or Generall Historie of Plantes* (1597, 1633). On looking at the sumptuously attired Gerard's engraved portrait on the title page, it is immediately obvious that the surgeon saw himself *not* as a tradesman, but as a confident *gentleman*.

Another organizational improvement was the regular provision of medicine chests for army and navy surgeons, often designed, stocked, and supplied by the City of London Worshipful Company of Apothecaries (founded 1617), to guarantee good quality, unadulterated drugs, first for the armed forces, then for anyone who chose to purchase them. The Apothecaries' Company also became the legally chartered licensing body for druggists, setting up apprentice training and qualifying standards. While apothecaries were not officially permitted to dispense medicines without an expensive physician's prescription, they became, practically speaking, the ancestors of the modern general practitioner, being much cheaper, easier to get hold of, and more numerous than university-trained MDs.

One remarkable man who just *may* have sat in on Harvey's Lumleian Lectures was the English surgeon John Woodall, whose *The Surgeon's Mate* (1617, 1639) was aimed specifically at sea-surgeons, and supplies us with a mine of information not just about nautical surgery but also about medicine. For the sea-surgeon – just as in the army – was also the physician and apothecary to an isolated, predominantly male travelling community. Indeed, Woodall's book might also be thought of as the first treatise on occupational ailments, for unlike the previous treatises on military wounds, *The Surgeon's Mate* deals with a wide sweep of practice, including bowel obstructions, hernias, fistulae, ulcers, fractures, fevers, and – as had been all too familiar to the *Mary Rose*'s surgeon of 1545 – venereal diseases. Especially fascinating was Woodall's statement that the curse of sea voyages, scurvy, could be kept at bay with "citrons" – lemons and other citrus fruits.

Woodall later went into civilian practice as senior surgeon at St Bartholomew's Hospital, where he must have met Harvey. And in the 1639 edition of *The Surgeon's Mate*, he makes something of an inroad into the then infant science of medical statistics, telling us that over 24 years at Bart's, he had performed just over 100 amputations, losing only 20 patients. As a devout man, however, Woodall always approached surgery with a deep sense of responsibility, stating "it is no small presumption to Dismember the Image of God".[10] As with John Gerard, the assured countenance of the richly attired John Woodall looking out from the title page of the 1639 edition of *The Surgeon's Mate* is *not* that of a tradesman, but that of a cultured, well-to-do gentleman.

Woodall's older colleague William Clowes, whose experiments concerning the toxicity of gunshot wounds we saw in Chapter 10, also described a heroic civilian thigh amputation sometime before 1588.[11] What terrible accident had necessitated this radical operation we are not told, but it was performed on a female patient, a "Mayde of Hygate", London. Clowes, who clearly knew his anatomy well, and the massive blood loss which the operation would occasion, prepared for the ordeal with care and acted with speed. Not only did the "Mayde" live, but was cured "within a very short time".

19. John Gerard, apothecary, author, and master of surgery. It is clear from this portrait that Gerard did not see himself as a tradesman, or his professional status as in any way subservient. From his dress and demeanour, one could easily mistake him for a courtier, bishop, or judge. (John Gerard, *The Herball* (1633 edn.), frontispiece. A. Chapman collection.)

New drugs were also appearing in Harvey's time, such as crude quinine, or "Jesuit's bark", from the South American cinchona tree, which would revolutionize the treatment of malarial fevers, including the indigenous English and European "ague". Appearing, too, were those enduringly popular mild and refreshing stimulants, tea and coffee, both the products, like quinine, of new intercontinental maritime trade. John Aubrey tells us that William Harvey became very fond of coffee, even before it had grown from a curiosity sold by apothecaries to the new sensation of the 1660s. Another sensation, also a product of the long-haul economic trade, was a brown or white crystalline substance that boosted one's energy and was put in coffee and all manner of confections, yet caused the teeth to rot – sugar. Refined sugar first became available in seriously large quantities in the seventeenth century, and its popularity went from strength to strength. When archaeologists compare the teeth of people who died in the fourteenth and seventeenth centuries, the impact of sugar upon the Western diet is obvious. Yet it was from this growing epidemic of bad teeth that the dental profession would be born in the eighteenth century.

Then, more controversially, there were the new mineral drugs, made popular (or infamous) by the "iatrochemists" or chemical physicians who, in the wake of that exotic Swiss alchemist–physician Paracelsus – openly admired by Woodall – had argued that the traditional plant-based drugs were feeble alongside the new chemical preparations: deliberate products of the chemical laboratory, not of mother nature. We saw in Chapter 10 how mercury compounds had the power to produce vomiting and copious salivation and were – misguidedly – believed to be a good purgative for victims of the new disease, syphilis. Then there were antimony preparations, which were believed (incorrectly, as we now know) to have purgative functions, causing perspiration (a diaphoretic) and vomiting. Could not diaphoretic antimony "break" a fever by "sweating it out of the patient"? (No!) Tartrate of potassium (nitre) and antimony could form a "tartar emetic": good, it was believed, for making a person throw up violently, to "purge" the stomach.

216

In the seventeenth century, Glauber's salt (sodium sulphate) would enjoy some popularity as a vomit and stomach purge. Likewise, "spirits of hartshorn", or an ammonia-based fluid obtained from deer and other animal horn (*sal volatile*) became popular for "clearing [purging] the head" and as an ingredient of cough medicines. Most of these Renaissance mineral preparations were, in reality, toxins and irritants, likely to do far more harm than good, especially if used regularly.

By the time that William Harvey died, at his brother Eliab's London mansion, in 1657, the medical arts as a whole were in a state of rapid evolution. While most diseases still eluded cure, and life for most people was still very short, profound changes were in progress. Central to these changes was the growing status of experimentation as a new benchmark of "truth", replacing the ancestral one of ancient wisdom. This reorientation in thinking, in evidence in the great medical schools of Padua and Leiden and in Harvey's own work, would become yet more prominent among his disciples in Oxford and London, the early fellows of the Royal Society, and in that man who, in 1664, would lay the foundations of neural anatomy: Dr Thomas Willis.

CHAPTER 13

The Neurologist and the Archbishop of Canterbury, Part 1

We saw in Chapter 12 how the young gossip, antiquarian, and biographer John Aubrey of Trinity College, Oxford, was privileged to have several conversations with William Harvey and to record many of the incidents of his life for posterity. Likewise, he was fortunate to get to know his fellow Wiltshire-man and Oxonian, the notorious political philosopher Thomas Hobbes, who also happened to be a friend of Harvey (and a token beneficiary in Harvey's will). Aubrey mentions in his *Brief Life* of Harvey that Hobbes had written of him "he is the only man, perhaps, that ever lived to see his own Doctrine [discovery] established in his lifetime", meaning that by the time of Harvey's death in 1657, the circulation of the blood had largely become an accepted idea – even before Malpighi's microscopic discovery of the capillaries.[1]

This change in the perception of Harvey's discovery from being "crack-brained" in 1628 to an accepted fact by 1657 had been neither quick nor easy. Collateral discoveries that fitted in with Harvey's theory helped, such as Gaspare Aselli's discovery of the lymphatic (from Latin *lympha* = water) ducts, published posthumously in 1627. Indeed, the lacteal ("milky") ducts were especially significant, as they conveyed the milky-white chyle from the intestines to the bloodstream. Aselli traced chyle vessels to the mesenteric glands, while other anatomists, such as Jean Pecquet and Thomas Bartholin, took them further. Bartholin's *Vasa Lymphatica* ("Lymphatic Vessels") (Copenhagen, 1653) established the basic

218

anatomy of the lymphatic system, and its relation to the pancreas and liver.

But the discovery of the lymphatic system fundamentally undermined the credibility of the old Galenic explanation for the digestive system and the supposed relationship of the intestines, liver, and blood. A whole new department of physiology was opening up, which fitted nicely with Harvey's circulatory work. As we shall see in this and the following two chapters, a cluster of discoveries relating to fevers, blood, air, and heart action would, by 1675, supersede the Galenic physiology, while at the same time also elegantly according with Harvey's work. All this came about not by rhetoric or bombast, but through a succession of elegant experiments and discoveries, the cumulative weight of which became indisputable.

But now let us look at an incident which saved an almost certainly innocent young woman's life, while at the same time bringing a fortuitous lustre to Harvey's disciples in Oxford, and the young Dr Thomas Willis.

THE HANGING OF ANNE GREENE

On the chilly morning of Saturday 14 December 1650, a 22-year-old country servant-girl from Steeple Barton, Oxfordshire, was hanged near Oxford Castle, following her condemnation in the Assizes. She was accused of aborting her foetus, the dead baby – "a lump of flesh" – having been found in a privy. Anne Greene claimed that while she was working in the house of Sir Thomas Read at Duns Tew, Oxfordshire, the elderly knight's grandson Jeffrey had been carrying on with her. She claimed, however, that she had not even realized that she was pregnant until she suffered a miscarriage – presumably in the privy – after doing strenuous manual work in the malt-house. For Anne was clearly a strong country lass, being described as "fleshie".

From the condemned cell, Anne was taken to the scaffold, a psalm was said (probably Psalm 51, sometimes nicknamed the "neck-verse" from its association with criminal trials), and a

219

confession of her lewd life was made, then she was "turn'd off the Ladder".[2] She hung for half an hour, and when cut down, a soldier stood on her body, pressing the butt of his musket into her throat, to make sure she was dead. Anne's body was then handed over to the medical faculty to be dissected, by Dr William Petty, Dr Thomas Willis, Dr Ralph Bathurst, Dr Clark, and other medical gentlemen.

Just before they made the first incision, according to Dr Robert Plot, the doctors "perceived some small ratling [sic] in her Throat", and realized that, against all the odds, Anne might still be alive.[3] What happened next, is a fascinating collection of circumstances; medico-scientific, devoutly Christian theological, a possible "out-of-the-body experience", and legal. They are reported in the curiously entitled *Newes from the Dead. Or a True and Exact Narration of the Miraculous Deliverance of ANNE GREENE* (1651) and Robert Plot's *Natural History of Oxfordshire* (1677, 1705). As soon as Professor Petty, the senior doctor, Willis, and their colleagues realized that their "cadaver" was actually alive, they did everything in their power to save her. The blow-by-blow description of what they did gives us a fascinating insight into seventeenth-century resuscitation techniques, as, for example, might have been tried on a drowned person.

"Cordials", probably rich in alcohol, were poured down her throat to act as a stimulant. Anne's legs and other extremities were rubbed to warm her up, her fingers initially being locked rigid until warmth and massaging got the blood flowing once more. Then she was given an "odorous" clyster, or enema, whereby warm aromatic fluid was injected into her anus with the intention of heating up her insides and occasioning a bowel movement: all aimed at getting her body warmed, loosened up, and moving once again. The account of Robert Plot and *Newes from the Dead* also tell us that Anne had a vein opened and was bled: yet another way of nudging her heart back into action. Finally, she was put into a warm bed, with another young woman to lie upon her and impart body warmth, like a human hot-water bottle. Cataplasms, or hot poultices, were also placed upon her breasts and sides, all with the explicit aim of imparting heat. Interestingly enough, however, there was no mention of the subsequent resuscitation technique of inserting a pair of bellows into Anne's mouth, and attempting to restart her respiratory cycle. The whole emphasis – still remarkably Galenic in its rationale – lay in trying to *warm* her icy body back to life.

And gradually, Anne regained consciousness. Fourteen hours later, she began to speak – hoarsely at first, understandably – drank some cold beer, and by 9.00 pm on Sunday evening "she laughed and talked merrily, looking fresh and of good colour".[4] When moves were later made to have her re-hanged, the doctors firmly interceded with the magistrates and judges to the effect that her recovery was a clear indication of the will of God. Her innocence appeared even more obvious when, three days after her hanging, Sir Thomas Read, the grandfather of the youth who got her pregnant and, it seems, Anne's main accuser and prosecutor, suddenly fell down dead. Once again, the hand of God was plain for all to see.[5]

As Anne appeared to have passed through the gates of death, only to be restored to life again by the will of God and the ingenuity of the doctors, she was asked what she had witnessed when on the other side. "What fine visions… what celestiall musick or hellish howling she heard; what spirits she had conversed with; and what Revelations she brought back with her".[6] Plot mentions that there were tales circulating that she had seen "a fine green *Meddow*" with a river "glittering like *Silver* and *Gold*", but these were dismissed as suggestions by "the *Women*".[7]

Disappointingly, on more formal enquiry, Anne recalled no "out-of-the-body" experiences whatsoever. Between leaving her clothes to her mother just before going to the gallows and being revived by the doctors, her recollection was completely blank. She did not even recall the singing of the psalm that immediately preceded her execution, although some two weeks later, as scraps of memory returned, she did remember seeing a man wrapped in a blanket – her executioner. It seemed that Anne's conscious mental life simply picked up from where it left off when she revived, with no recollection in between: "like a Clock whose weights have been taken off a while, and afterwards hung on again", as the author of *Newes from the Dead* put it.[8]

No one doubted that Anne had a soul, that it had probably left, and then returned to her body. Rather, the trauma and shock of her ghastly experiences had benumbed her recollection: "As we often see it fares with men that are buzz'd in the head with drink, or transported with madnesse, who, though they seem sensible enough of every present object that moves them, yet after they recover can but [recall] little of what they did or said before."[9]

The gentlemen of the university, fascinated with her case, generously raised money to help set her up with some security once she had returned to her village in triumph, complete with the coffin in which she had lain as a trophy. Even so, her astute old dad, sensing the new celebrity status of his daughter, put her on show, charging admission to people who came from miles around to see the merry young woman who had been dead and was now very much alive!

The medical gentlemen who had been God's instruments in the revival of Anne Greene were also disciples of Dr Harvey: circulationists, dissectors, and experimenters – men of the "new medicine". Surely, their success was more than mere coincidence. Who, since the days of wise old Hippocrates, had put life back into a long-hanged man or woman? So Anne's resuscitation turned out to be excellent publicity for the doctors as well.

Anne became something of a legend in her lifetime, and she is far from forgotten even today in Oxford, as a sort of folk heroine. She stimulated the composition of rhymes and stories, some of which appear in *Newes from the Dead*. One rhyme was still doing the rounds in *c.* 1870 and was published by George Valentine Cox under his "Anatomy School Stories".[10] Elements in the rhyme suggest its composition was after 1685, when the infamous and incompetent hangman Jack Ketch was at work: "Jack Ketch" in the eighteenth and early nineteenth centuries being the generic nickname for the public hangman. Anne was described thus:

Anne Greene was a slippery queane,
in vain did the Jury detect her.
She cheated Jack Ketch,
and then, the vile wretch,
'scaped the Knife of the learned dissector.

But as Christian gentlemen, committed to the saving of life, Dr Petty, Dr Willis, and their friends seem to have been more than delighted to lose their cadaver, and to send a fit and healthy Anne, complete with a nice bag of money, laughing merrily back home to Steeple Barton.

DR THOMAS WILLIS OF OXFORD: PIONEER OF
NEUROLOGY

When he was busily trying to revive Anne Greene, in 1650, Thomas Willis was 29 years old and a rising star in English medicine. He had been born in Great Bedwyn, Wiltshire, in 1621. The son of a farmer, or perhaps connected to the minor gentry, his father, also Thomas, reputedly had an association with St John's College, Oxford. His mother's name was Rachel, and the family moved to North Hinksey, just outside Oxford, in 1630, when Rachel, a North Hinksey girl, inherited land in the village, shortly before she died in 1631. For young Thomas it was only a mile or so down the road to Edward Sylvester's famous classical school in central Oxford, where he acquired his training in Latin and Greek.

Willis left school at 15, to go 200 yards down St Aldates to Christ Church, which was and still is both a college of the university and the seat of Oxford's cathedral. Willis entered Christ Church as a "poor scholar", being "batteler" to Thomas Isles, one of the professorial canons of the cathedral, to whom, according to John Aubrey, he was related.[11] As a "batteler" Thomas was expected to do a variety of ordinary jobs for his master, such as wait on him, and in return, the young man received a full, free education. This was a time-honoured way by which generations of young men from modest backgrounds might rise up to careers of distinction –

undermining the myth that Oxford and Cambridge only educated the elite. (Canon Isles's memorial tablet, with its Latin inscription, can still be see in the Lady Chapel in Christ Church Cathedral today. It even mentions his "star" pupil, Thomas Willis.)

Young Willis and Canon Isles seem to have got on well. Being a devout young man of high church Royalist parentage and inclination, it was probably Willis's desire, in due course, to become an Anglican clergyman himself. Yet the English Civil War broke out in 1642, and in 1645 the triumphant Parliamentarians both executed the Archbishop of Canterbury, William Laud, and began the abolition of the Church of England as a national institution, which inevitably put paid to any intentions Willis may have had regarding a clerical career. (In 1660, however, both the church and the monarchy would be restored, and Willis would become both physician to and friend of a later Archbishop of Canterbury, as we shall see.)

Between 1644 and 1646, Thomas Willis served as a soldier in a Royalist university regiment, on the side of King Charles I against the Parliamentarians. But as a clerical career was no longer possible, he turned to his second love: medicine. Strong circumstantial evidence suggests that this probably came about through his having assisted Canon Isles's wife (probably Martha, his second wife) in her own dispensary. Martha Isles was "a knowing woman in physique and surgery and did many cures", and according, once again, to Aubrey, she would "oftentimes have him [Thomas Willis] to assist her in making of medicines".[12] Mrs Isles was not alone among women in her passion for medicine, for many ladies, often clerical wives, taught themselves medicine from books, especially if they were Latin literate, and at their own expense ran a charitable medical service for the poor from their husbands' vicarages or canonries.

It seems, then, that it was assisting Mrs Isles with her bottles, stills, and medicaments that led Willis to think of medicine as a profession. Exactly what formal training he acquired in a Royalist university under siege and in a country torn apart by civil war, however, is uncertain. Willis's Christ Church was also under siege, for having quit London, King Charles I had moved his shattered court to Oxford and was residing within the royal foundation of Christ Church, in the Deanery. The Parliamentarian gunners took especial delight in firing their artillery at Christ Church Hall when they knew the king was likely to be dining, and the cathedral still possesses a large iron cannonball which was shot through the roof. (The ball is about six or seven inches in diameter; I have lifted it myself, and it was clearly no love-token from Oliver Cromwell!)

Willis's growing passion for experimental medicine and anatomy may also have been influenced by William Harvey, who, as the king's physician and friend, had become warden, or head, of Merton College, only 200 yards from Christ Church, and where Queen Henrietta Maria held her besieged court. There were also around Oxford other men who we know were admirers of Harvey, such as Dr Ralph Bathurst of Trinity College and Dr (later Sir) Thomas Clayton, and they were also friends of Willis. Clayton, it seems, had been Oxford's Regius Professor of Medicine, and recommended Willis for the degree of Bachelor of Medicine in 1646, largely on the ground of his unswerving loyalty to King Charles I and the high church Royalist cause.

It may strike one as outrageous today that a man should be awarded a major medical degree on the grounds of political and ecclesiastical loyalty. For by 1646, Willis's medical knowledge had been picked up on a seemingly ad hoc basis: from Mrs Isles, from private study, and probably from having attended occasional lectures and done some hands-on dissection. Oxford had no "medical school" worth the name at that date and nothing remotely resembling Padua, Bologna, Montpellier, Leiden, or Paris, although it enjoyed the ancient privilege of conferring Bachelor's (BM) and Doctor's (DM) degrees in medicine to men who, on *oral* examination, seemed to know what they were talking about. Yet that is how the system worked, and in Willis's case, it worked spectacularly well, for the 25-year-old turned out to be not only a "natural" doctor and biomedical scientist, but one of the greatest medical discoverers of all time.

In the meantime, however, he had to earn a living, although we do not know what rental income he may have been receiving from his inherited North Hinksey properties, his father and his second wife

Elizabeth (Thomas's stepmother) both having died in the "camp fever" (typhus) epidemic of 1643. Willis started to receive private patients in his ancient and now long-demolished "Peckewater Inn chamber"[13] or college rooms in Christ Church, which he also set up for chemical experiments. We also know that he travelled to nearby Abingdon and other county towns on market days to take patients, probably in an inn or public house – a perfectly respectable practice at the time, especially for a young doctor. In addition, he visited private patients in their own homes, as was almost certainly the case with the woman of Eaton, a village just outside Oxford. The poor lady seems to have been suffering from sudden changes of mood and behaviour, which Willis recorded in his surviving case books and ascribed to fluid movements within her brain. Among his other achievements, Willis was said to be the first doctor to describe (but not to name) bipolar syndrome.[14]

His private practice shows Willis to have been a brilliant, instinctive clinician, who, although "learning on the hoof", had all the right instincts, asked the right questions, kept detailed case books, and displayed a great sense of responsibility towards his patients. And in conjunction with all this medical science, Thomas Willis was inviting dispossessed Anglican clergymen to celebrate illegal Book of Common Prayer services in his rooms. In 1657 he married Mary, the daughter of the Very Reverend Dr Samuel Fell, the former Dean of Christ Church, who had been evicted by the Puritans. The Fells had become icons for the opponents of the Cromwellian regime: Dean Fell was imprisoned by the Parliamentarians, while his wife Margaret, son John, and daughter Mary had to be physically thrown out of the Christ Church Deanery because they flatly refused to obey Puritan demands and leave when ordered to do so. Only the king, Fell argued, could order them out. Willis was obliged to give up residence in Christ Church, however, when he married (only the Dean and Canons, *not* the lay dons, were able to retain their rooms after marriage). So Thomas and Mary moved into Beam Hall, Merton Street, only a minute's walk from Christ Church's back entrance, and set up practice there. Willis also, in conjunction with some medical colleagues, took over the run-

226

down Angel Inn a little further down the street, which they ran as a very successful private hospital, ideally placed for the arrival of coaches conveying invalids between London and Bath.

In their new and spacious Beam Hall home (which still stands), the Willises increased their invitations to renegade Anglicans during the final Puritan years, and covert Royalists crowded in to hear a dispossessed bishop preach or to celebrate the Eucharist. I have often wondered about the apparent irony of celebrating illicit Anglican services either in college rooms or in a house that also served as medical research premises, but in which the solemn words of the Eucharist might have been spoken in the presence of bottles of chemicals, distillation retorts, and furnaces, while half-dissected body parts might even have lain under sheets nearby. Yet Willis would have seen nothing strange in this, for his deep Christian faith was held in conjunction with a driving scientific curiosity, and a passion to discover how the human body was constructed and how its ills might be rationally cured.

Thomas Willis's star truly rose in 1660, when the Puritan regime ended and King Charles II was restored and welcomed home as king, some 11 years after the Parliamentarians had tried and executed his father. Willis was given the Oxford Sedleian Professorship of Natural Philosophy (science), while Mary's brother and Thomas's friend John Fell replaced his deceased father Samuel as Dean of Christ Church in 1660, and later as Bishop of Oxford.

During all this time, from the mid 1640s to becoming Sedleian professor at the age of 39 in the Restoration year, Willis had been maturing into a physician, anatomist, and experimental scientist of outstanding brilliance. He had, among other things, become associated with that Wadham College-based "Club" of experimental scientists under the aegis of the Reverend Dr John Wilkins, Warden of Wadham, which, in 1660, would be chartered as the Royal Society of London (he would be elected a fellow in 1663). Yet before his brain researches immortalized his name in medical circles, he had begun to study how chemical actions taking place in the human body might lie at the heart of a variety of maladies, and fevers in particular. All of this was accomplished within the "Harveian" tradition: one emphasizing meticulous physical research, dissection, vivisection, and controlled and quantified experiments, and taking it as axiomatic that the blood circulated around the human body under the systolic, or contractive, force of the heart, while at the same time respecting the ancients.

20. Dr Thomas Willis, aged 45, by David Loggan. (Willis, *Opera Omnia*, ed. Gerard Blasius (Amsterdam, 1682). (Christ Church Library, © Governing Body of Christ Church, Oxford.)

FERMENTATION, FEVERS, AND CHEMISTRY

Chemistry was Thomas Willis's original medico-scientific passion: in particular, the chemical changes taking place within the human body. But he saw medical chemistry as part of a wider understanding of chemical action in the world in general. As we shall see in Chapter 15, Oxford, between *c.* 1650 and 1675, was to be one of Europe's foremost "centres of excellence" in chemical research. Playing a leading role here was Dr Wilkins's informal "Club" of like-minded friends: the proto-Royal Society. Among them, as well as Thomas Willis, were the Honourable Robert Boyle, Robert Hooke, Ralph Bathurst, John Locke, Seth Ward, Christopher Wren, Richard Lower, and, slightly later, John Mayow. The friends even clubbed together to build a chemical furnace and equip a laboratory in Wadham.

There was no shortage of other private laboratories in Cromwellian and early Restoration Oxford, including those which Willis set up first in his Christ Church rooms, then in Beam Hall and in the Angel Inn hospital. Willis was coming to think that traditionally "mysterious" illnesses, such as sudden fevers or mental illness, were caused not by the winds, bad smells, the planets, or witches, but by *physical* changes in the patient or the environment. Fundamentally, those changes were *chemical* in nature. Thomas Willis's skills as a manipulative laboratory chemist were well known, and the ubiquitous biographer John Aubrey would refer to him as "our Chymist": the old word indicating a highly skilled chemical manipulator, good at distillations, evaporations, separating out substances, making medicines, and such. When one sees these skills displayed in conjunction with those of an adroit anatomical dissector and a perceptive clinician, one begins to grasp why this largely self-taught doctor became so significant.

In the mid and late 1650s, both in his Christ Church rooms, then, after marriage, in Beam Hall, Willis was exploring the nature of *substance*. He was a member of that "iatrochemical" fraternity mentioned at the end of Chapter 12: men who were thinking beyond the four Elements of antiquity – Earth, Water, Air, and Fire – towards a new concept of chemical *process*. The founder of this tradition, in the first half of the sixteenth century, had been that exotic Swiss alchemist–doctor Paracelsus, with his new chemical "principles" of reaction that lay at the heart of all chemical change: mercury (volatility), sulphur (inflammability), and salt (stability). The Middle Ages had discovered a range of chemical phenomena not known to the ancients, such as explosives, and powerfully corrosive mineral acids which could destroy all metals but gold.

Paracelsus's clarion call to the world had been the pre-eminence of the chemical laboratory in the discovery of new drugs and the developments of new treatments, as we saw with mercury, antimony, and others in Chapter 12. Yet in Paracelsus there was a great deal of speculative philosophy and quasi-chemical mysticism. Paracelsian iatrochemistry had then been brought back down to earth by the great Flemish chemist Johannes Baptista van Helmont who, like Vesalius, was a native of the region of Brussels. Van Helmont, who died in 1644 aged 64, would be inspirational to the rising generation of medical chemists in his concept of chemical action as process, not mystery. Chemical action, he saw, was driven by key forces, such as heat, solution, evaporation, and such, which could be controlled by a skilled "chymist" to unravel the innermost secrets of nature and to cure diseases.

In addition to van Helmont's influence, however, chemical and physical ideas were beginning to change in response to the revived "Christian atomism" of the French Roman Catholic priest–scientist Pierre Gassendi. Contra the atheistic or agnostic atomism of Lucretius of c. 60 BC, who argued that the world around us has no especial meaning, being simply the product of chance groupings of swirling atoms, Gassendi said one could see meaning if one saw the atoms as guided by God to form all manner of things, from planets to plants to human beings. Gassendi's Christianized atomism would be inspirational for many devout "chymists" and physicians, most notably Willis's friend Robert Boyle. The French philosopher René Descartes, the declared admirer of Harvey, would also give fresh inspiration to chemical, medical, and physical thinkers, for while Descartes was not an atomist, he nonetheless saw *mechanism* and mechanical action – not humours and the classical four Elements – as lying at the heart of everything, including the human body.

The young Willis, like most of the early Oxford chemists and doctors, including Dr (later Sir) William Petty – who was due to have led the dissection of Anne Greene – were all basically Helmontian and, in varying degrees, Gassendian and Cartesian in their approach to chemistry and medicine, which brings us to that body of researches by Willis which would lead to his first major internationally acclaimed and much reprinted *Diatribae duae medico-philosophicae, quarum prior agit de fermentatione, altera de febribus* ("Two medico-scientific treatises, the first on fermentation, the other on fevers"), 1658–59.

In these treatises, Willis dealt with two common processes – fermentation and fever – and saw a connection between them. Suspending for a moment what we now know about microbial action and gases, what both fermentation and fever display is an apparent generation of heat in the absence of any source of flame. Both lead to seemingly mysterious changes of substance, which appear to be corollaries of the heat. How, for instance, do hops, malt, yeast, and water become beer in the vat? Likewise in cheese-, butter-, and wine-making, all of which change the basic nature of the ingredients, while spontaneously warming the vessels in which they are contained. As everyone living on a farm, a smallholding, or in an insanitary, congested town knew all too well, snow always melted when it fell on a dunghill, for dunghills somehow generated their own internal heat. Some "chymists" buried vessels requiring sustained, gentle heat in a dunghill in order to achieve a slow chemical reaction. So was heat *not* an "element", but the product of a chemical process?

Fevers likewise generated their own mysterious heat. As every doctor knew, a person might be hale and hearty at breakfast, uncomfortably warm at midday dinner, then in the grip of high fever at 6.00 pm. Where did the heat come from? Willis suggested in *Diatribae* that fever was related to fermentation and perhaps derived its heat from a similar chemical process. As the blood was the warm fluid that the heart drove to every part of the human

231

body, fever was most probably caused by the blood fermenting and generating excess heat.

In our modern post-bacterial and post-viral age, we now know that Willis was incorrect, although he was wrong for the right reason. Fevers are the visible manifestation of highly complex physiological reactions taking place in the patient's body, *not* caused by some malevolent or mysterious agency. While Willis still had no real conception of what caused different types of fever, he left some very perceptive observations on the natural courses run by typhus, typhoid, puerperal, and other fevers.

Thomas Willis's *Casebooks* and *Oxford Lectures*[15] supply us with a fascinating collection of astute observation and experiment-based clinical studies in which Willis, the devout Christian, nonetheless saw disease as a natural process amenable to rational study – as did Hippocrates. By Willis's time, there was in existence a growing body of meticulous anatomical and physiological knowledge, while a range of substance-analysis techniques were available to the seventeenth-century "chymist" of which the Greeks had been ignorant.

One medico-chemical discovery, which Willis subsequently announced in *Pharmacopoeia Rationalis*, 1674, was the diagnosis and description of what he styled diabetes *mellitus* (Latin, honey-sweet). While diabetes was an already known medical condition, it was Willis, in the *Pharmacopoeia*, who identified the sugar component. Skilled "chymist" that he was, the simple laboratory technique that enabled an "on the spot" diagnosis was a warm summer's day when a dish of sugary urine might attract flies and bees. (We shall return to diabetes in Chapter 26.)

One young man, the equivalent of a modern postgraduate research assistant, who worked with Willis during the preparation of the *Diatribae* in the 1650s was Robert Hooke, also of Christ Church. As a biographer of Hooke myself, I believe it was during his time with Willis that Hooke acquired his conspicuous skills as a laboratory "chymist", an anatomical and physiological dissector, an experimenter, and – with reluctance – a vivisector. And all in the tradition of William Harvey and the circulation of the blood.

ARTHUR COGA AND THE SHEEP: EXPERIMENTS
WITH BLOOD AND CIRCULATION

Once Harvey's circulation theory had been announced, it opened up all manner of physiological and experimental possibilities. If, for example, any vein or artery gave access to *every* part of the body within a minute or so – the heart pumping 70-odd times per minute – then was it not possible to use that rapid circulation to convey substances to the internal organs? Or to transfuse blood from one living creature to another?

In 1656 an experiment was performed, probably in the gardens of Wadham College, Oxford, by men who were part of Dr Wilkins's "Experimental Club". The young [Sir] Christopher Wren injected a warm mixture of opium dissolved in sack wine into the (possibly crural) vein in a dog's back leg, the purpose of the experiment being to test how a drug might act if administered by a non-oral route. The dog fell into an almost instant stupor: the drug acted much more rapidly than if the same fluid had been lapped up orally. (It soon made a full recovery.) The result of this experiment chimed in with Harvey's circulatory theory, for had the drug simply stayed in the blood of the leg and been absorbed there, as Galen's theory taught, then it would not have rendered the dog unconscious.[16]

This experiment made clear that the extremities of the body are directly connected with the brain, through the cardiovascular system. By 1656, no anatomist doubted the fact that it was the *brain*, and not the *heart*, which controlled both voluntary and involuntary bodily movement. (It would, however, be another 200 years before medical knowledge had advanced sufficiently to enable Alexander Wood of Edinburgh to use a hypodermic syringe with a very fine hollow needle to intravenously inject drugs into human patients.)

Another possibility which Harvey's circulatory theory opened up was the transfusing of blood between living beings. Transfusion, vein to vein, was a very different thing from the much older, ineffective practice of giving blood from a vigorous young person to a sick elderly one as a *drink*, in an attempt to "rejuvenate" them.

To understand the logic behind therapeutic transfusion in the 1650s and 1660s, however, we must lay aside our knowledge of blood biochemistry, and think more in classical and biblical terms. While Marcello Malpighi in Bologna and Anton van Leeuwenhoek in Delft were coming to realize that blood was not the homogeneous red fluid that it appears to the naked eye when it is examined under the microscope, most people took it as axiomatic that different living creatures had their own special types of blood. After all, Genesis taught that life "is the blood",[17] with all human and animal forms deriving their unique features from their blood. Brave men had strong and fiery blood; sheep had timorous, pacific blood; mad people had disturbed blood; and fish and reptiles cold blood.

What would happen if you gave a madman lamb's blood (or, as the witty Samuel Pepys FRS said, what if "the blood of a Quaker be let into an Archbishop")?[18] The creative possibilities of blood transfusion first became clear after the Montpellier professor Jean-Baptiste Denys supposedly cured a Parisian lunatic by transfusing sheep's blood into him. On 23 November 1667, the newly formed Royal Society of London undertook its own transfusion experiment. The willing victim was one Arthur Coga, a drunken and debauched Cambridge graduate who had apparently been employed in an ecclesiastical capacity by John Wilkins.

Whether Thomas Willis was present at the transfusion is not clear, although it was his Oxford protégé, Dr Richard Lower, who acted as transfusor, while his other protégé, Robert Hooke, by now curator of experiments at the Royal Society, almost certainly assisted. A pipe was set up to allow a direct flow of blood from a living lamb directly into Coga's upper arm vein. It was estimated that half a pint, or 281 cc, of sheep blood flowed directly into Coga's body. The blood had clearly not coagulated en route, for when the pipe was removed from Coga's arm, sheep blood flowed from it.

Arthur Coga sailed through his ordeal without a hitch, and on 12 December he returned for another 14 fluid ounces (398 cc) of lamb's blood, after which he addressed the Royal Society, saying how well he felt. By all the laws of physiology, he should have died from a violent allergic reaction after his first transfusion, for sheep blood is wholly incompatible with that of a human being. And he should *never* have survived a second dose.

I have discussed Coga's case with numerous medical friends, and all agree that he should have died. So why did he not? A new line of thinking opened up during the question session following a lecture on medicine and the early Royal Society which I delivered to the Royal Society of Medicine, London, in spring 2014. A medical gentleman asked me "Do you know the systolic blood pressure of a sheep?" No, I did not, and neither did he, nor any of his colleagues in the audience. Yet his question suggests a reason for Coga's survival. If a sheep's blood pressure is *lower* than that of a human, and the transfusion was being performed with a simple flow through a pipe, then the laws of hydrostatics suggest that Coga's blood might have inadvertently been *transfused into the sheep*.

I have since been informed by a veterinary physiologist at the University of Lancaster that a sheep's blood pressure is around 100–110 mmg of Hg (mercury) systolic, over 75–85 mmg diastolic.[19] "Normal" human blood pressure is ideally around 120 systolic and 80 diastolic, but as Coga was a heavy brandy drinker, it is likely that his was considerably higher. Yet the tied-up sheep was distressed, with its heart pounding, so who got whose blood is something of a guess. Sadly, the Royal Society account of the transfusion experiment tells us nothing about the fate of the sheep. Did it, perhaps, collapse in a drunken stupor as Coga's brandy-enriched blood entered its veins?

These early blood transfusion experiments, conducted in complete ignorance of human, let alone animal, blood-typing, were horrendously risky undertakings, and as the French and English experiments were repeated elsewhere in Europe, there were fatalities. Very soon, they had to be stopped. In addition to the purely physical risks involved, animal-to-human experiments opened up major religious and ethical questions. Was there not something *blasphemous* in trying to re-engineer the mind, body, or habits of a human being made in the image of God, by imparting animal characteristics? Eating cooked meat from a lamb was one thing, but vein-to-vein transfusion of the warm blood of a living lamb was altogether different. Ethical questions associated with the use of living animal matter in human medical experiments would arise on future occasions, most notably, as we shall see in Chapter 20, when Dr Edward Jenner successfully used cow-derived matter as a preventative for smallpox. They still do today, especially in the area of genetic engineering, and the reusing of animal parts in human surgical procedures, as when pig valves are transplanted in human heart surgery.

All of the above could only have followed in the wake of William Harvey's circulation discoveries and the new physiological perspectives that they opened up. While circulation theory's therapeutic potential was but marginal in the seventeenth century, it triggered the growth of a critical mass of medical discoveries that would usher in the post-1850 therapeutic revolution.

In his admiration for Harvey, his part in the revival of Anne Greene, his work on fevers, fermentation, and chemistry, and his inspiration of an "Oxford School" of medical experimenters, Thomas Willis played a vital role. Now we must examine that body of researches by Willis which, after *c.* 1660, would lay the foundations of modern neuroscience, bequeath to the world the term "neurologie" or neurology for the scientific study of the brain and nervous system, and which he would dedicate to his friend the Archbishop of Canterbury.

CHAPTER 14

The Neurologist and the Archbishop of Canterbury, Part 2

When Thomas Willis entered Christ Church, Oxford, in 1636, fresh from Edward Sylvester's School, to be a batteler to Canon Isles, the college had long been home to a bachelor clergyman don of extensive medical interests and erudition, who had a European reputation (but only under his pseudonym) as an authority on what we would now call "depressive illnesses". Whether Willis ever met or conversed with this clerical gentleman, or simply exchanged respectful bows, is a moot point, although he was 19 years old when the clergyman died in 1640.

THE REVEREND ROBERT BURTON: ANATOMIST OF
MELANCHOLY

Robert Burton's death, on 25 January 1640, is not without paradox. The young Robert Hooke, who came up to Christ Church in 1653, told his friend and biographer John Aubrey, sometime around 1680, that he, Hooke, once occupied the same Christ Church rooms "that was Mr Burton's, of whom 'tis whispered... he ended his dayes in that chamber by hanging him selfe".[1] Other contemporary writers too, such as the acerbic Antony Wood, spoke of Burton as having hanged himself: "he sent up his Soul to Heaven thro' a slip about his Neck".[2] Yet in the seventeenth century, suicides were generally buried in unconsecrated ground, whereas Burton appears to have been interred within the Christ Church Cathedral precincts. What is more, in the north transept of the cathedral stands a magnificent funerary monument to Burton, complete with portrait bust, the Latin inscription which he composed himself, and horoscope, all exquisitely carved – a mark of respect not generally accorded to a suicide, even a famous one. (A fine portrait of an almost smiling Burton, by Gilbert Jackson, hangs in Brasenose College, while his face appears yet again within the allegorical frontispiece of the third edition of *The Anatomy of Melancholy* (Oxford, 1628).)

Nowhere does Burton's personal name appear on his memorial. Instead is the pseudonym by which he was already famous: Democritus Junior. Now medieval and Renaissance academics loved playing Latin word-games, and this, I suspect, is what happened here. For while there is a marble bust of Burton in the centre of his memorial, painted to resemble the real man in life, the inscription reads *Paucis, Notus, Paucioribus, Ignotus*, or "Known to few, unknown to fewer". The inscription ends, equally paradoxically, with *Cui, Vitam, Dedit, et, Mortem*, or "To whom [Melancholy] brought both life and death." Does this mean that while his acclaimed *Anatomy of Melancholy* (1621) certainly brought his pseudonym "Democritus Junior" fame and some money – hence life – the melancholic affliction also brought him death? Does "death" here mean darkness of the soul and misery to be endured, or actual death: suicide? Either way, melancholy was a passion which dominated his life.

Some indications suggest, however, that Burton's melancholy – a "cruel disease... which crucifies the soule in this life"[3] – was episodic, for some elderly Christ Church dons who had known Burton before 1640 told Antony Wood when he was researching Burton's life for his *Athenae Oxonienses*, published in 1691 that he could be jolly good company and a merry companion, when the mood took him.[4] By the time of his death in 1640, brought about by whatever means, natural or unnatural, Burton was 63 years old. He had lived in Christ Church for the previous 40 years, ever since becoming a student (senior member) of Christ Church and Chaplain of the House after completing his undergraduate studies at Brasenose College, Oxford, and then being ordained as an Anglican priest. He also held the Christ Church living of St Thomas's Church, Oxford, as well as ecclesiastical benefices in Lincolnshire and Leicestershire. But he lamented his failure to rise higher in church or state, seeing himself as "left behind, as a Dolphin, on shore, confined to my Colledge, as *Diogenes* in his tubbe".[5]

Robert Burton had a serious interest in medicine and collected medical books, and in his masterpiece he chose to treat of a specific form of mental illness – melancholia, or depression – in the same way that a medical man might explicate the secrets of a human body: by conducting an *anatomy*, or dissection, upon it. One of the fascinating things about *The Anatomy of Melancholy* is the vast range of erudition which Burton brings to bear upon it, with extensive quotations from both ancient and modern writers.

In his pathology of melancholy, Burton follows the traditional humoral line: black bile was at the root of the disease, while it was also influenced by the movements of the planets which, as we have seen, were believed to affect the four bodily humours. Saturn was the planet believed to be the astrological "patron" of the disease: Saturn's slow motion around the zodiac the very antithesis of vitality, and, in Greek mythology, Kronos (Saturn) was a mad god who even ate his own sons. (The Spanish artist Francisco Goya produced a horrifying re-creation of this incident, in a painting of *c.* 1820, now hanging in the Prado Museum, Madrid.)

Burton, in keeping with the wisdom of the age, also saw melancholia as a thing to which scholars were especially prone, due to the solitary and often introspective nature of their occupation. He was sound in his advice, for nothing was worse for a melancholic than to let himself be drawn into deep solitary brooding, which would only make things worse. Instead, Burton advises constant activity and diversion as a way of lifting heavy spirits. "Be not solitary, be not idle" are his valedictory words of advice to melancholics at the very end of his *Anatomy*.

Burton considered that melancholics suffered from one or more basic types of the disease and even had them depicted – accompanied by their correct astrological symbols – upon the title page of the *Anatomy*. They were Love-lorn (Inamorato),

239

Religious (Superstitiosus), Raving (Maniacus), and Hypochondriac (Hypocondriacus) melancholy. All were obsessive, inclining the sufferer to shut himself away and brood.

Much of the *Anatomy* deals with particular ways in which a melancholic might pursue healthy diversion, and these in themselves give us a fascinating insight into seventeenth-century thinking about depression. Some of the suggestions might fall into the category of what we would now call "occupational therapy". Included are changes of air, travel, exercise, diet, mirth, various diversions, and music: the latter being an agency by which young David, playing his harp, brought some relief to the increasingly unstable King Saul in the Old Testament. Plato too had been very much aware of how different types of music could alter a person's mood.

Burton then cites a whole range of herbs that have influences upon the human body, to the effect that they, by causing the humours to move or change, might expel melancholy. He cites Galen, Dioscorides, and others about the power of plants such as bugloss, which "is hot and moist [contra the cold of black bile], and therefore worthily reckoned up among those herbs which expel melancholy and exhilarate the heart".[6]

Burton's *Anatomy* is not an easy book to read. It is, rather, a stunning cornucopia of diverse information about melancholy, drawn from classical and modern writers, and his English text is frequently interrupted by long quotations, mainly in Latin. Nor is Burton attempting what might be called an original neurological study, for he does not tie up behavioural, temperamental, and emotional states with the actual anatomy of the brain in the way that Thomas Willis would. But his study, in its scale and scope, furnishes us with the first in-depth analysis, literary, philosophical, and medical, of a single mental condition.

THOMAS WILLIS AND HIS "CIRCLE"

Over the late 1650s, much of Willis's attention seems to have been focused upon laboratory research, although he was constantly expanding his clinical expertise as an increasingly successful (and wealthy) practising physician. This had laid the foundation for his work on fevers and fermentation in the *Diatribae* of 1658–59, which were to win him fame across Europe. By the early 1660s, however, and now holding the prestigious Sedleian Professorship of Natural Philosophy [Natural Science] in Oxford University, with which he had been presented at the Restoration in 1660, Willis was turning his attention to the blood supply to and the nervous appendages of the human brain.

The brain was a mysterious organ, especially when seen in relation to its complex nervous appendages which ran – like Harvey's cardiovascular physiology – to every part and extremity of every living body. The brain's physiological chemistry too appeared a puzzle in 1660. Why, for example did the heart drive warm blood up the carotid and vertebral arteries into the head, and the jugular and other veins return it to the heart, while the cortex itself in dissection seemed cold and largely bloodless? How did some of this blood transform into the watery "animal spirit" (cerebrospinal fluid) found in the brain ventricles of dissected cadavers? All of that was dwarfed by the *huge* mystery of how the brain enabled us to perceive, respond, think, and imagine.

An incident took place sometime before 1664, which Willis described in his *Cerebri Anatome* ("Anatomy of the Brain") of that year, which would turn out to be transformative in our scientific understanding of the brain. Willis was dissecting the cadaver of a man whom he appears to have known in life, when he was struck by the presence of a long-standing blockage of the right carotid artery in his neck. The carotid had become *osseam* ("bony") or *lapideam* ("stony") hard and long since incapable of conveying blood.[7] Yet the right vertebral artery, which runs close by the carotid, had swollen two or three times in size, as though to compensate for the failed carotid.

By traditional physiological reasoning, the man should have been either dead long since or should have manifested signs of major brain-function impairment, such as an "apoplexy" or stroke, for the right hemisphere of his brain would not have been receiving its full and necessary blood supply. This is what Galen's anatomy

241

taught: when blood passed up the carotids, it entered the *rete mirabile* or "wonderful network" of vessels, half of which nourished the left brain hemisphere and half the right, and even the swollen right vertebral artery would have left part of the brain blood-deficient. But as Willis made clear, the man had displayed no signs of brain dysfunction and had died from a stomach problem: a large *schirrus* [scirrhous] or hard, ulcerated tumour of the abdominal mesentery – a cancerous tumour, most likely, though it was probably not recognized as such in the early 1660s.[8]

It was this problem of the failed carotid which led to Willis's recognition of the presence and true function of a circular artery at the base of the brain – an artery into which both carotids and vertebrals, right and left, pumped blood, and which in turn distributed to both hemispheres equally, via a series of smaller branching vessels. Hence, if one carotid or vertebral artery failed and others grew in size and compensated accordingly, then the "circle" would continue to receive its full blood supply, and from it, the rest of the brain.

Key features of this complex arterial system supplying the brain had already been described by several anatomists, including the Paduans Gabriele Falloppio, Giulio Casserio (who had taught Harvey), and Johann Vesling, along with the Swiss Johann Jakob Wepfer. So the arterial *structures* would have been familiar to the very well-read Willis. His originality lay, therefore, *not* in "discovering" that anatomical structure which subsequently immortalized his name – the "Circle of Willis" – but in explaining its correct physiological function. And this came about partly through experiments with injecting black ink into dissected arteries, veins, and brains, enabling him to trace blood pathways, partly through vivisection experiments on dogs and other animals involving the application of ligatures to arteries (reminiscent of Harvey's work on the circulation), and partly through trying to make sense of what had happened in the case of the man whose closed and "stony" carotid had in no way affected his mental or coordination capabilities.

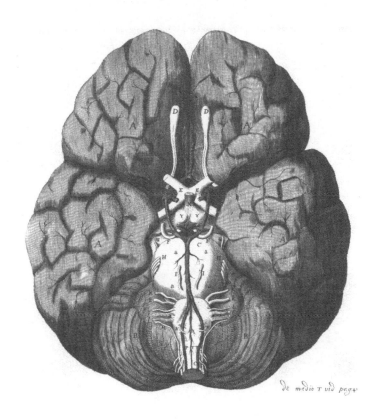

21. The base of the human brain showing the "Circle of Willis" and its adjacent
blood vessels. Drawn by Willis's friend and fellow Oxford professor, Sir
Christopher Wren, from Willis's *Cerebri Anatome* (1664). (Christ Church
Library, © Governing Body of Christ Church, Oxford.)

While the circular blood vessel at the base of the brain (which
his Oxford friend, astronomy professor, and later architect of
St Paul's Cathedral, Sir Christopher Wren, would illustrate in his
superbly drawn plate in *Cerebri Anatome*) would immortalize his
name, Willis made many more contributions to our understanding
of the brain and nervous and blood circulatory systems. In many
respects, Willis was anatomy and physiology's "next step" after
Harvey. He would not merely coin the term "neurologie" in 1664;
he would also found the clinical discipline itself.

He did this in two particular ways. Firstly, he made detailed dissections of sections of the interior of the cortex, delineating specific structural components of the brain and noting the colour changes in the different parts. Secondly, he assiduously traced the nervous appendages of the brain from the interior of the cortex to the chest, abdomen, and extremities of the body. For the term "neurology", which we today associate primarily with the study of the brain, derives from the Greek *neuron*, or nerve. Willis, therefore, quite correctly saw the brain and the nerves as one extensive yet integrated *system*.

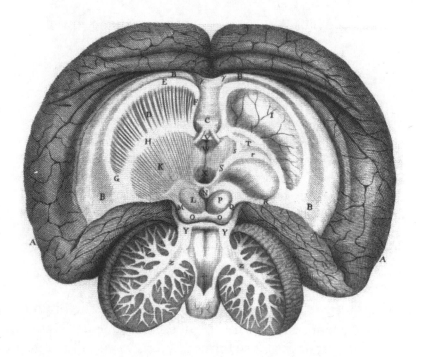

22. Dissected human brain, showing the delicate interior structures of the cerebrum and cerebellum and internal white matter. (Willis, *De Anima Brutorum* (1672). Christ Church Library, © Governing Body of Christ Church, Oxford.)

Central to Willis's legacy to medicine was the concept of localization of function within the brain: that different zones within the two brain hemispheres perform different functions. His studies include the external convolutions or *gyri* of the outer brain, which he saw as connected by fibres to the interior white matter of the corpus callosum. As the human brain has far more external *gyri* or convolutions than do the brains of animals – birds and fish having virtually no convolutions – Willis associated them with the ability to learn, humans naturally having a vastly greater intellectual capacity than birds, reptiles, or beasts. Then there were those two structures, the cerebrum and the cerebellum. Willis saw the cerebrum as the control centre of movement and the seat of the human rational soul, and the cerebellum as the "Mistress of the involuntary functions". Then there were the corpus callosum and corpus striatum, the medulla, and the fornix.

In Willis's *Cerebri Anatome,* the brain is an internally complex control centre, the different external and internal zones possessing their specific functions, facilitating an internal dialogue within the brain itself – thought, volition, imagination, and such – and conveying "instructions" to all the organs of the body. Many pages of the *Cerebri Anatome* were devoted to tracing the brain's nervous and related appendages. Willis made a major contribution, for instance, in his reclassification of nerves, most especially those of the cranium, and most of all in the study of what he styled the *vagus* ("wandering") nerve, so named because it wandered throughout many bodily structures. It was this nerve, traced in human cadavers, and experimented upon vivisectionally in dogs, that led to his recognition of its crucial role in controlling cardiac, arterial, and vascular functions, and how its dysfunction, through whatever cause, could lead to trauma, cardiac irregularities, and the constriction of blood vessels.

In an age still ignorant of electrical action beyond that of friction-generated sparks, what did Willis see as the force that *powered* the nervous system? He spoke of the classical "animal spirit", but interpreted it in a new and more physical way. To Willis, this light and fast-moving spirit (or whatever it was) was not stored in the brain ventricles, as the ancients thought, nor did he think of the nerves as resembling pipes. The spirit was a sort of subtle fluid, perhaps like the force seemingly emitted by a magnet, or the sparks that could be generated by rubbing an amber (or "electrum") rod in the dark. This is just an analogy, for in 1664 neither Willis nor anyone else had any real idea of what activated the central nervous system.

It was the movements of this spirit, as it acted upon anatomical structures, that produced both normal and abnormal mental and physical activity. Abnormalities in its function could cause mental disturbance, madness, and spasmodic behaviour, as well as normal and healthy voluntary and involuntary functioning. As with the anonymous man whose abdominal tumour led Willis to discover a blocked carotid and to elucidate his "circle", he clearly liked to "know his corpses", as it were, while they were alive.

In a number of his explanations of localized cerebral function, Willis attempted to establish a connection between traits in the known living person and features in their post-mortem brains. He also suggested, from the dissection of the brains of dead babies born to epileptic women, that a hereditary component was probably involved, and that epileptic seizures were caused by the erratic movement of blood and spirit in the brain. Brain malfunctions also lay at the root of strokes ("palsie"), senility, and the behaviour of "natural fools" or what we would call persons with learning difficulties. The brain was the key to the body.

Even musicality lay in brain anatomy, according to Willis. People with an ear for music when living were found to have soft cerebellums when their brains were dissected, whereas those "destitute of the Faculty of Musick have hard cerebellums".[9]

Few of Willis's proposed links between brain, body, and conscious functions stand up today, for his only research tools in the 1660s were sharp eyes, a scalpel, meticulousness, and great patience. Yet Thomas Willis's significance derives not from any specific physiological connections which he may have suggested in *Cerebri Anatome* (1664), *Pathologiae Cerebri* (1667), his surviving "Casebooks", "Oxford Lectures", and other works, but in the *approach* which he established. In the same way that his *Diatribae* established a concept of fever and disease being caused by chemical agencies (which he attempted, often incorrectly, to replicate in the laboratory), so his dissection- and observation-based model for brain function would lay down the procedural foundation for the emerging science of "neurologie".

But let us turn to what was, most probably, one of Thomas Willis's most enigmatic dissections.

DEATH BY LIGHTNING IN 1666

In the seventeenth century sudden death, especially if not occasioned by something obvious such as drowning, a bullet, or a sudden fall, could be baffling, as most deaths came about as a result of protracted diseases, such as fevers and infections. So when a healthy undergraduate student was struck "stone dead" by a lightning bolt while boating with friends on Oxford's river Isis on a warm spring day, 10 May 1666, Dr Willis was keen to dissect him to see what state his organs, and especially his brain, might be in.

As with the case of Anne Greene 16 years before, we have detailed written accounts of the incident, one of which was published in the new *Philosophical Transactions* of the Royal Society.[10] The unfortunate student in question is not named, but we know that he and his boating companions were from Wadham College. I therefore examined college records, and found the only Wadham man to have died on that day was 17-year-old Samuel Mashbourne, a "gentleman commoner", or fee-paying student, from Northamptonshire.[11]

We also know that, quite independently of Mashbourne's death, the Reverend Dr John Wallis, Savilian Professor of Geometry, was carefully monitoring the progress of the storm, timing the intervals between the flashes and the thunderclaps with his watch, from the high vantage point of the Bodleian Library tower.

Mashbourne and his friends had just cast off from Medley, on the Isis, he being standing up, probably working the punting pole, when a sudden thunderbolt struck him down. One of his student companions, Ahasuerus Rigemorter (whose father was a Dutch doctor in London), was stunned, knocked overboard, and driven like a post into the mud, yet luckily recovered. Mashbourne's body was taken back to college, to lie in state in Wadham Chapel prior to burial. It seems to have been at this juncture that Dr Willis became involved.

At a time when electrical phenomena were only vaguely understood, a death by lightning posed all manner of puzzles, both in physics and physiology. One of the current explanations for lightning was that of the Polish "chymist" Michael Sendivogius, who suggested that it might be the result of a chemical explosion in the upper air: a thing which seemed convincing in the seventeenth century due to the stink of "sulphur" that follows a great lightning storm. We now know this to be caused by ozone. Yet why was Mashbourne's body wholly intact, with only superficial burning to his neck and down his chest (the latter being caused by metal waistcoat buttons)?

The dissection took place by candlelight, which may suggest that Willis and the large company of Oxford doctors and surgeons present had "borrowed" the body for an hour or so. Samuel Mashbourne's corpse was opened and all of his internal organs were found "healthy" and fresh, so what had killed him? His brain was then removed, using a standard technique that left the face unscarred for the final pre-burial farewell: the scalp would be cut through down to the bone from the left to right ear, through the hairline. The flesh would then be peeled down, front and back, to reveal the dome of the skull. We are told that a surgeon dexterously made a shallow cut around the skull with a saw, enabling the top to be taken off like a cap, revealing the pia and dura mater membrane linings and the brain, which could then be lifted out for examination.[12] When the dissection was over, the "skull-cap" would have been replaced, the flesh drawn tight once more and stitched within the hairline, leaving the face and forehead unmarked.

Dr Willis and his friends came to no clear conclusion about exactly *how* Samuel had died, and it would not be until the eighteenth- and early nineteenth-century discoveries in electrical potential and its connection with the brain, nerves, and muscles that it would become possible to establish that. (This electrical shock potential was used in reverse by Dr Andrew Ure in the 1810s in an attempt to "shock" cadavers back to life, which would go on to inspire Mary Shelley's *Frankenstein*, 1818, as we shall see in Chapter 21.)

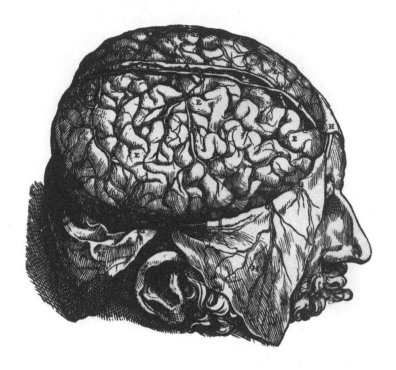

23. The human brain after the dome of the skull has been sawn away to reveal the cortex and convolutions, with their dural membrane wrappings pealed back. The brain would then have been lifted out of its chamber in the lower skull as a necessary preliminary to dissection. This is how Thomas Willis would have first seen Samuel Mashbourne's brain after the surgeon had sawn around his skull to remove the dome in May 1666. (Vesalius, *De Fabrica* (1543), Book VII. Christ Church Library, © Governing Body of Christ Church, Oxford.)

The well-documented account of the death of Samuel Mashbourne is historically significant for a number of reasons. It occurred at a time when European scientists were coming to think of thunder and lightning in physical and chemical terms, which explains why a mathematical scientist like John Wallis was studying them. It also tells us much about the "dissection culture" of the seventeenth-century Oxford and London medical men, who were especially keen to tie up phenomena in the physical world with responses in the human brain and body, especially in rare phenomena such as death by lightning, as part of a *physical* causation model of health and disease. This willingness to "desecrate" the dead, and to search out the subtlest physical aspects of the body, was not seen to be in any significant conflict with a deeply held Christian faith: a point to which we will return anon.

One wonders, however, what Mr and Mrs Mashbourne thought about the dissection of their dead son? Or whether they even knew about it? As was the custom of the day, Samuel's body was probably never taken back to his Northamptonshire parish churchyard, but was most likely buried in the Wadham burial ground, off the cloister and between the college chapel, library, and kitchen.

FATHOMING THE WORKING OF THE MIND IN SEVENTEENTH-CENTURY ENGLAND

Since classical times, a range of figures, including Aristotle, Avicenna, Chaucer, Montaigne, and Descartes, have wrestled with what goes on inside our heads and how our minds, bodies, and souls work together. I would like, however, to examine in this section the ideas of three significant Oxford men, because each had a singularly enduring influence. One was Thomas Willis, to whom I will return at the end of this section. The other two, both household names as political philosophers, based their whole rationale of human action upon certain psychological presuppositions. One became in his own day and remained long thereafter a byword for abomination; the other had an incalculable impact upon British social thinking for the next 200 years, while his ideas were destined to underpin both the American Declaration of Independence in 1776 and the US Constitution. The first was Thomas Hobbes, an old Magdalen Hall graduate; the second John Locke, a student, or don, of Christ Church.

As I have discussed Hobbes in another book,[13] I will only mention him briefly here, but in Part I, "Of Man", of his monumental

Leviathan (1651) he set out a psychology based upon mechanistic reductionism, as a preliminary to his wider agenda of explaining the origins of political society. To Hobbes, man was a machine, both in body and in mind, and he used analogies to springs, levers, and gear trains in his explanations of how we react. Basically, Hobbes's human mind was a sort of computer preprogrammed to achieve specific ends: to extend its own existence for as long as possible and to attain "felicity", or happiness. Death was the supreme evil. Man was no more than an extremely versatile animal, unique in the creation, for we humans had a power of language that vastly outstripped that of any brute. As words, to Hobbes, were "wise men's counters", or calculating units (as in an abacus calculator), they enabled humans to undertake much more complex trains of thinking than could a dog.

The human machine is driven by two contrary forces: "Vainglory", or an urge to dominate, and "Fear", or a dread of pain and destruction. All social thinking is no more than an endless process of calculation, where we compute the odds of every situation: will I dominate or will I suffer? Idealism is nonsense, and self-sacrifice for others plain madness. We are but machines that acquire data through our senses, file it away for future reference, then act upon it in the fight for survival. This model for human perception and action led Hobbes to formulate a radically egalitarian politics – based on our all being in the same mess – and to offend both Parliamentary Puritans and Royalists in equal measure: Puritans because, in Hobbes's thinking, mankind was *not* guided by an "inner light" theology: Royalists because his mechanistic egalitarian analysis was an affront to the divine right of kings.

How far Hobbes actually *believed* in his system, and how far he was enjoying flying an outrageous intellectual kite, is a matter of debate. Hobbes the man seems to have been a genial and remarkably generous individual, who included among his friends William Harvey, several Dukes of Devonshire, and even King Charles II. Yet Robert Boyle, the great chemist, once turned round and left a London bookshop upon hearing Hobbes's voice coming

251

from around a corner, while the devout Thomas Willis had no time for him. Irrespective of whether one loved or loathed Thomas Hobbes's analysis of human motivation, what is undeniable is that he set up an argument that his successors felt obliged to respond to, one way or another. And one man who did produce a psychology (and a political philosophy) that perhaps owed at least something (unacknowledged) to Hobbes was John Locke.

Both Hobbes and Locke were classical scholars and accomplished "Grecians". Yet where Hobbes's wide-ranging inclinations led him to a passion for the purity of geometry and mathematics, those of Locke drew him to chemistry, experimental science, and medicine – and political philosophy.

Born in 1632, making him 11 years younger than Willis, Locke was the son of a Somerset gentleman who had served as a captain in Cromwell's Parliamentary Army: on the opposite side to Willis's Royalists. As with many men of this period, experimental science captivated both of them, and Locke, the Westminster School boy, then Christ Church teacher of classics, even found time to take a Bachelor of Medicine degree at Oxford in 1675. He became physician to his political ally Anthony Ashley Cooper, First Earl of Shaftesbury, and in 1668 doubtless saved the earl's life when he was suffering from a liver abscess, probably a "suppurating hydatid" liver abscess. After consulting with more experienced medical colleagues beforehand, the still technically unqualified Mr Locke used a small silver pipe, which he inserted through his patient's side, to drain away the infection. It worked![14]

While it is true that Locke did not take a formal medical degree until seven years after successfully treating the earl, his lack of precise paper qualifications was less of an obstacle to being taken seriously as a doctor than it would be two centuries later. This was still an age in which a learned gentleman and an Oxford or Cambridge don (or, in the case of Willis's medical inspirer Mrs Isles, a cathedral canon's lady) could take patients in good faith and be held in respect by qualified doctors, provided the person was reputed to know what they were doing and did not behave like a showy quack.

No one could deny John Locke's standing as a scientific gentleman, despite his daytime job as a Christ Church classics tutor. He had been actively involved in chemical research since at least his early twenties, had attended Thomas Willis's medical lectures in Christ Church, had participated in the dissection of corpses, was a fellow of the Royal Society, and would even be elected to one of Christ Church's two medical studentships, or tutorships. (In Christ Church the dons have always been called "students" rather than "fellows", as is the case in other Oxbridge colleges.)

The political changes during and after the reign of King Charles II, the Duke of Monmouth's abortive grab for the Crown in 1685, and the hated and unsuccessful Roman Catholicizing policies of King James II, however, obliged the radically minded Locke to leave both Oxford and London, and flee for safety to Holland. There he perfected an "empiricist" philosophy of the human mind which he published in *An Essay Concerning Human Understanding* (1690). In this very substantial *Essay* he begins in "The Epistle to the Reader" by pointing out how Boyle, Sydenham, Huygens, and Newton, along with many other great discoverers of the age, had amply demonstrated that experiments and scientific instruments (such as the telescope and microscope) were transforming natural knowledge in a remarkable way. Locke's very modest personal claim was, as he put it, to serve "as an under-labourer in clearing ground a little and removing some of the rubbish that lies in the way to knowledge".[15] By that, Locke meant clarifying the then often confused philosophical framework within which science was then practised.

Intuitive, philosophically deductive, or "abstract" agencies were simply not part of what Locke saw as the experimental, empirical agenda of creative science. In this respect, he was fundamentally at odds with the massively influential Cartesian philosophy, widespread across Continental Europe. In Locke's thinking, there were no preconceived or "innate" ideas, either in the universe itself or in the mind of the beholder. At birth the mind was the classic *tabula rasa*, or blank sheet, and what wrote upon it was direct, physical experience. This brings us to what might be called

Locke's psychology. In spite of his extensive medical knowledge, the analysis of mind and action which Locke presents in his *Essay* is not neurologically tied in with any real or supposed structures in the cortex. It is, rather, about ideas in themselves, where they come from, and how they develop with relation to perceived or imagined reality.

A steady stream of impressions from the outside world, caused by the motion and impact of physical particles, is conveyed through the five senses from birth onwards, laying the basis of what might be called *mind*. These impressions constitute "simple ideas", and they leave their record upon the blank *tabula rasa*. These simple ideas are then "filed away", as it were, to form memories, or if one put it in computer terms, a *database* upon which we might draw.

From the combining of these memory-recorded simple ideas – such as number, movement, hard, and soft – we then go on to construct what one might call reflective concepts, such as *"phantasm, notion, species, or whatever it is which the mind can be employed about in thinking"*.[16] Locke was by no means the first European thinker to wrestle with the problem of how we know and perceive: Descartes and even Galileo, among others, had dealt with it in one form or another, and fundamental to the whole perception debate was the old problem about primary and secondary qualities. Primary qualities were those about which we were all likely to agree, whereas secondary qualities generally required some sort of judgment.

Every conscious person falling downstairs would agree that they were *falling* and that the stairs, and finally the floor, were *hard* (motion and hardness being primary qualities). Yet were the stairs *brown* and the floor below *red* (colour being a secondary quality)? A falling blind person would not know the colours, yet he or she would never doubt the falling and hardness sensations. What is more, colours can change depending on a variety of factors, such as the level of illumination. If we encountered four ginger cats brushing past our legs in a semi-darkened space, we might think that the cats were grey. Yet we would not doubt that we had made contact with four substantial, moving objects.

I give the above examples in the hope of conveying some of the concepts involved in Locke's analysis of how the human mind handles its perceptions. It is a very physically based model, rooted not in abstract concepts, but in what Locke believed to be common-sense reality: common sense deriving from *sensus communis*, sensory experiences common to all human nervous systems.

One can see how a man with Locke's passion for experimental and "hands-on" data as a way of getting at the "truth" of the natural world could articulate such a way of thinking. When taken further, one can also see how Locke's physical, communally verifiable approach to reality could migrate from the laboratory to psychology, and on to ideas about society and politics. His "primary quality" and "simple ideas" building blocks of the mental process were profoundly egalitarian, as all minds worked essentially the same, and both our thought processes and our models for a just society operated through the association and complexification of ideas. In the eighteenth century Newtonian gravity astronomers, French materialist philosophers, subjective idealist philosophers such as Bishop George Berkeley and sceptical David Hume, American Revolutionaries, and British reforming politicians all drew on Locke in one way or another – as did working men of science.

Yet let us return from the psychological implications of the new science and consider how Willis came to see how the psyche, the mind, and the soul might relate to the brain.

ARCHBISHOP GILBERT SHELDON, DOCTOR WILLIS,
AND THE SOUL

As indicated above, it may seem strange that a man like Willis could be an enthusiastic experimental scientist and a dissector of dead men and women, *and* a devout Christian who ranked not only clergymen, but also bishops and archbishops among his personal friends. It may strike people today as even more peculiar that many clergy, of all ranks, were keenly interested in science and sometimes working men of science in their own right. If one checks the fellowship lists of the new Royal Society, one will see that Gilbert Sheldon,

Archbishop of Canterbury, was elected to that august fellowship in 1665. In those days, this did not mean that Archbishop Gilbert FRS was a *working* scientist; rather, he was a friend and promoter of the new "experimental philosophy". But very importantly, he was a *friend* of science and *not* an opponent.

Archbishop Gilbert was joining a company of episcopal colleagues – John Wilkins, Seth Ward, and Thomas Sprat – who either were, or soon would be, fellows of the Royal Society *and* bishops of the Church of England, and a few cathedral deans. Gilbert's later successor Archbishop John Tillotson, FRS (whose wife Elizabeth was an amateur physician in the tradition of Willis's Christ Church inspirer, Martha Isles) would confer a Lambeth MD degree on his old friend Robert Hooke, FRS, in 1691.

What Willis saw as giving him an open ground upon which he could dissect the bodies and brains of cadavers without serious religious qualms was his concept of the human soul and its relation to the brain. There was nothing new about this, and previous researchers, such as René Descartes and the French Roman Catholic priest–scientist Pierre Gassendi, had already published in this area. Their model was that of dualism, whereby the immortal soul and the transient body operated closely in harmony in life, but separated – one decaying, the other going to God – after death. In life, Descartes had actually suggested that the human soul resided in the pineal gland in the brain, while Father Gassendi – Copernican astronomer, theorist of early atomic physics, physiologist, and cathedral dignitary – had proposed a "two souls" model for humans. It was this Gassendian soul which especially interested Thomas Willis, for Willis suggested that humans had both a "corporeal" and an "immortal rational" soul. The corporeal or bodily soul was instinctual, enabling our bodies to respond to their environmental circumstances, as do animals; the "immortal soul", however, was unique to humans, giving us mental, moral, rational, and spiritual capacities. At death, it flew to God, while the instinctual corporeal soul was simply extinguished with the last heartbeat.

Willis's position on the brain, soul, and God relationship was laid out unequivocally at the start of *Cerebri Anatome*. In Samuel

Pordage's translation of Willis's *magnum opus* of 1681, in the "Epistle Dedicatory" to Archbishop Sheldon, he says "... I resolved to unlock the secret places of Man's Mind, and to look into the living breathing Chapel of the Deity" – or the anatomy of the brain. Willis posits no particular theology of the mind–brain relationship, but his whole emphasis is to refute the "School-house of Atheism" by showing the elegant harmony between not only medical but *all* scientific research and the Christian religion; a position which he occupied alongside that of his chemist friend Robert Boyle, and their wider circle in Oxford and the early Royal Society, as we shall see in Chapter 15.

Willis's approach to science and faith was that of the emerging, distinctly English, indeed *Anglican*, "natural theology" tradition, epitomized in the Oxford experimentalists and the Royal Society. It saw itself as supplying a vigorous counter-argument to what was seen as Roman Catholic "superstition" on the one side and Puritan "enthusiasm" on the other. Instead, it drew attention to the elegant, *rational* symmetry of the divine creation – extending from planetary orbits to brains – the balanced providence of the mind of God, and humankind's ability to "trace God's hand" in nature. It also fitted in nicely with the perceived "balance" and beauty of the Church of England, in contrast to perceived Catholic "excesses" and Nonconformist "ecstasies". Most of these English scientists tended, like Willis, to be members of the Church of England and on an ideal footing to enjoy warm friendships with scientifically minded archbishops.

Very importantly, Willis's whole physical approach to brain function fundamentally challenged prevailing – albeit increasingly discredited – accusations of witchcraft, demonic possession, extreme religious enthusiasm and similar superstitions, by seeing these manifestations as caused by fluid or chemical changes taking place within the brain. Was not the true God a deity of reason, light, and love, rather than of despotic fury?

Religion notwithstanding, Willis's work was fundamental to the future development of neuroscience. What he had begun in *Cerebri Anatome*, he developed further in *Pathologiae Cerebri, et Nervosi Generis*

Specimen. In quo agitur de Morbis Convulsivis et de Scorbuto ("Pathologies of the brain, and an instance of a nervous type [of disease]. In which are discussed convulsive diseases and scurvy"), 1667, and in *De Anima Brutorum* ("On the Soul-Function of Animals"), 1672. In these works, he extended the range of his dissections of the brain, nerves, bodies, and cardiovascular systems of a wide variety of animals, which he performed comparatively alongside those of humans. In *De Anima Brutorum* he explicated his corporeal and rational sentient souls idea (beasts lacking the latter), while in "Caput" 1 of *Pathologiae Cerebri* he developed his explosion or gunpowder theory of muscle action: *spiritus… expansos, pulveris pyrii ritu. veluti ignitos explodi* ("as happens with gunpowder, the spirits [i.e. vital power] when expanded explode as if they have been ignited").[17] In 1667, no one understood how gunpowder exploded, yet great force was generated seemingly from nothing; so did muscles function likewise?

So could muscular action, like fevers, fermentation, and brain function and dysfunction, all be somehow *chemical* in their operation? This style of thinking becomes more comprehensible when one considers the way in which chemistry, and especially medically related chemistry, were developing at this time: first in Continental Europe, and then within that brilliant scientific "dynamo" world of mid seventeenth-century Oxford, and the early Gresham College, London-based Royal Society.

CHAPTER 15

Breathing and Burning: Cardiology, Chemistry, and Combustion

*I*n the early twenty-first century, people tend to take for granted the rapid advancement of medical knowledge, as new cures and treatments almost routinely follow upon one another. Yet while real *cures* were still sadly thin on the ground in 1680, knowledge of how our bodies work had changed beyond recognition in little more than a generation. It would be these discoveries, relating to the functioning of the cardiovascular and lymphatic systems and of the brain, nerves, and muscles that would lay down that foundation of experimentally verifiable data upon which later discoveries, and then rational *cures*, would come to stand.

Central to that saga of discovery would be the solving of the ancient mystery of what is our breath? How does it relate to the air, what happens to that air once it is inside our bodies, and what is that vital body heat and motion which Doctors Petty, Willis, Bathurst, and Clark successfully stimulated in the cold "corpse" of Anne Greene? This is where chemistry and physiology really first came together, not in the experimental pharmacology of Paracelsus, van Helmont, and their iatrochemist followers, but in medicine's first, unexpected encounter with pure chemical research.

THE BREATH OF LIFE

It all began in Italy in 1643, when the physicist and engineer Evangelista Torricelli was trying, and failing, to devise powerful

259

pumps for the Grand Duke of Tuscany, to raise water to a greater height than 30 old Italian feet. Being an instinctive experimentalist, Torricelli realized, from his old teacher Galileo's hydrostatic researches, that air must have weight, and that the resulting air pressure could only support a thirty-foot column of water in a pipe. But as mercury is 14 times denser than water, he began to experiment with long, vertical mercury-filled tubes, open at the bottom yet sealed at the top, only to find that no matter how much he modified the parameters, he could only sustain about 29 Italian inches of mercury in a tube.

Inadvertently, Torricelli had invented the "mercury barometer", discovering that the weight of the atmosphere is only sufficient to support around 30 feet of water or 29 inches of mercury in a top-end-sealed vertical tube at sea level. If one filled a 40-inch-long top-sealed tube with mercury, before upending it, the mercury still fell to about 29 inches. So what was in the sealed area between the mercury and the top of the tube? As researchers across Europe began to experiment with similar tubes, they found that it contained *nothing* whatsoever. This became immortalized as the "Torricellian vacuum". Aristotle had posited that vacuums could not exist, for nature was "full". But all of a sudden, it was discovered, one could make a vacuum on demand.

By the 1650s, members of Wilkins's Oxford Club were experimenting with vacuums, and discovering that changes in air pressure, as indicated by the changing level of the mercury, could be used to predict the weather. Eminent Irish physical and medical experimenter, Robert Boyle, who was currently living in Oxford, wanted a bigger and more versatile vacuum, and commissioned his current research assistant and future lifelong friend, Robert Hooke, to devise a "Pneumatick Engin", or vacuum pump, in 1659. It was Robert Hooke who had originally assisted his Christ Church mentor Dr Willis with the laboratory work for his *Diatribae* on fermentation, but when that was complete Hooke began to work with Boyle, at Boyle's spacious lodgings at Deep Hall, on Oxford's High Street.

24. Robert Hooke. No authentic portrait is known to exist, although we do possess detailed "pen portraits" from men who knew him well. The present portrait – in the style of Hooke's old art teacher, Sir Peter Lely – was composed from these descriptions by the artist Rita Greer, *c.* 2005.

Hooke's "Pneumatick Engin" suddenly made it possible, using a piston and pump, to evacuate a large glass sphere over a foot in diameter, inside of which a carefully devised experiment had previously been set up. Astonishing revelations followed thick and fast. Candles went out in a vacuum. Sound could not cross a vacuum, yet light could. Ignited gunpowder would not explode *in vacuo*, yet it would slowly fizzle and smoke. Cold-blooded creatures with simple hearts went into hibernation for a time *in vacuo* and could often be revived, yet animals with four-chambered hearts, such as cats, died and did not revive when air was admitted. What, therefore, was the relationship between air, respiration, and combustion? Over the next 15 years, the Oxford researchers, with their friends at the Royal Society and in Gresham College, London, after 1660, would begin to change the entire chemical and physiological landscape of Europe.

Why would gunpowder fizzle and slow burn in a vacuum, whereas candles and normal combustibles would not react at all when, *in vacuo*, intensely hot points of sunlight were focused on them with a burning glass? "Nitre", or saltpetre (potassium nitrite, with its triple oxygen bond, KNO_3 as modern chemists call it), had been the vital ingredient of gunpowder since it was first invented around 1260. Why was this snow-white crystalline salt, which appeared identical to common salt, so "fiery", whereas common salt was "fixed" or inert? And even more baffling, why did it retain its "fieriness" even in a flame-smothering vacuum?

Many European "chymists", such as van Helmont and Boyle, had investigated the peculiar "fieriness" of saltpetre before Hooke perfected the "Pneumatick Engin" and were already associating the words "nitre" and "nitrous" with inflammability. But it was that ground-breaking series of researches by Boyle, Hooke, and John Mayow, first in Oxford and then in London, using the "Engin" or air pump and thus burning combustible materials in air-restricted experimental environments, that led them to the conclusion, by 1665, that "air" was not the simple Element of Aristotle.

From his experiments on charcoal, in *Micrographia* (1665),[1] Robert Hooke suggested that atmospheric air contained a component that was especially fiery – "aerial nitre" – along with a part that was not. Did not wood suddenly burst into flame when heated to a point where the "aerial nitre" set about devouring it, whereas if the air-supply were deliberately restricted, the heated wood simply toasted to charcoal with no devouring flame?

25. Robert Boyle, Robert Hooke, and the air pump of 1659, set in the Oxford laboratory of John Crosse. Rita Greer, *c.* 2008, based on a drawing by A. Chapman.

Neither Hooke nor any of his friends had any concept of an atmosphere made up of chemically distinct gases, such as oxygen and nitrogen. That would have to wait another hundred years, for Lavoisier, Scheele, Davy, Dalton, and others. The seventeenth-century Oxford experimenters, rather, thought in terms of "air" as *elastic* and as possessing different *allotropic* properties – such as fieriness or inertness – depending largely upon "physico-mechanical" pressure differences. Even so, their published and well-known researches, both with air pump and with other laboratory experiments, would fundamentally undermine the classical doctrine of the four Elements. It was becoming increasingly implausible to think of "air" and "fire" as unchanging Elements, or primaries. Rather, they appeared to be the products of chemical and physical *processes*; and this new thinking would guide one of Robert Hooke's young Oxford disciples, Dr John Mayow of Wadham and then

All Souls, before Mayow's tragic death in 1679, probably from tuberculosis, at the age of 38.

DR JOHN MAYOW: AIR, FIRE, BLOOD, AND LIFE
TESTED IN THE LABORATORY

John Mayow came up to Wadham College, Oxford, at the very end of the Reverend Dr John Wilkins's brilliant Wardenship, at a time when his Experimental Club was on the verge of migrating to London at the Restoration to become the Royal Society in 1660. Like his older Christ Church colleague and friend Dr Richard Lower (Willis's pupil and assistant), John Mayow was descended from Cornish gentry.

In common with virtually all the great Oxford experimental doctors, Mayow never pursued a formal course of medical studies in the way that William Harvey had done. Instead, he read books, attended lectures, dissected cadavers, and learned much on the hoof, as it were. Like Boyle, Willis, Hooke, Lower, and Locke, he turned out to be a born doctor and experimentalist who, when not in his laboratory college rooms, took fee-paying patients. Yet while the others subsequently applied for, or had conferred upon them, Bachelor's and Doctor's degrees in medicine, John Mayow took an Oxford DCL (Doctor of Civil Laws) degree, as the statutes of All Souls College required, for he officially held a jurist's, or a lawyer's, fellowship in that College.

Being a classical scholar and an MA "in good standing", Mayow easily obtained a formal licence to practise medicine after taking his DCL in 1670. In the seventeenth century, Oxford and Cambridge doctorates, in divinity, civil laws, or medicine, were interchangeable. They proclaimed the holder was a deeply learned gentleman who might, if required, turn his hand to healing an ulcer, advising on a point of law, or (if ordained) preaching an erudite sermon.

Taking the air pump, combustion, and "aerial nitre" discoveries of Boyle as his starting point, Mayow began to conduct experiments inside water-sealed glass vessels. He found that if a burning candle, or a burning-glass-ignited piece of camphor, were sealed inside an upturned glass jar with a dish of water at its base, then a point was reached where all burning ceased. At this point, he noted, the water in the dish below had risen up to occupy one-thirtieth part by volume of the entire glass. Then when he enclosed a vigorous living mouse inside a similar air-locked apparatus, the mouse died at a point when it had breathed one-fourteenth of the air volume within the glass.[2]

Very clearly, what Mayow had discovered was that both combustion and biological respiration depended upon the same "nitro-aerial spirit" component in common air. Perhaps it was the same invisible fiery stuff that facilitated the rapid combustion (explosion) of gunpowder in normal air and which *might*, somehow, be contained within saltpetre crystals, thereby enabling it to fizzle, or slow burn, *in vacuo*?

John Mayow presented these remarkable findings in his Latin – intended for an international readership – *Tractatus Quinque Medico-Physici* ("Five Medico-Physical Treatises"), 1674. These five "Treatises", or sections, covered (1) the nature and properties of saltpetre and the "nitro-aerial spirit"; (2) respiration; (3) respiration of the foetus and egg; (4) muscular action; and (5) the disease of rickets. Having discussed the combustive properties of the "nitro-aerial spirit", he then goes on to examine its physiological functions, and especially, its role in respiration, or what it does in the lungs, and its relation to body heat, and why arterial blood is lighter in colour than venous. Mayow's *Tractatus* is one of the first great treatises on body chemistry, demonstrating how the meticulous conduct of experiments, combined with astute clinical observations, could tell us not only how fluids moved into specific organs within the living body, but also the vital role played by chemical action in the overall life-process.

Very little of this would have been imaginable before Torricelli's realization of the existence of the vacuum, and more significantly, Boyle's and Hooke's custom-built "Pneumatick Engin", with its ability to produce precisely controlled experimental environments within which to design and execute experiments involving air, flame, and blood. This was probably the first time in the history of

265

medicine that a piece of engineering physics made possible a series of dramatic breakthroughs in the life sciences, and a wonderful example of the cross-fertilizing potential of the experimental method, not only within but *between* disciplines.

One initial observation Mayow made was how venous blood changed colour when in contact with ordinary atmospheric air, in Chapter VIII of the First Treatise, which deals with the role of "nitro-aerial spirit" in animal (and by extension, human) respiration. He mentions that if a vial of venous blood were stood upright with its top open to the air, the surface blood would cease to be dark, but would become "scarlet and florid", although the blood beneath remained dark. If one then exposed this darker blood to the air, that in turn "will often a short time become ruddy".[3]

Perhaps influenced by an experiment performed sometime between *c.* 1666 and 1670 by Boyle and Hooke and reported to the Royal Society in August 1670, Richard Lower too experimented with putting fresh blood into the glass sphere of an air pump and noted the results. When Boyle and Hooke had put a pot of warm sheep's blood, fresh from the slaughterhouse, into their new and even more powerful air pump, and then evacuated the air, they observed "... the Blood was so Volatile and the expansion so vehement, that it boyled over the containing glass". Whether the pot contained venous or arterial blood is unclear, but it was clear that the "boyling", or effervescence, was caused by the rapid escape of air bonded into the blood in life, and kept there in part by normal air pressure, yet escaping once *in vacuo*.[4]

Mayow himself found that if warm arterial blood were placed in a "void place" or vacuum, "it will expand in a remarkable way and rise in an almost infinite number of bubbles", as he reported in the *Tractatus*.[5]

By 1674, Mayow was to bring in Harvey's circulation theory full circle, in many ways. He achieved this through a series of elegant laboratory experiments, for he was not an especially avid dissector and vivisector, unlike Harvey, Willis, and Lower. Laboratory experiments led him to conclude that when inhaled air passed through the lungs, before ascending through the left-side chambers of the heart to enter the aorta, it absorbed "nitro-aerial spirits". These spirits in turn caused some kind of invigorating reaction, making the blood, as Mayow styled it in the Willis tradition, "ferment", expand, and change colour from dark to scarlet. So whereas Malpighi with his microscope had demonstrated the venous–arterial transit in the capillaries, and Willis had elucidated the brain's blood supply and the existence of compensation mechanisms when a major vessel failed (such as the carotid artery), John Mayow had supplied key ingredients in forming the bigger picture of what happens during respiration.

As Mayow was all too aware by 1674, however, the advance towards a more complete knowledge of the physiology and function of the heart as an organ, as well as the observation of blood colour changes in the lungs, stood upon the recently published discoveries of his Oxford friend Robert Hooke, and "the illustrious [Richard] Lower".

ROBERT HOOKE AND THE DOG

Before examining Richard Lower, the founder of cardiology, mention should be made of Robert Hooke's frankly horrendous experiment conducted upon a dog in November 1664, and its subsequent repetition on other dogs before the Royal Society in 1667. The big question under investigation was the still imperfectly understood relationship between the heart, lungs, air, and blood. In this post-Harvey age, every doctor knew that the blood passed through the lungs, but exactly what happened to it there, mechanically and chemically, was still a mystery. Did we respire as the result of what we now call a "bio-chemical" process, activated by the inhaled air itself? Or did we do so, as "some Eminent Physicians" claimed, because the windpipe and lungs possess an innate, spontaneous vital respiratory motion separate from any chemical interaction with the air?[6]

On 9 November 1664 Hooke reported the vivisection experiment of "insouflation", which clearly distressed him, to the Royal Society. The chest of a living dog was cut open to expose

267

the heart and lungs. When the animal's windpipe was immobilized, just below the epiglottis, its lungs and heart went into a pre-death spasm. Yet when a pair of ordinary bellows were inserted and blew air in, then the cardio-pulmonary respiratory cycle normalized. Stop the bellows, and spasm returned; restart the blowing, and things normalized once more. Very clearly, the key to the cycle was the "blast" of air, as Hooke styled it, and not any innate pulmonary motion. It was also found that when a scalpel point was used to make small punctures in the lungs and heart chambers, small spurts of blood made it possible to discern the phases of the cardiac systolic and diastolic and lung expansion cycles.

Scientifically, the experiment was extremely informative, and was not only published in the society's *Philosophical Transactions* in November 1667, but also included as an essay in Thomas Sprat's (later Bishop of Rochester) *History of the Royal Society* (1667).[7] Yet when the Royal Society requested the kindly Hooke to repeat the experiment before the whole society, he begged to be excused. As he had told his friend Robert Boyle in November 1664, no known opiate could relieve "the torture of the creature".[8] Only after the expert vivisectors Dr Edmund King and Peter Ball had hopelessly botched the experiment, however, was Hooke, assisted by his friend Lower, induced to provide a conclusive demonstration before the Royal Society on 10 November 1667. The pieces in the heart, breathing, and blood circulation puzzle were falling into place, one after the other.

Let us, though, spare a thought for the animals who made all these world-changing discoveries possible, as well as for the feelings of those well-intended Christian gentlemen who, while inspired by scientific curiosity, took no pleasure whatsoever in doing what they had to do to living creatures.

RICHARD LOWER, *TRACTATUS DE CORDE*, AND THE FOUNDATION OF CARDIOLOGY

Lower's work on the action of the heart was, in some ways, an inheritance from his teacher Thomas Willis, for both men, along with Hooke, had puzzled over blood colour changes in the late 1650s, when working on Willis's *Diatribae*. All of this was an inheritance from the big picture of the circulation established by Harvey, which left many crucial questions still to be solved by the succeeding generation. Was the colour change from dark venous to scarlet arterial blood due to the blood's spontaneous "effervescence" as it shot out of the left ventricle of the contracting heart, to begin its arterial circuit? And what actually *powered* the blood circulation? Harvey had recognized the crucial importance of the systolic phase in the heart cycle and had correctly identified the function of the cardiac "clacks", or valves, in regulating the blood flow, but it was still far from clear whether the heart operated by a spontaneous blood "fermentation pressure" action (like the overspilling generated by shaking a fizzy drink), or by a mechanical pump-like action. All of these questions would be addressed by Richard Lower in those researches which would make his 1669 *Tractatus De Corde* ("Treatise on the Heart") one of the foundation stones of scientific cardiology.

We saw in Chapter 13 how Lower played a leading role in the post-Harvey experiments in blood transfusion in 1666 and 1667. While much of his "bio-chemical" thinking was formed during the late 1650s when he was assisting Willis in his fermentation researches, Lower came to recognize limitations in his old teacher's explanations, and then – most respectfully – to advance beyond them. As Willis's pathological interests moved increasingly towards fathoming the nature of brain function in the early 1660s, Lower was able to move into his old master's vacated field of heart and blood studies, to elucidate the cardiac action.

The *Tractatus* is divided into five main chapters. The first three deal with the anatomy, structure, and function of the heart. Chapter 4 deals with Lower's blood transfusion experiments, amplifying what was reported in the *Philosophical Transactions*, while Chapter 5 looks at the chyle, the blood, and aspects of digestion.

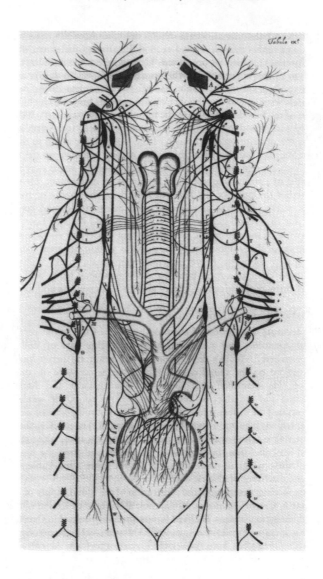

26. The autonomic (self-acting or non-volitional) nervous system, controlling processes such as digestion, heart action, and respiration, traced by Willis and his pupil Richard Lower. In particular, they were impressed by the extent and complexity of the vagus ("wandering") nerve system extending from the brain to all parts of the body. Lower's drawing also delineates the great structure of nerves that run from the brain to the heart to control the precise cardiac phases necessary to ensure the circulation of the blood. (Willis, *Cerebri Anatome* (1664). Christ Church Library, © Governing Body of Christ Church, Oxford.)

270

Among the most significant insights contained in the *Tractatus* was the demonstration that the heart was a *muscle*, and that the force that drove the blood from the heart and around the body was essentially mechanical rather than chemical. The heart's action was one of pumping rather than serving as a chamber for some sort of chemical action in the blood. Without this crucial realization, cardiology as we know it could never have come into being, nor modern cardiac surgery. Probably during his work with Hooke on the bellows-insufflation of a dog's lungs, published in 1667, Lower realized that it was the mixing of air with the blood during its pulmonary passage within the heart–lung cycle that caused it to change colour from dark to light. This would, as we saw, be taken up by John Mayow in his "nitro-aerial" laboratory experiments.

While not concerned with the action of the heart, Lower's *Dissertatio de Origine Catarrhi* ("Discourse on the Origin of Catarrh"), 1670, explored other aspects of fluid movements and pathological conditions, inspired by his work with Willis. Here, Lower was able to correct the time-honoured classical doctrine that catarrh was a sort of overspill of fluid from the brain ventricles into the nose.

Why was all this ground-breaking pathological and chemical work going on in Oxford during the third quarter of the seventeenth century? Oxford was a modest cathedral city, and while it was home to Europe's second-oldest university (after Paris), Oxford University was not large by the great Continental university standards. Nor did it have a formal medical school or medical teaching establishment (as opposed to ancient degree-granting rights), and it would not have a hospital for another century. Its doctors, chemists, and scientists, as typified in Dr Wilkins's Wadham-based Experimental Club and the Christ Church physiologists, pursued their researches not because it was their job to do so, but because it was their passion.

Oxford had some surprising advantages that would not necessarily be obvious to us today, unless we knew where to look.

OXFORD'S ENTERPRISING APOTHECARIES

For a city with a population of only a few thousands in the 1670s, Oxford appeared to be surprisingly well served by apothecaries, pharmacists, and chemical manufacturers. On a quarter of a mile of the High Street alone, modern scholars, such as my former research student Carole Brookes, have identified the well-documented premises and business activities of over 14 such individuals, and some 40 others residing elsewhere within the then small walled city.[9] The clear leader of their fraternity was John Crosse of Deep Hall, on the High Street, where Boyle lodged and had his laboratory between 1654 and 1668. Surviving will inventories suggest that even the poorest of these apothecaries left estates of around £320 – a very respectable sum in c. 1670 – whereas Crosse, apothecary, Oxford city businessman, and "hotelier", with fingers in many lucrative commercial pies, may have left an astonishing £9,200.

Just as it is simplistic to think of surgeons as glorified tradesmen, so the same is true of apothecaries. They might have been trained by apprenticeship, yet were often very well-read, scientifically involved, and felt no qualms about conversing with university gentlemen on easy terms. Not only did they talk with university gentlemen, but the more enterprising of their fraternity actually *educated* them as well. For just as Mrs Isles ran some sort of laboratory and compounded medicines in her husband's Christ Church canon's house, as testified by the undergraduate Thomas Willis, so it was enterprising apothecaries who first offered formal courses in chemistry for university gentlemen. By 1659 we know that both John Clerk and Peter Stahl (brought over from Germany by Boyle) were offering commercial, pay-on-the-nail courses in their private laboratories – instruction, chemicals, and apparatus provided. The diarist Anthony Wood recorded attending Clerk's course, where he and his friends did "exercise themselves in chimicall extracts", and where John Locke, the 27-year-old Christ Church don, was sometimes noisy and a nuisance.[10]

The men attending these courses would not be undergraduates, but senior members or private gentlemen, who, like the 27-year-old Anthony Wood, had interests in science and medicine. Practical teaching appears to have been the order of the day, however, with an emphasis on how to do things in a laboratory, rather than discussing Paracelsus's or van Helmont's theories.

Two newly concocted substances fascinated the mid seventeenth-century Oxford chemists: *aurum fulminans* ("banging gold") and *aurum vitae* ("living gold"). The first was a highly unstable gold compound – probably a complex gold chloride – which was found to be touch-sensitive. Put a few crystals in a spoon, balance a heavy coin on the spoon, and gently tap. Bang! The coin is blasted up to the ceiling. How could *percussion* generate fire, contra the "like produces like" of Aristotelian thinking? Here was fire without prior flame! One Oxford "chymist" who discussed these gold compounds was the Reverend John Ward, whom we will meet in the next chapter.

It is *aurum vitae*, "living gold", however, that most interests us in a medical context, where it was assumed that a medicine made from the noblest of metals would possess universal powers of healing. My former research student Michael Osborne, who tried to reconstruct these gold compounds using chemical equation calculations, concluded that *aurum vitae*'s only conspicuous medical property was most likely that of a violent and stomach-damaging emetic! I am indebted to him for computing a possible chemical reactive structure of these compounds.[11]

None of the chemical, physiological, and anatomical researches discussed over the last three chapters were conducted as an official part of Oxford University business. All were performed in college rooms, or private labs, houses, and dissecting rooms, for they were not a statutory part of the official essentially book-based degree-granting machinery. All of these researches represented the intellectual passions of private gentlemen: men who were driven by a desire to know how God had put both the world and living creatures together. They were *Grand Amateurs* of science in the noblest sense: *Grand*, because they addressed themselves to the great, cutting-edge issues in medicine and science; and *Amateurs* from the Latin *amat* – lovers of scientific knowledge.

It would not be until 1683 that Oxford University would acquire its first, institutional, custom-built laboratory, dissecting room, and museum of specimens and curiosities. This came about via that curious old member and benefactor of the university, Elias Ashmole. Ashmole had come into possession of one of the most spectacular collections of natural historical, medical, ethnological, and historical objects in Europe, built up over 60-odd years by the two John Tradescants, senior and junior. Ashmole came into his possession by means that might not stand up to too close a scrutiny. While incapable of bequeathing a sufficient sum for an endowment, nonetheless he succeeded in having his name attached to the institution that housed the collection: the Ashmolean Museum. This 1683 building, in Broad Street, Oxford, is now the Museum of the History of Science, and stands next door to Sir Christopher Wren's magnificent academic Theatre, paid for by the dedicatee of Thomas Willis's *Cerebri Anatome*: Archbishop Gilbert Sheldon, FRS

In the great stone-vaulted, fireproof basement – which now holds the museum's medical instrument collections – was the great *Officina Chemica*, or laboratory, with an adjacent cool chamber for anatomical dissections. The first keeper and professor of chemistry had been Dr Robert Plot who, as one might recall from Chapter 13, had also published an independent account of the hanging of Anne Greene. With Ashmole's Museum and Plot's professorship and keepership, Oxford University had acquired its first scientific and medical research and teaching institution after 1683. Yet the already well-established English tradition of private gentlemanly research would continue in tandem, right down to the Victorian age two centuries later. It had been that ingenuity- and curiosity-driven tradition, in the hands of qualified doctors, enterprising apothecaries, classical scholars, ingenious private gentlemen, a few ladies, and lots of clergymen, that had laid the foundations for, and carried forward, that rich saga of English medical discovery from William Harvey onwards.

And one outstanding part of that tradition had been the clergyman–doctor–scientist, to whom we shall now turn.

CHAPTER 16

John Wesley's Primitive Physick and
the British Priest–Physician

*M*any years ago, when I first encountered Wesley's *Primitive Physick, or, An Easy and Natural Method of Curing Most Diseases* (1747) I was struck by the apparent incongruity of a major religious leader writing what I discovered to be a major bestselling and very well-informed medical book. Since that time, however, I came to realize, as we have already seen in previous chapters, that medically inclined clergymen were by no means rare and many were in active practice well into the nineteenth century. By the time that Parliament passed the *Medical Act* in 1858, the number of formally trained medical men had become much more abundant and professional boundaries far more clearly defined, while the medical vicar, and often his wife, were still part of the English landscape.

Following the closure of monastic medical services at the Reformation, a new market came into being for cheap, accessible English-language medical books, and even Latinate ex-monks, such as Andrew Boorde, and other classically educated clergy took to writing them and to practising medicine. Their motive for doing so could be Christian compassion, the need to earn a living, intellectual curiosity, or all three together.

Yet a good century before John Wesley published *Primitive Physick*, there was a young Oxford scholar with a passion for science who was destined to become vicar of Holy Trinity Church, Stratford-upon-Avon. He is generally remembered as the man who recorded various local Shakespeare anecdotes, yet he also mixed with medical men, dissected corpses, performed chemical experiments, and studied theology, seemingly without any sense of incongruity.

THE REVEREND JOHN WARD, MA:
EXPERIMENTALIST AND SHAKESPEARE ANECDOTE
COLLECTOR

It is likely that Ward's tangential Shakespeare family connection played a part in his subsequent renown, for his extensive and detailed diary became a proud possession of the Folger Shakespeare Library in Washington DC. Nearly a century earlier, however, in 1839 and when still in the possession of the London Medical Society (now the Royal Society of Medicine) Library, extensive extracts from Ward's diary were published in book form. Microfilm copies of the whole manuscript diary are now easily available, such as the one in the Library of the Museum of the History of Science, Oxford.

Ward's diary, covering the period 1648–79, is a fascinating and informative source, providing a valuable insight into the mind of an ordained medical and scientific *cognoscente*. Ward, the son of a Royalist officer, seems to have begun the diary when he was a 19-year-old student at Magdalen Hall, Oxford (now Hertford College), though he was subsequently admitted to Christ Church. There, most likely, he became part of the Thomas Willis medico-chemical circle, along with Richard Lower, Robert Hooke, and John Locke, and, elsewhere in the university, he probably knew Robert Boyle, Christopher Wren, and the young Robert Plot. It is also likely that he had contact with John Wilkins and his Experimental Club in Wadham College, and with William Petty and the doctors involved in the 1650 Anne Greene story. For as we have seen already, Oxford was a veritable hotbed of experimental science and medicine between *c*. 1650 and 1670.

Divinity, however, was to be Ward's formal course of study, and he became an Anglican priest, and after the Restoration served as vicar of Stratford-upon-Avon from 1662 to his death in 1681, aged 52. Yet Ward, the bachelor academic scientist, when not dutifully serving his flock in Stratford, enjoyed return visits to his friends in Oxford, as well as in London, where he could walk the wards of St Bartholomew's Hospital and inspect cases with his doctor friends. When in Oxford, he usually stayed with his apothecary friend Stephen Toone on the High Street, enjoying the delights of Toone's and other friends' laboratories, accompanying Dr Conyers and other medical chums on their patient rounds, and often writing what he saw in his diary. After Ward took his MA degree in 1652, he often stayed in London, at The Bell Inn, Aldersgate, "to be near Chyrurgeons' [Surgeons'] Hall" in Monkswell Street, so that he could attend lectures and dissections. Ward also recorded that the eminent physician Dr (later Sir) Charles Scarborough "had a wooden man, wherein the muscles, and all their motions, were represented". A fascinating teaching aid.[1]

The living of Holy Trinity, Stratford, carried with it a handsome stipend of £105 per annum, along with a good parsonage house, and from the surviving evidences Ward appears to have been a conscientious parish priest. At the same time, he obtained a licence – presumably on the strength of his Oxford MA degree – to lawfully practise medicine. Dr Charles Severn, who produced the 1839 edition of Ward's diary, states "a combination of the two professions [the church and medicine] in one individual being, [was] at that era, not unusual; the bishop of the diocese possessing the power of granting to the clergy licences to practise medicine, as well as to evangelize their flocks".[2] This is further confirmation of a situation which we saw in earlier chapters. (Stratford-upon-Avon already had qualified medical practitioners by Ward's time. Shakespeare's daughter Susannah had married Dr John Hall, of Hall's Croft, Stratford, though Hall's posthumously published *Select Observations on English Bodies* (1657) was essentially conservative and Galenic in its approach.[3])

How much actual *physicking* of his own patients Ward did is not clear, though his diary is packed with accounts of illnesses, operations, clinical opinions, and experiments that he witnessed or assisted with in Oxford and in London. Cancer has always exerted a fascination on the human imagination, and Ward's Diary makes several references to the disease. "1666. Mrs Townsend, of Alveston [near Stratford] being dead of cancer, Mr Edes and I opened her breast in the outward part, and found it very cancerous; itt [*sic*]... yielded two porringers full of a very yellow substance..."[4] I will stop here!

Mrs Townsend had been a woman of amazing courage and toughness, having apparently endured, fully conscious, a mastectomy operation at the hands of Mr Clerk of Bridgnorth and Mr Leech of Stowbridge before Messrs Ede and Ward anatomized her dead body. One of her surgeons admitted that, during her ultimately fruitless operation, she bore unflinchingly pain that would have killed him, the surgeon! Ward recorded: "the Dr. said hee [*sic*] had read that women would endure more [pain] than men, but did not beleeve itt till now."[5]

Clearly, the Reverend Mr Ward could not resist a good post-mortem, and was always eager to see what might be inside a dead body, and even to try, or to record from colleagues, chemical tests on the same: "A woman opened at Oxford, who died of a kind of dropsie... Dr Conyers took out of the woman's bellie three buckets full of water, and afterwards went to distill a good deal of it."[6] But Conyers was surprised to find that the ensuing distillation process only produced "three or four spoonfulls" of water, the "rest of it settled turnd [*sic*] to a kind of slime or mucilage." Then there was the post-mortem on Mr Gwinne "of our house" (Christ Church?), whose dead body showed no obvious mortal signs, although his "heart was exceedingly large, allmost as large as the heart of an ox, but not perisht att all".[7]

What makes Ward's diary such fascinating reading is the sheer amount of curious and sometimes bizarre medical accounts and attempted treatments which he either witnessed or tried out over the years. For example: "My Lord Cherburie's eldest sonne died with drinking at Ludlow and was found dead in his vomit and blood." Attempts were made to revive him, however, somewhat reminiscent of Anne Greene. Sneezing powder was put up the young nobleman's nose, while he was cupped and scarified – attempts to encourage the blood flow – but all to no effect, he being "perfectly dead".[8]

278

Ward records a rather improbable autopsy, related by his Oxford friend Dr Conyers, of a man who had died of love in London. When his body was opened up "they found (at least, they fancied they did) the impression of a face upon his heart".[9] Then there was the account of the woman in the agonies of childbirth, whom the visiting Lady Puckering philosophically reminded that it was all from "her grandmother Eve": to which the suffering woman responded "I wish the apple had choak'd her" – which made Lady Puckering laugh![10]

John Ward was a highly skilled practical "chymist" and "adept". Whether he possessed his own laboratory in Oxford, London, or, more likely, at his parsonage in Stratford, or used the labs of Toone and other apothecary and physician friends, is not clear, but his writings say quite a lot about preparation techniques. He possessed, or had access to, a sophisticated laboratory furnace which enabled him, using dampers and controlled air-intake, to obtain four levels of heat. In this pre-Bunsen-burner age, such a brick-built, charcoal-fired furnace would have enabled Ward to smelt and fuse metals, slowly toast a mineral compound in a crucible, and perform all manner of distillations and separations at a variety of temperatures. Ward also left detailed instructions on how to prepare the dangerous *aurum fulminans* ("banging gold") and the "medicine" *aurum vitae* ("living gold") mentioned in Chapter 15.

The diary abounds in curious medical and medically related morsels which Ward records, perhaps uniquely. For example, the English physician Dr Thomas Sydenham ("the English Hippocrates"), whom Ward admired, affirmed that medicine should be taught not from books in universities, but by apprenticeship: one might as well "send a man to Oxford to learn shoemaking as practising physick".[11] Then, he "had heard... to be a certain truth" that women who were "scolds" invariably had "blew lips" (cyanosis?);[12] while the hospital porter told him that Mr Thomas Hollier, surgeon to St Bartholomew's, performed 30 lithotomy, or bladder stone, operations in one year, and did not lose a single patient.[13] (This was the same Hollier who, in 1658, had successfully performed the lithotomy operation on Samuel Pepys.)

Combining his clergyman's insights with those of the medical man, Ward cites the Gospel story of the miraculous cure of 10 lepers, only one of whom bothered to thank Jesus afterwards: "so itt is with physitians' patients, they promise fair [much] till they are cured, but they never so much as come back and thank you".[14]

From this sampling of the diary of John Ward, we must now turn to a medically minded clergyman with a very different approach.

JOHN WESLEY AND SIMPLE MEDICINES FOR THE COMMON MAN

John Wesley's *Primitive Physick* (or "Simple Remedies") must be one of the great medical bestsellers of all time. The British Library Catalogue, a legal statute deposit library, records some 35 English editions between 1747 and 1847 alone, while there were quite separate American editions and, most likely, others printed elsewhere across the world. Not only did *Primitive Physick* carry the imprimatur of one of the most influential Christian teachers since the Reformation, but it contained a treasury of useful tips, being, as its subtitle optimistically states, *An Easy or Natural Method of Curing Most Diseases*. From English farm cottages to settler homesteads across the American plains, from the Australian outback to the jungles and grasslands of Africa it travelled – wherever Methodist missionaries went, and human suffering was to be found.

John Wesley and John Ward had a variety of things in common. Both were Oxonians, and Wesley was also a Christ Church graduate. Both came from educated families, both became Anglican clergymen, and they shared a deep sense of service. Wesley and Ward also were interested in contemporary science, and saw its discoveries as revealing the hand of God.

There were major differences. Never was John Wesley an aficionado of post-mortems and dissections, nor does he seem to have taken an especial delight in the detailed business of medical experimentation. His driving force was the spread of the Christian gospel, not fathoming the nature of disease. To Wesley, the source of disease, plainly and simply, was human sin, and the key to avoiding its worst effects was to transform one's life and bring oneself into conformity with gospel teaching. While the devout John Ward would have agreed in general with this argument (Ward dying some 22 years before Wesley was born), the two men might have disagreed about the potential of scientifically based medicine as a truly curative force in its own right.

Wesley believed, judging from the contents of *Primitive Physick*, in what might be called "the good old cures". While sin afflicted the world, God had mercifully granted some relief to humankind in the art of medicine. Some of this relief might even be fairly recent in origin, such as the application of a mild electrical stimulus, as administered by one of the hand-cranked "shocking" or electrostatic machines currently being demonstrated by the popular scientific lecturers of the day, such as John T. Desaguliers.

For the "Palsy" or stroke Wesley recommends: "Be electrified, daily, for three months, from the places wherein the nerves spring, which are brought to the paralytic part. If parts beneath the head are affected, the fault is in the spinal marrow: if half the body, half the marrow is touched." Physiologically shaky, perhaps, even by the best knowledge of 1747, but along the right lines, with its identification of paralysis with the nerves, spine, and perhaps – shades of Willis's *Cerebri Anatome* – the brain! What fascinated the eighteenth-century medical "electricians" was the strange power which the "electric fluid" possessed to make palsied limbs jerk and tingle by inducing muscular spasm.

Most of Wesley's other cures, however, are much more "low-tech" and suitable for the cottage kitchen. For (as we will soon see with the Reverend Edward Stone) had not God in his mercy placed the cure near to the places where a given disease was most likely to occur, so that what the wise healer needed to do was to read nature's "signatures" and signs?

Wesley and Ward would have most clearly disagreed on the subject of dissections, vivisections, and laboratory and cadaver-related experiments. Through these – to Wesley – futile activities, medicine had simply lost its way. As when "men of philosophical [scientific] turn" became dissatisfied with the age-old cures and began to examine "the human body in all its parts; the nature of the flesh, veins, arteries, nerves, the structure of the brain, heart, lungs, stomach, bowels… Men of learning began to set experience aside to set physic upon hypothesis… theories increased, simple medicines were more disregarded" until ultimately "physick became an abstruse science, quite out of the reach of ordinary men". It was Wesley's desire to return physic to the people.[15]

Primitive Physick contains no big arguments or ideas, beyond Wesley's above-mentioned preliminary observations regarding the folly of medical men of a "philosophical turn" and the primacy of traditional wisdom. There is no contents list: the "cures" are simply listed in alphabetical order, from "Abortion (TO PREVENT)" to "Wounds". The book opens with advice on the "Manner of using the Medicines" and their "Plain Easy Rules", and ends with a discussion of popular, approved, medicines of the day, including how to make that eighteenth-century cure-all, "Daffy's Elixier".

It should be noted that Wesley was not a teetotaller, for the Daffy's recipe required all the ingredients "to be bruised and put into a quart of best brandy". No wonder Daffy's was such a favourite! Nor was it to be taken merely by the spoonful. In the preliminary "Plain Easy Rules" section, Wesley came down heavily on alcoholic excess: "Strong, and especially spiritous liquors, are a certain, though slow poison" (Rule 6), by which he was referring to the contemporary and deadly craze for gin-drinking. Yet even "Malt liquors are extremely hurtful to tender persons" (Rule 9), and even tea and coffee are "hurtful to persons who have weak nerves" (Rule 10). Most of Wesley's "Rules", however, relate to the traditional common-sense virtues of temperance, exercise, cleanliness, and a peaceful mind.

Here follows a short sample of Wesley's recommended remedies. "A Bruise. Immediately apply treacle spread on brown paper (Tried)." "A Burn or Scald. Immediately plunge the part in cold water: keep it in an hour if not well before." "Cancer of the Mouth. Boil a few leaves of succoury, plantain, and rue, with a spoonful of honey, for a quarter of an hour. Gargle." "A Slow Fever. Use the cold bath for two or three weeks daily." "Legs, Sore and Running. Wash them in brandy and apply elder leaves, changing twice a day." "Leprosy. Wash in the sea often and long. Or, drink half a pint of celery whey…" "The Pleurisy. Take half a drachm [4 grams metric] of soot…" "To one Poisoned. Give one or two grains [120 mg] of Verdigris [a green copper compound, probably copper carbonate or acetate]. It vomits in an instant." "The Sciatica. Is certainly cured by a purge taken in a few hours after it begins," "To Clean the Teeth. Rub them with ashes of burned bread." "To Cure the Toothache. Be electrified through the teeth. (Tried.)" "A Fistulous Ulcer. Apply wood-beteny, bruised, changing it daily." "Putrid Wounds. Wash them morning and evening with warm decoction of agrimony." And so on!

All these are what might be seen as good, old-fashioned "cottage cures". Yet as mentioned above, while sceptical about "philosophical" doctors, Wesley was a big fan of electric stimulus treatment – as with "Toothache" above. In his concluding "Medicines" section at the end of *Primitive Physick*, Wesley lists 29 illnesses, ranging from "Deafness" to "Menstrual obstructions" to "Ophthalmia", all of which can supposedly benefit from "Electrifying".

It is hard to estimate the impact of *Primitive Physick*, especially on people who could not afford, or get to, a doctor, but judging from the number of editions published over more than a century, one might safely say that it was massive, sustained, and international.

I must take issue with certain modern-day people who try to claim that John Wesley was a sort of prophet for "natural healing", or what is now called "alternative medicine". In the eighteenth century, academic physicians, with their use of mercuric and arsenical compounds, often unwittingly did fatal damage to their patients, as they tried to wrestle with ailments which we now know were entirely beyond the therapeutic capacity of the age. What Wesley was saying, in effect, was leave these dangerous chemicals and experimental concoctions alone and simply try the God-given old natural remedies (his beloved "electrifying" apart). Things are profoundly different today, when a vast body of scientifically tested techniques, from artificial heart valves to radiotherapy, are available in our hospitals, and can *really* cure illnesses against which celery root, rue, whey, soot, or verdigris would have been entirely ineffective. There is only one realistic "alternative" to modern scientific medicine: death.

The country vicar who paved the way for aspirin

The Reverend Edward Stone was a much more scientifically and experimentally inquisitive "medical" clergyman than John Wesley. Born of comfortable yeoman farmer parentage at Lacey Green, near Princes Risborough, Buckinghamshire, in 1702, Stone was one year older than Wesley. He studied at and became a fellow of Wadham College, Oxford, taking his MA in 1727, before becoming rector of Drayton and Chipping Norton in the northern part of Oxfordshire, serving as Justice of the Peace, and marrying his patron's daughter. It would be interesting to know if Stone and Wesley ever met, as both were Oxford contemporaries.

In eighteenth-century England, the "ague", or "quakes", was a common affliction, especially for people living near boggy or marshy places, of which there were many in 1750, before the great land drainage and reclamation schemes of the late eighteenth and nineteenth centuries. Today we associate malaria with the tropics, and particularly with the mosquito (the mosquito connection was first suggested by the papal physician Giovanni Maria Lancisi in 1717, and scientifically demonstrated in India by [Sir] Ronald Ross in 1898), yet there are mosquito species that can thrive in colder climates. And thrive they did, until land drainage destroyed their habitat.

Edward Stone's interest in malarial agues began at Chipping Norton about 1756, when he was suffering from quaking fits himself. Whether or not he had heard of the benefits of willow bark from locals, he does not say, for he appears, as he testified to the Royal Society in 1763, to have been led to try willow bark by the providential belief that God in his mercy put the remedies for "natural maladies... not far from their causes".[16] And willows "delight in moist or wet soil where agues chiefly abound".

Associated with this traditional way of medical thinking was the theory of "signature". Stone found that willow was very bitter to the taste; and so was cinchona, "Jesuit's bark", from South America, which contained crude *quinina* or quinine: the best-known and most expensive prophylactic for the "ague". Had God "signed" with a bitter taste those plants that could break a malarial fever?

Guided by this traditional, quasi-theological thinking, Edward Stone began to do quantitative experiments. Collecting the stripped bark from pollarded willows, he dried it in a bag by a baker's oven, and then pulverized it. When he had his next ague fit, Stone dosed himself with 20 grains, old apothecary's weight (1.296 grams), of the cinnamon-coloured powder every four hours, between the paroxysm cycles, and immediately felt better. Increasing the dosage, he discovered that 1 drachm (4 grams) was best of all.

Stone now began to try powdered willow bark on his sickly parishioners. Dosing some 50 individuals over five years, he enjoyed an amazing success rate: only a post-chaise coachman had a relapse, because he had failed to take his willow bark as instructed. When one-fifth part of quinine was added to the willow bark, even the worst cases were "totally routed".

After five years of successful quantitative experiments, Stone communicated his results to the Royal Society's *Philosophical Transactions* in 1763, under the title: "An Account of the Success of the Bark of the Willow in Cures of Ague".[17] One cannot help but be impressed by Stone's meticulous experimental technique, and the only way in which he falls short of a modern clinical trial is in his failure to try a "double-blind", or to give one group of sufferers the willow bark, and another a dummy drug or "placebo", to test how each responded. (The first such "double-blind" clinical trial was made in 1747, at sea, by Dr James Lind, FRS, who tested the efficacy of lemon juice against various placebos as a treatment for scurvy.[18])

No one is claiming that Edward Stone "discovered" aspirin, but he initiated a process of discovery. By the 1830s, however, rapid advances in the new science of organic chemistry enabled French and German scientists to isolate salicylic acid (from *salix*, the Latin for willow), which was the active ingredient in Stone's willow bark. Then salicylic acid was also found in the meadowsweet plant (Latin *spiraea*), so when Friedrich Hoffmann announced his new salicylic-acid-based compound in 1899, it took its name from its plant of origin: *a spiraea*, or *aspirin*. We will return to aspirin in Chapter 23.

STEPHEN HALES, SYDNEY SMITH, AND OTHER MEDICAL CLERGYMEN

By the time that Wesley and Stone were writing in the eighteenth century, the scientific and medical clergyman was an established part of the English landscape, many of these gentlemen being Fellows of the Royal Society. Take the Reverend Stephen Hales, FRS, a graduate of Corpus Christi College, Cambridge, perpetual curate of Teddington, Middlesex, vicar of Farringdon, Hampshire, and an avid chemical and medical experimentalist from his undergraduate days onwards. He turned down a prestigious canonry of Windsor so as to have more time for his Christian ministry, philanthropy, and scientific experimentation. Hales's passion was for exact measurement, and he applied this both to his researches into "airs" (or what would later be identified as gases) and to cardiovascular physiology. How much he may have "physicked" his parishioners for their ailments is not clear, but with such a scientific gentleman in the vicarage, it was probably to him that one might have resorted for a purge or a soothing salve.

All the evidence suggests that Hales had a very thorough knowledge of contemporary practical medicine, having attended lectures and dissected corpses in both Cambridge and London. As a fellow of the Royal Society, he would have enjoyed easy social access to the leading physicians and surgeons of the day. In 1739, he was awarded the prestigious Copley Medal of the Royal Society – the international "Nobel Prize" of the eighteenth century – for his work on the nature, causes, and treatment of bladder stones. One can see the influence of the Boyle, Hooke, and Mayow approach in his thinking about the chemistry of "airs", that of Harvey and

Lower in his cardiovascular researches, and perhaps that of Sir Isaac Newton in his wider physical understanding. In many ways Stephen Hales was the very archetype of the English country vicar scientist: driven by an intense, informed desire to know how nature worked on the one hand, and a genuine sense of Christian compassion and duty to his flock on the other.

His reputation was firmly and internationally established in 1727, when, at the age of 50, he published his monumental *Vegetable Staticks*, or on what might be called biochemical "functional equilibrium". How did all the physical and chemical parts fit together in a living system to enable it to function properly? The work described numerous experiments with "airs" (gases), pressures, volumes, and chemical reactions. Hales's primary goal was to see what "airs" he could extract, quantify, and test from a wide variety of living things. While he saw all of these explosive (hydrogen), invigorative (oxygen), combustive (coal gas), and suppressive (probably carbon dioxide) products as part of the *structure* of living, and sometimes mineral, things, he still regarded them all as *allotropic*: variant states of one "air". It would be later eighteenth-century chemists who would finally recognize them as chemically unique *gases*.

Hales's contributions to chemistry were very significant. He emphasized the importance of exact quantification in biological and chemical research. He also devised many techniques for extracting and collecting gases, which are still used in refined forms in modern laboratories: such as developing Mayow's water-trough technique for collecting gases by displacing volumes of water. Hales collected what we now know as oxygen by toasting saltpetre in an old gun barrel, and hydrogen by putting iron filings into sulphuric acid. He also became a pioneer of plant chemistry and respiration, noting that at night a plant kept beneath a glass vessel gave off different invisible "airs" [carbon dioxide] from those emitted in daylight [oxygen]. He was a pioneer, too, of the study of phototropism, or the way light affects plant growth.

Hales further suggested that the strange lights seen dancing above swamps at night were probably caused by the inflammable airs (marsh gas) emitted from rotting vegetation, and *not* by fairies.

Likewise, the wandering "spirits" seen hovering above newly filled churchyard graves on a dark, moonless night were probably caused by nothing more than airs rising up from the decaying corpse.

In 1733, Hales published his *Haemastaticks* [blood measurements]; *or, an Account of some Hydraulick and Hydrostatical Experiments made on the Blood and Blood-Vessels of Animals*: a truly ground-breaking body of research into the volumetric capacities and blood-throwing actions of living hearts. It was the next stage forward towards the complete elucidation of the cardiovascular system, after Lower's *De Corde* (1669) and Mayow's *Tractatus Quinque* (1674). One of the gems of *Haemastaticks* was his quantitative study of what we now call "blood pressure".[19]

To this end, Hales inserted the adapted brass-fitted end of a nine-foot glass tube into the crural (lower abdominal, leg) artery of an immobilized old horse. The blood shot four feet up the tube, then rose an inch or two with each pulse, up to 8 feet 3 inches, in accord with the systolic and diastolic actions of the horse's heart. He repeated the experiment on other animals, also using the jugular vein. In *Haemastaticks* he went on to compute the internal volumetric capacity of the heart,[20] and thus demonstrated "blood pressure" as a shock-wave of blood coming from the systolic, contractive, phase of the heart, and travelling around the arterial and vascular system. It would not be until the 1880s, however, that the first inflated air-cuff "sphygmomanometers" were developed to measure a patient's blood pressure without an incision.

The Reverend Sydney Smith is best known to history as a literary wit, ecclesiastical careerist, social "lion", humourist, and raconteur. He was also a conscientious and courageous parish priest, a keen amateur doctor, and something of a pioneer of rheumatology.[21] While a New College undergraduate and Oxford gentleman-about-town, he had listened to lectures on chemistry and probably witnessed dissections. But when a virulent fever struck his rural parish of Foston, Yorkshire, in 1816, the 45-year-old Smith did not flee. Instead, he battled with it, and saved the lives of many of his agricultural labourer parishioners with the drugs which he purchased and the remedies he tried upon them. He also appears to have obtained an archbishop's licence to practise medicine.

In addition to the epidemic crisis of 1816, it is clear that the local folk came to his vicarage door to seek relief from their aches and pains. Smith invariably had a variety of bottles of medicine to dispense to them. The sophisticated London wit, one gathers, was also the kindly, jolly country parson, for these bottles of medicine were often given comic names: such as the popular "Rub-a-dub". All were given out free of charge, as was the case with most parson-doctors. He was particularly significant in his treatment of the sore joints and muscles of the hard-working country folk, for he had made sets of "rheumatic armour", or specially shaped metal hot-water bottles, which could be tied on to a foot, arm, or elsewhere that was causing a person pain. Gentle heat can be very soothing for patients with joint and rheumatic problems.

In this chapter I have only scratched the surface of the English parson–doctor, for there were plenty more known to history: the brave Reverend William Mompesson, whose quarantine measures in his Derbyshire parish of Eyam stayed the bubonic plague epidemic of 1665–66; the Victorian qualified physician cum Congregationalist minister astronomer William Rutter Dawes of Ormskirk, Lancashire; and Elizabeth Tillotson, wife of the Archbishop of Canterbury, who brewed her medicinal potions in Lambeth Palace in the 1690s. In Chapter 18 we will meet the Reverend Dr Francis Willis, the pioneer eighteenth-century "Mad-doctor", or what we would now more politely style "psychiatrist".

An essential feature shared by all these medical practitioners was a sense of the sacred duty of care towards their patients. In the eighteenth century in particular, this coming together of care and progressive scientific medicine was most clearly manifested in the founding of new hospitals, asylums, and medical schools.

CHAPTER 17

The Duty of Care: New Hospitals,
Charities, and Medical Innovation

*I*t was the eighteenth century that invented our modern concept of charity. The great monastic-based hospitals, almshouses, and colleges of the Middle Ages had been charities, though most had been founded and liberally endowed by individuals: bishops, noblemen, or rich aldermen. Many had been swept away at the Reformation. The eighteenth-century charities, however, took on a new complexion, often being corporate rather than individual benefactions, especially in Great Britain. Even when a charity had been founded by an individual – such as Oxford's Radcliffe Infirmary – it was the corporate activity of more moderately well-off private subscribers who kept the ball rolling on a daily basis, as well as public-spirited unpaid trustees who managed the original endowment: all for the benefit of the poor and the advancement of learning.

There were all sorts of charities – often designated "societies" – for medical aid of various kinds; overseas and domestic Christian missions; the relief of old folks; food, clothing, and coal for the poor; orphanages and free schools; and, by the 1780s, charities to benefit the distant stranger in the burgeoning slave-trade abolition movement. Charity concerts, balls, tea parties, theatrical productions, and all manner of events were being hosted by 1780, the price of a ticket for which would help give relief to a poor, sick, homeless, or enslaved person.

Several factors led to this flood tide of charity which is still part

290

and parcel of Western, and especially Anglo-American, civilization today. Firstly, there was an approximate doubling of Great Britain's population between Gregory King's demographic tax-based computation in 1688 and the first proper national census in 1801 – from 5.5 million to nearly 9 million souls in England and Wales. There were more people to care for than ever before.

Secondly, the technological transformation of agriculture, accompanied by the massive expansion of overseas trade, banking, stockbroking, and mercantile wealth, and, most of all, a veritable explosion in industrial output, had not only generated unprecedented wealth, but spread it across whole swathes of society – from the London merchant-princes of the East India Company, to the country vicar investing £50 in a new local coal mine, to Richard Arkwright, the Preston barber and small shopkeeper, whose business acumen would transform him into a Lancashire cotton millionaire, to become *Sir* Richard by his death in 1792.

Thirdly, there was that growing concern with *improvement*: coming not from some vaguely idealistic "Enlightenment" notion, but from something much more down to earth and powerful: the Christian idea that those who had been blessed with talents and abundance had a duty to those who just could not make it. Cartoons of obese parsons notwithstanding, this was also the age of Wesley's and Whitfield's Methodist revival, conscience searching, chapel building, Bible reading, and abundant hymn writing and singing. While not everyone who chose to give away money was necessarily religious, the broader public culture of eighteenth-century Great Britain certainly ensured that over the century vast sums of private money went to endow "good works" enterprises. Many of these had a significant medical dimension.

A NEW TIDE OF HOSPITALS: LONDON

One inescapable feature of this "good works" movement was the wave of new hospitals, refuges, and dispensaries established entirely for the benefit of the poor, disabled, or defenceless. In addition to its ancient hospitals of St Bartholomew (1123), St Thomas (pre-

1215), and St Mary of Bethlehem – Bedlam – (1247), eighteenth-century London saw both the expansion of existing resources and the establishment of five entirely new general hospitals, plus some new specialist institutions. All were necessary, for by 1800 the growing metropolitan sprawl was approaching a million people, or about one-tenth of the nation's population.

27. Bethlehem Hospital ("Bedlam"), in Moorfields, London. This was the new model mental hospital, laid out in gardens, the fabric designed by Robert Hooke for the Governors in 1674–76. (Original engraving, *c.* 1680, A. Chapman collection.)

The new London hospitals included one which still bears its city financier–speculator founder's name: Guy's. Thomas Guy had made a killing in the city, especially by shifting his stock just before the notorious South Sea Bubble burst in 1720. Yet this city "fat cat", like so many others over the centuries, right down to today (and I have met a good few), had a deep sense of duty. In 1721

292

he founded and endowed an institution for "incurables" suffering from various diseases. It was destined to grow, like its south London neighbour, St Thomas's, to become one of Europe's great hospitals for the poor, and one of its great teaching and research centres. In common with similar foundations, treatment has always been free of charge to the patient – and still is.

Then there was the Westminster Hospital (1719), to cater for the burgeoning overcrowded suburbs in London's "theatre district" of the new West End: a district teeming with brothels, cheap pubs, slums, thieves' kitchens, and gaming dives such as those immortalized in William Hogarth's low-life moral engravings of the 1730s and 1740s. If any hospital needed an accident and emergency unit, one suspects that it was the Westminster; and many an enterprising medical student must have learned volumes by keeping his eyes open. Other new foundations followed: St George's (1733) and The London (1740): all on the strength of charitable giving and philanthropy.

Several specialist charitable foundations were also established in eighteenth-century London, such as the Lock Hospital, (1746), which catered for persons with sexually transmitted diseases. Over the decades countless destitute "ladies of the night" found a bed, succour, and some sort of treatment within its walls.

In an age with only the most rudimentary birth control, childbirth and unwanted children were making increasing medical demands as the population swelled, both by people migrating to London from elsewhere in Britain to seek employment and from "indigenous" metropolitan births. This led to the Middlesex Infirmary, founded in 1745, and re-established as the Middlesex Hospital in 1757: another foundation for the sick poor, now incorporating London's first "lying-in" (maternity) hospital. Then there was the very pointedly named Hospital for Married Women, (1751). Most married women had their babies at home, and it was generally only the unmarried ones who sought the refuge of a lying-in hospital. Four "lying-in" hospitals were founded in London by 1770.

By the 1730s, London's streets were so full of homeless children that a rich, kind-hearted, and strongly Anglican merchant sea-

293

captain and successful London businessman, Thomas Coram, was inspired to found and endow England's first proper orphanage, just north of the city, in "Coram's Fields". Giving a home, a simple education, and a pre-apprenticeship training to countless thousands of children over the years, usually brought soon after birth by their desperate, often unmarried, mothers, the Foundling Hospital became London's most fashionable charity. Not only was it patronized by London's most fashionable ladies, but musicians and artists of the calibre of George Frederick Handel and William Hogarth put their talents at its service.

Although we will be looking specifically at mental health care provision in Chapter 18, it should be noted that back in 1674 the governors and trustees of the cramped and notorious "Bedlam" Asylum in London had commissioned Dr Robert Hooke to design a new, spacious Bethlehem Hospital for 120 patients on a fresh site in Moorfields, just outside the city walls. Dr Hooke, as we know, had been Thomas Willis's assistant and was an accomplished architect, and very much in demand. The new "Bedlam" was a glorious piece of architecture, similar to a French château, set in beautiful gardens, with airy and humane accommodation for the patients. Sadly, however, by the early eighteenth century patient demand had led to appalling overcrowding.

In 1751, London's new asylum, St Luke's, was opened, it being the conviction of Dr William Battie, the physician, that with humane care the mentally ill could be restored to health no less than the fevered or broken-limbed.

Alongside the formal hospitals, with their wards and beds, another type of medical institution came into being in the eighteenth century to deal with the sick poor: the dispensary. These were outpatient institutions, supported by charitable subscriptions, the first of which came into operation in 1773, and by 1800 there were some 16, dealing with around 50,000 cases per annum. Yet even before 1773 George Armstrong had founded his Dispensary for Infant Poor, a pioneering paediatrics outpatient clinic, and in 12 years is said to have treated an incredible 35,000 sick children.[1] A dispensary was a bit like a free "medicine shop" or clinic. The poor could drop in, be examined, given medicine or dressings, and even receive house visits from the attendant apothecary or physician.

The booming of dispensaries and charity hospitals tells us there was a fast-growing concern for the indigent sick, and the polite classes perceived it their duty not merely to sing hymns and respectfully listen to sermons, but to put their faith into practice and do something about the problem on a practical, hands-on level. While this was taking place within London and its suburbs, what was happening across the counties?

NEW HOSPITALS ACROSS GREAT BRITAIN

By the early eighteenth century, England had by far the biggest and fastest-growing body of comfortably off people – the "middling sort", as they were called – in the history of civilization. These extended from village tradesmen who owned their own premises and a few acres of freehold land to the middling country gentlemen or successful businessmen in the shires. What defined them was their relative independence: their ability to do things without seeking permission from a superior, and their possession of a financial "cushion" between themselves and adversity, be that cushion £100 or £100,000.

The men of this class would generally be able to vote for Parliament, sit on local councils and parish church or chapel vestries, pay taxes, and employ one or more servants. But they were most definitely *not* aristocracy; nor were they poor. They included members of the church, medical, and legal professions, and collectively generated much of the nation's wealth. This 10 per cent of the population were the "movers and shakers" of Georgian Britain, the role model for their relatives in the 13 American colonies, and the envy of the educated classes of a much more autocratically governed Europe. And their ranks were growing all the time.

These literate and thinking people were also the driving force behind charities and "good works" enterprises, as much in Edinburgh or Newcastle upon Tyne as in London. They founded

295

and maintained the 28 or so general hospitals that were set up in eighteenth-century provincial Britain, and asylums, dispensaries, and other institutions providing comfort for the distressed. While Continental visitors to Great Britain often commented that the English "poor" were not as destitute nor as cowed as their French counterparts, there was real poverty in Great Britain: in the one-roomed farm labourer's cottage, the pitman's overcrowded urban brick box, and the teeming, reeking slums of St Giles's parish in London. Such problems were often made worse by cheap gin-drinking and alcoholic poisoning, until Parliament began to levy prohibitive taxes on those "distilled liquors" condemned by John Wesley and others. A farm labourer who made £20 per annum on which to keep his family was doing well, and the rest of the rural and urban workers would not be doing much better.

These circumstances moved people to acts of kindness, conscience, and Christian compassion. What is impressive about medical charity in the eighteenth century is the sheer scale of it, as new general and specialist hospitals went up across the land. These include the Edinburgh Royal Infirmary (1729), Bath Royal Mineral Water Hospital (1736), Bristol Royal Infirmary (1737), Winchester Infirmary (1737), York County Hospital (1740), Royal Devon and Exeter Hospital (1741), Bath Infirmary (1742), Northampton General Hospital (1743), Newcastle Infirmary (1751), Manchester Royal Infirmary (1752), Gloucestershire Royal Infirmary (1755), Norfolk and Norwich Hospital (1771), and Norwich Eye Hospital (1822).

The two English university cities also acquired their hospitals. Oxford's Radcliffe Infirmary came out of an enormous benefaction left by Dr John Radcliffe in 1714. It provided medical scholarships to his old college – University College – plus a fund which was to provide his *alma mater* with a scientific library (Radcliffe Camera), an astronomical observatory, and a hospital. The well-managed funds accruing to the Radcliffe Trust are still being disbursed today.

28. Dr John Radcliffe (1652–1714) of Oxford, with a patient and assistant, *c.*
1690. In the background stands the future Radcliffe Camera Library, 1737,
and to the left the Radcliffe Infirmary, 1770, both benefactions by Radcliffe.
(Pen and ink cartoon, A. Chapman, 1978.)

The Infirmary that would bear his name was initiated in 1758, and
its magnificent buildings, wards, and gardens opened on St Luke's
Day, 18 October 1770, at a cost to the Radcliffe Trust of £4,000.
Its patient base was to be the sick poor of Oxford city and county,
and the magnificent allegorical picture of the hospital, from the
Oxford Almanack of 1760, says it all. Here is the shade of the long-
dead Dr Radcliffe, resplendent in academic gown, accompanied by
a smiling Asclepius, the classical god of healing, welcoming the
halt, lame, and blind into the new Infirmary. The picture combines
all the learned medical motifs of the age: a smiling and munificent
benefactor, along with all the classical trappings of architecture and
mythology, inviting the poor to come and obtain relief – all on the
Saint's Day of St Luke, the Christian Gospel writer and "beloved
physician". Although the John Radcliffe Hospital has now been

297

relocated to a large site outside Oxford city centre, one can still see the names of the early subscribers and trustees, the "great and good" of the Georgian county, painted on panels inside the elegant Board Room of the original Infirmary. Its architecture, the subscribers' lists, and the visual images speak of that combination of local pride, learning, and genuine concern for the poor which underpinned all of these charity hospitals.

29. The benign shade of Dr John Radcliffe welcomes the crippled, sick, and broken into his new charity Infirmary in Oxford, finally opened on St Luke's Day, 18 October 1770. (*Oxford Almanack* plate, 1760. Christ Church Library, © Governing Body of Christ Church, Oxford.)

Cambridge's Addenbrooke's Hospital, originally created by a £4,500 donation from the late Dr John Addenbrooke, fulfilled exactly the same role both for Cambridge University and the city and county poor, in 1766. In Oxford and Cambridge Universities in particular, the Regius Professors of Medicine (both chairs created by King Henry VIII over 1545–46) had a technical supervisory role over their new hospitals, as well as over their regular medical staff.

It was in these burgeoning metropolitan and county hospitals for the poor in the eighteenth century that the subsequent staffing pattern of the English hospitals would begin to emerge. While the hospitals might employ an apothecary, a young "junior" doctor or surgeon, nurses, and attendants on a regular basis, it became a mark of professional prestige for a senior medical man to attend the hospital on certain days, walk the wards, and perhaps teach, in a consultant capacity. This he did either free of charge, or in return for a very modest stipend.

One may wonder where, if all these new hospitals were for the poor, did the better-off classes go when sick? They stayed at home, where a bedroom would be turned into a sick-room or even an operating theatre; servants would be trained to act as nurses or professional nurses employed; and the doctor or surgeon would make his private house calls. This tradition took a long time to die out. In the late 1960s I had a conversation with an elderly gentleman who told me that when a boy, he recalled an operation being performed on an anaesthetized household member lying on a well-scrubbed table, by a surgeon visiting his parents' house in the country!

THE HOSPITAL AS A "MUSEUM" OF DISEASE

In the new charity hospitals housing and treating the sick poor, the patients themselves often served an educative and even an experimental purpose for medical and surgical students and qualified practitioners alike. If a hospital had 30, 40, or more beds, and its patient catchment was London or an English county and its town, then a student might find a fascinating cross-section of human illness all under one roof: scarlet fever, pneumonia, a bad compound fracture, cancerous lumps, cataracts, a complicated hernia, childbirth complications, severe burns, and perhaps even madness.

Strangely, though, even after getting their fine new hospitals in the 1760s, neither Cambridge nor Oxford University used them to establish a clinical medical school until the late nineteenth century.

299

What was more normal, in fact, and continued even after the 1760s, was that a "Probationer BM or DM" (registered candidates for the degree of either Bachelor or Doctor of Medicine), having attended the requisite academic lectures and read the prescribed books, would go off to London, and perhaps Paris, to walk the wards of the big hospitals and get the opportunity to see at first hand as wide a spectrum of cases as possible. There he would watch leading physicians and surgeons in action, treating a variety of often frightened charity patients, in that continuity of medical theatre that extended back to Vesalius's Padua and Galen's Rome.

Ironically, however, although London hospitals were proliferating in the eighteenth century, the city had no university. Not until the foundation of University College in 1826 and King's College in 1829 (their hospitals founded 1834 and 1840 respectively) did London possess university-level academic institutions. Instead, most of the London hospitals gradually evolved their own formal teaching departments, and students qualified by passing the examinations of the Royal College of Physicians, the Company of Surgeons (or the Royal College of Surgeons after 1800), and the Society of Apothecaries.

The other place for a medical student to head towards was Scotland. All four Scottish universities, St Andrews, Edinburgh, Glasgow, and Aberdeen, enjoyed the privilege of conferring MD degrees (the BM and DM are Oxford peculiars). With the growth of population in the eighteenth century, Edinburgh and Glasgow in particular developed world-class medical schools. These comprised a conglomerate of university teachers, private medical teachers (as we shall see in Chapter 21), and new hospitals: Edinburgh Royal Infirmary (1729), Aberdeen Royal Infirmary (1739), and Glasgow Royal Infirmary (1791–94).

Trinity College, Dublin also awarded degrees and taught young doctors, and Ireland acquired its own charity hospitals with patient catchments in Dublin, St James's (1727), Cork Infirmary (1744), and, originally supported by a local musical society, the City of Belfast (1797).

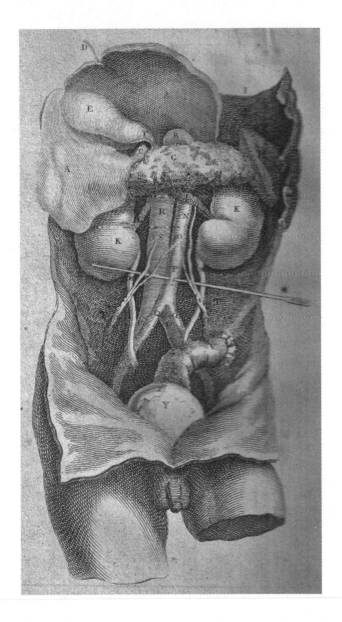

30. Dissected human body, front view, showing the viscera: liver (A), gall bladder
(E), pancreas (G), kidneys (K), iliac arteries (Q), part of the colon leading
to the rectum (V), and bladder (Y). The long needle passes through the
beginning of the inferior mesenteric artery. (Andrew Fyfe, *A Compendium of
the Anatomy of the Human Body* (Edinburgh, 1807), Vol. II, Table XXII. A.
Chapman collection.)

301

From a teaching perspective, a hospital was a museum of *living* diseases, so most of these old and new hospitals built up museums of preserved specimens. The seventeenth-century realization that soft tissue could be preserved in sealed jars of alcohol led to a veritable explosion of bottled body parts, and skeletons and muscular tissue that had been carefully dried and varnished.

St Bartholomew's Hospital, London, still has a wonderful historical museum, as do both the London and the Edinburgh Royal Colleges of Surgeons: places which some people find creepy – with their 200-year-old-plus bottled foetuses, brains, and livers, bleached white by the light – and others absolutely fascinating. In particular, I know the Royal College of Surgeons' Hunterian Museum in Lincolns Inn Fields, London, and have lectured there on several occasions. It is, along with its magnificent library, a medical history cornucopia, as is Edinburgh.

But who were the men who practised and taught in these new hospitals, and what did they achieve? And who was John Hunter, FRS?

TEACHERS AND DISCOVERERS IN THE EIGHTEENTH-CENTURY HOSPITALS

Scholars have long debated why the British population almost doubled in the eighteenth century, and various explanations have been advanced. The broadly scientific transformation of agriculture, diminishing famine and increasing the supply of grain, meat, and dairy products, certainly helped lay the foundation for a better-nourished population, although less so in Ireland. The mass manufacture of cotton cloth from the 1760s onwards provided cheaper, cleaner, and more easily washable fabrics than linen, while the burgeoning coal mining industry made warmer and more comfortable homes available to many.

One major factor was the impact of that technique known as "immunization" against smallpox, popularized in England in the 1720s by Lady Mary Wortley Montague, then living in Constantinople. It had long been known that if one caught smallpox and survived, one gained immunity for life. In the immunization technique, a healthy person deliberately took a (hopefully) weak dose of the disease from a current sufferer in an attempt to gain lifelong immunity. It involved an enormous risk, and many died, though conversely, many survived – and lived to produce more children: creating, perhaps, an exponential population growth. Then in 1798, Dr Edward Jenner developed the vastly safer technique of *vaccination*, to which we will return in Chapter 20, and which helped to make possible the population explosion of the nineteenth century. Yet what happened in the eighteenth century?

The London teaching hospitals, old and new, made innovations in many areas: in accurate pathological and clinical descriptions of diseases, in new, pre-anaesthetic, surgical techniques, in the study of occupational diseases, and even in the development of new drugs.

In 1785 Dr William Withering, FRS, a Shropshire-born Edinburgh MD who was physician to the new Birmingham General Hospital, published his clinical trials in which he successfully treated 156 patients for "dropsy", or the swelling of the legs due to fluid retention. (This is caused by a weakened systolic, or heart contractive, force, now known to be related to an irregularity in the action of the atrial valve.) On the suggestion of a local folk-healer, Mother Hutton, he began to experiment with foxglove, or digitalis, which does indeed have a beneficial effect on the contraction and rhythmic action of the heart.[2]

Yet one of the great bedside clinicians of the eighteenth century, who died in 1801 aged 91, does not appear to have held any hospital appointment, but to have drawn his vast clinical experience from bedside consultations in private homes. This was Dr William Heberden, FRS, a Cambridge graduate and active teaching fellow of the Royal College of Physicians. Heberden was not an experimentalist, but belonged, like Thomas Sydenham, Richard Mead, and John Radcliffe before him, to the Hippocratic tradition of carefully studying and recording the disease process in living people, and learning by experience when best to intervene with a drug, bloodletting, or other treatments. He left classic descriptions of angina, arthritis, and the stages through which tuberculosis passed, as well as other diseases, in his posthumous *Medical Commentaries on the History and Cure of Diseases* (1802).

The "museum of illness" aspect of a great hospital or college was also significant for the clinical work of Percivall Pott, of St Bartholomew's. Among other things, he realized early on that tuberculosis, probably the biggest single killer of the age, was not just a disease of the lungs: it could also cause joint deformity, and affect the spine – "Pott's disease". Working as he did a century before the discovery of bacteria, Pott did not know the causal agent of tuberculosis and how it proliferated and attacked different parts of the body. Yet he made and published the crucial empirical observations and saw the vital connections: the very stuff of scientific progress.

Pott also made the vital connection between certain irritant substances and cancer. He noted in 1775 that the incidence of cancer of the scrotum in chimney-sweeps was particularly high. Young boys were being forced to climb, often naked, up chimneys to sweep out the soot. The delicate scrotal folds could harbour and slowly absorb the carcinogenic soot, especially if the boy was rarely, if ever, bathed.

In December 1735, the French-born Huguenot surgeon of the then new Westminster and St George's Hospitals, and sergeant surgeon to both King George I and II, Claudius Amyand, FRS, reported what we now recognize as the first, inadvertent, "appendectomy". Amyand told the Royal Society, in October 1736, that he was at the time operating on an 11-year-old boy who had a serious hernia. The lad's name was Hanvil Anderson, and Amyand gave him a special mention because of the immense courage that he had displayed during his 30-minute unanaesthetized ordeal. On the operating table, his scrotum – already perforated by a fistula – seems to have burst open, and poor Hanvil's intestines poured out. Then, as Amyand attempted to push them back, it was noticed that a "Pin incrusted with stone" was sticking out through his *Appendix Coeci*. Amyand cut it off, ligatured the gut, and stitched young Hanvil up again.

304

The brave boy survived and made a rapid recovery, apparently without complications, and was discharged from the hospital one month later, simply wearing a temporary truss to "confirm the Cure". Amyand was clearly a pioneer of gastroenterological surgery and stated that while previously working as a military surgeon, he had saved the lives of men who had been shot in the stomach: quite an achievement with no anaesthetics or antiseptics.[3]

Yet probably the most significant and long-lasting influence of the eighteenth-century clinical teaching tradition came from the Scotsman John Hunter.

JOHN HUNTER FRS

Hunter was born in 1728, the younger son of a modest farmer from South Lanarkshire, Scotland. His elder brother William had trained as a doctor at Glasgow and Edinburgh universities before going south to seek his fortune as a physician in London. John was not a university man, however, but followed his brother to London to become his assistant, having learned some anatomy and surgery beforehand. Before long, however, younger brother John would outpace Dr William.

In many ways, it was the British Army that was to be the making of John, for he was commissioned into the Medical Department as a surgeon to serve in the Seven Years War. Following a battle on the island of Belle-Île, northern France, in 1761, Hunter encountered a group of French soldiers who had musket balls lodged in some of their soft tissue, though the bullets had fortunately not severed any major blood vessels. They had been living rough with no medical treatment, yet Hunter noticed that their wounds were healing naturally, whereas soldiers with similar wounds brought fresh into the field hospital were in a much worse way. Could this be because the well-meaning surgeons had made matters worse by poking around in the wounds to remove the bullets, or even amputating the limbs?

This led Hunter to propose a "conservative" or minimal interference technique with stable gunshot wounds. Just leave the bullet in, clean the wound, bandage, and let nature take its course. He received much criticism for this minimally invasive method, but poking around with unsterile fingers and probes is likely to do a wound more harm than good, causing infection, and necessitating amputation. Hunter had no concept of bacteria, but his experiences in the field gave rise to his future researches into the nature of blood, infection, tissue growth, and healing.

Over the next 30 years, before dying suddenly from a heart attack following a heated meeting with colleagues in St George's Hospital in 1793, Hunter was to pursue several lines of research which would have a profound impact upon Western clinical thinking and practice thereafter. He was a brilliant observer, clinician, experimentalist, and practical surgeon.

His army experience also sparked his interest in sexually transmitted diseases. In his day, physicians disputed whether gonorrhoea and syphilis were caused by the same or by different pathogenic agents. Hunter, the born experimentalist, devised what he hoped would be a defining experiment. He injected a healthy subject – some sources suggest Hunter injected himself – with pus from a gonorrhoea patient and observed the ensuing development of the disease. The experimental patient developed not only gonorrhoea but syphilis as well, probably due to cross-contamination resulting in doubly infected pus. This led Hunter to conclude that both diseases shared a common pathogen, whatever it was. Unfortunately, Hunter was wrong, and as later doctors would discover, primarily using high-power microscopes in the nineteenth century, each disease had its own unique pathogen. But he was wrong for the right reasons.

After leaving the army, Hunter also became a pioneer of transplant surgery. When working with a London dentist, he developed a technique whereby teeth taken either from a cadaver or from a living donor (usually a young poor person willing to sell their teeth) could be immediately transplanted into the still-bleeding sockets of a rich patient. The technique sounds cringingly painful, as bad teeth were extracted and matching donor teeth – molar, incisor, and such – *hammered* into the new cavity. The curious thing was, the teeth sometimes stayed in the mouth and healed, and sometimes they fell out. Why? This led Hunter to formulate ideas about tissue transplantation. He concluded that the pivotal factor in healing was maintaining an abundant healthy blood supply, in addition to a minimally invasive approach in the treatment of wounds.

His decades of research into this topic would be presented in his posthumous *Treatise on the Blood, Inflammation, and Gunshot Wounds* (1794). Without any knowledge of cells, or bacteria, Hunter attributed health and the good healing of flesh to a vigorous blood supply. On the basis of decades of clinical experience of all manner of wounds, he came firmly to the opinion that simple cleaning and dressing, with minimal tampering, was the best road to healing all breaks of the skin, both small and great. Like the medieval and Paduan physicians and surgeons, Hunter believed that "first intention", or pus-free, healing was both the most natural and the quickest way for any kind of wound. He was absolutely right, for minimal tampering minimizes infection and the risk of complications.[4]

By the time of Hunter's death in 1793, the healing arts had reached a crucial juncture. The anatomy of the human body, along with that of many animals, had been delineated with remarkable detail, and the basic functions of the cardiovascular, respiratory, cerebral, and other systems outlined. Surgical techniques had achieved a remarkable sophistication, and were poised, with the discovery of anaesthetics and antiseptics after 1846, to fundamentally transform the power of the healer. Infection was still largely a mystery, yet the value of cleanliness was coming to be recognized on a purely empirical level. Increasing prosperity, combined with a largely Christianity-derived sense of duty and philanthropy, had set in motion a stream of hospital and care facility foundations, whose impetus would continue down to the present day.

And while "madness" still remained a puzzle, the eighteenth century would lay the foundations of our understanding and humane care of the mentally ill.

CHAPTER 18

"Remember Poor Tom o' Bedlam": Dealing with the Mad

*I*n 1698 the pioneering "tabloid" journalist, Ned Ward, was running a humorous series of 18 monthly-instalment articles called *The London Spy*. The series proved so popular that they would be reissued in book form in 1703. In this first golden age of the English gutter press, following the lapse of the 1662 publication-censoring Licensing Act in 1695, which gave birth to our free press, Edward Ward wrote a series of fascinating and revealing articles about the sights and scandals of contemporary London. Deliberately mischievous and often mocking in tone, the *Spy* articles allegedly described the instructive perambulations around London of Ward and a gentleman friend. "All human life is here" – from churches to brothels, from learned societies to low boozing kennels. Having visited, and sent up, the august Royal Society (where some years before Robert Hooke had examined insects with microscopes), Ward and his friend "remov'd from Maggot-mongers Hall [the Royal Society] to survey Madman's College".[1]

This was "Bedlam" Hospital, Robert Hooke's grand fabric having opened in 1676. Here, for a small charge, one might be admitted to view the lunatics, the entrance money officially being used to meet running costs, though a good bit disappeared into the pockets of the keepers. Ned Ward and his friend were introduced to a curious array of unfortunate folk, including highly educated persons, a deranged alcoholic, another who commanded a legion of eagles, and a depressed academic. The last was "a scholar of St.

John's College, in Cambridge, who was possess'd with melancholy, was very inoffensive and had the liberty of the gallery".[2] In the "women's apartment" there was a quick-tongued lass who embarrassed some young women visitors with her racy quips.

The whole institution was a true "bedlam" of "rattling of chains, drumming of doors, ranting, holloaing, singing and rattling", calling to mind Don Quixote's "Hell in an uproar",[3] all in a charitable institution, with custom-designed and built humane accommodation, only 20-odd years old. The building, designed by Hooke, the scientist–architect, had been built at the cost of a breath-taking £17,000, with a magnificent architectural frontage 540 feet long. It reminded some of King Louis XIV of France's new Palace of Versailles, leading the wits to quip that what was good for a French king was only meet for an English lunatic!

So what had gone wrong? Why was a palatial château designed to humanely accommodate 120 men and women in 1676 beginning to fill up by 1698, and, judging from Hogarth's engraving in *The Rake's Progress* of 1732–33, had by then become seriously overcrowded? In the rapidly expanding metropolis, with an ever-growing catchment of mentally ill people beyond, demand had already overwhelmed the facilities that "Bedlam" could supply. But why?

"POOR TOM'S A-COLD": HELPING THE "INSANE" IN STUART AND GEORGIAN ENGLAND

"Poor Tom's a-cold... Poor Tom hath been scared out of his good wits... Five fiends have been in poor Tom at once", says Edgar, in Shakespeare's *King Lear* (1606).[4] By 1698, "Tom o' Bedlam" had long been a stock character in English parlance, and a recognizable figure in popular ballads, plays, and tales. According to folklore, he was a licensed beggar, allowed out of the original medieval asylum to beg with his pitiful litany "Poor Tom's a-cold".

Whether such beggars were even licensed by the charitable governors is a matter of scholarly dispute. More likely, they were sane professional beggars, also known as "Abraham men", done up in rags and skilfully muttering, pathetically eye-rolling, or doing "idiot" capers. There was nowhere better to do them than outside a fashionable city church on a Sunday morning, as the congregation of rich merchants and their ladies were coming out, having heard a powerful sermon on the parable of Dives and Lazarus.[5]

If Poor Tom had already been a stock character in Shakespeare's day, when London had a population estimated at around 225,000 persons, how much greater must the demand for psychiatric help have been in Ned Ward's time, when it was approaching 575,000, let alone in 1800, when it was over a million? Who were the people needing such comfort, treatment, or accommodation? They included traumatized discharged soldiers and sailors unable to gain admission to the new specialist naval and military hospitals at Greenwich (1694), and Chelsea (1702); the depressed – like Ned Ward's Cambridge scholar; those who had run amok and from whom the public needed protection; those broken by life's woes; and those tragic victims of tertiary syphilis, or "general paralysis of the insane". These would include people like Hogarth's character Tom Rakewell, whose dark chancres, in the last Bedlam scene of *The Rake's Progress*, suggest the syphilitic source of his madness.

What could a doctor actually *do* for a deranged person, be it in 1650 or 1850? It depended on individuals, circumstances, and doctors, as we shall see later in this chapter. But as the physiological causes of mental illness were only vaguely understood, except that it was brain-related (at least no one was taking Shakespeare's Tom o' Bedlam's "Five Fiends" seriously by 1750), what were the treatment options for those poor Bedlam folk who could not afford a private mad-house or individual medical attention?

Simple custody and restraint to prevent damage to oneself and others was the first line of "therapy". Vomiting, bloodletting, and heavy purging could help shift malign humours, though on a more pragmatic level, an exhausted patient was not likely to be as rowdy or violent as beforehand. Then there were John Wesley's favourite cure-alls: for "Lunacy" "try [a] decoction of agrimony, sweet ivy, shaving and chafing the head. Or, be electrified. (Tried.)": For "Raging Madness", he advised pouring cold water on the patient's head. Broadly speaking, all of these might be considered as "shock", "purge", and "exhaustion" therapies. On a crude level they worked, for depending on how they were applied, they could silence or induce a reaction. But under no circumstances could they *cure*. The "madness" or "melancholy" either diminished naturally or the patient remained ill. These time-honoured "treatments" were still in use in 1815, when Dr Thomas Monro, physician to the Bedlam, outlined them to a Parliamentary Committee investigating private madhouses.

The profoundly isolating and reality distorting nature of mental illness invited a strong response from the physician, whether this meant trying to silence the patient or to provoke some kind of reaction, however crude the method used. It was not until German organic chemists synthesized chloral hydrate in 1832 that things began to change, for chloral hydrate, it was discovered in 1869, could bestow a good night's sleep upon a frenzied or insomniac patient, with no addiction or obvious (in 1869) nasty side effects, as was the case with opiates. Chloral hydrate made an asylum a quieter and more easily managed institution, although a quiet patient was not by any standards a *cured* or *peaceful* patient.

Much horror and indignation would later be expressed at how overcrowded eighteenth-century Bedlam treated its patients. Yet we, with our efficient psychotherapeutic drugs and modern technologies, can all too easily forget the sheer sense of *helplessness* which doctors, nurses, and other staff must have felt when faced with the needs of several hundred "ranting, holloaing, singing, and rattling" Bedlam patients. On the other hand, in the eighteenth century – as in the Middle Ages – "kindness cures" also came to be tried, and results discussed.

MAD-DOCTORS AND MADHOUSES

Before looking at (hopefully) benign practices, one must be aware of the fact that mental illness in particular opened up all sort of opportunities for appallingly abusive relationships, for an "insane" person had, by definition, lost their own right of self-governance and invariably fell into the power of others. And these "others"

311

could come in all shapes and sizes: from grasping relatives trying to gain access to a mentally disturbed person's property, to corrupt asylum officials more keen on pocketing the watches or purses of a committed person than helping them, to negligent or incompetent doctors, to, perhaps worst of all, the cynical proprietors of private "madhouses", who were generally more anxious to fleece the relatives of "crazy Aunt Jane" than to try to help her get better.

We know of these abuses not only from official reports, such as those that instigated Parliament's passing of the Act for Regulating Private Madhouses in 1774, and from the scandal that erupted following the exposure of criminal neglect in the York Asylum in 1815, but even from the patients themselves. Many highly intelligent, literate, and even literary persons suffered psychiatric episodes and left harrowing accounts of their treatment, either within or outside "madhouses".

Take, for example, the published and reprinted narratives of Hannah Allen, a deeply devout and providence-seeking late seventeenth-century English widow; or, a century later, the wool merchant Samuel Bruckshaw, whose pamphlet *The Iniquitous Abuses of Private Madhouses* (1774) led to searching questions being asked at the highest level. Then there was William Belcher's *Address to Humanity... and a Sketch of a True Smiling Hyena* (1796) concerning the formal declaration of a lunatic and the seizure of his property. A name that was to ring down the next two centuries in the context of biblical scholarship was the London-based Scotsman Alexander "The Corrector" Cruden, whose *Concordance* to the Bible (1737) is still breath-taking in its lexicographical scope. All of their narratives are published in full in Allan Ingram's *Voices of Madness*,[6] and these are only the ones who committed their harrowing experiences to paper and found a publisher.

THE BEGINNINGS OF COMPASSIONATE CARE

Humane, diligent, and generous men, and often their wives, both medically qualified and unqualified, did practise compassionate care for the mentally ill. It was the growing problem of providing effective care at Bedlam that led to the Cambridge-educated Dr William Battie, MD, being appointed physician to the new London charity asylum, St Luke's, after 1751. As we saw in Chapter 17, Battie was of the opinion that mental illness was a natural disease just like any other and that instead of its victims simply being locked up, purged, vomited, bled, and exhausted, it should be studied scientifically. Consequently, it was hoped, rational and effective cures could be found. And one of Battie's "successful" patients was the famous poet, Christopher Smart. But as we saw above, mental illness is not like a fever or an inflamed wound, which might either respond to treatment, self-heal, or simply kill the patient outright. For "madness" comes in many different forms, its sufferers are often incommunicative, and it can drag on for years, until some other malady eventually kills them. So where does one start? It is true that in 1621 Robert Burton had produced a classic patient's-eye case-study of "Melancholy", but those conditions now clinically defined as schizophrenia, psychosis, and paranoia could all too easily present themselves as a chaotic jumble of symptoms, defying rhyme or reason, especially if the person were also perceived to have criminal tendencies or simply to be a "natural idiot".

Good intentions, therefore, often came to very little, especially when attempting to treat the "insane poor" in charity asylums in London, and then, by 1800, in Manchester, York, Lincoln, and Liverpool. What one might call "compassionate care" was by its very nature time-consuming and, irrespective of whether it was paid for by an individual or a charitable institution, expensive. Such care was most likely to work within self-selecting patient groups, such as the committedly religious, the genteel, and the educated: with people who shared basic values which provided natural avenues of conversation between doctor and patient, and between patients.

The essential logic underpinning this approach was that as man is made in the image of God, he must therefore be inclined to rational thinking and conduct. While the physical degeneration of the brain might lead to conditions such as "silliness" (dementia) in old age, most mental illness was simply, it was hoped, a temporary estrangement from a natural reason which could be restored. The patient's natural reason could have been upset by passionate love, obsessive behaviour, irrational fears, or unchecked delusions, and the way to bring about a cure was by rest, peace, calm, and sane discourse, within a secure and pleasant environment. It might even – as in the Lincolnshire institution run by the Reverend Dr Francis Willis (of whom more will be said shortly) – include healthy physical activity, such as digging the soil. Such a regime predicated the space and resources of a private, fee-paying establishment, though as we shall see, in the nineteenth century John Conolly successfully tried it in a public asylum.

One man who found peace and relief, though not a full cure, through this regime was the poet William Cowper. Educated at Westminster School, Cowper subsequently entered the private institution of the devout Leiden-trained physician Dr Nathaniel Cotton at St Alban's. Later, however, he was taken, presumably as a paying guest, into the rectory of the Reverend Morley Unwin and his wife Mary, in 1765. When Morley Unwin died suddenly from a fall from his horse, Cowper developed a close friendship with his devoutly evangelical widow. He also got to know the evangelical clergyman and former slave-ship captain, the Reverend John Newton – composer of the hymn *Amazing Grace*.

Cowper's occasional mental instability continued even after his Christian conversion, yet he appears to have benefited enormously from the peace of the Unwins' household at Olney, Buckinghamshire, and from conversations with men and women of shared religious views – a far cry from Bedlam.

There are other cases of clerical "mad-doctors" who took in paying guests. With their wives, they must have faced enormous strain within their households by having one, two, or more "mad" gentlefolk living with them, yet they felt it to be part of their Christian vocation. Without doubt, however, the best-known and most successful of these was the Reverend Dr Francis Willis, DM, Oxon.

The Reverend Dr Francis Willis and King George III

An incident took place in the autumn of 1788 which was to occasion a fundamental rethink concerning "mad-doctoring". In October 1788, His Majesty King George III, then in his fiftieth year, began to exhibit strange and upsetting changes of behaviour. Although he had been distressed at the loss of Britain's colonies in America a decade before, King George's manner to date had marked him out as a model monarch. He was highly intelligent – though sadly, this did not always show in his portraits – conscientious, devoutly Anglican, and happily married to Queen Charlotte. He maintained a private state-of-the-art astronomical observatory, laboratory, and instrument collection near Kew Park, enjoyed the company of scientists and artists, was an accomplished harpsichordist, and was a discerning art connoisseur whose agents bought up art works from European auctions, effectively creating the modern Royal Collection. George was also popular among his people, turned Windsor Great Park into a model farm, and, in his guise of "Farmer George", enjoyed attending country shows, dressed in plain brown fustian, casting a knowing eye over pigs and sheep, and chatting to ordinary farmers.

Then suddenly George started to talk to imaginary people. He chattered on, often incomprehensibly, for hours on end, cried, suffered from insomnia, had acute bowel problems, and all of a sudden, would act quite irrationally. Sir George Baker and the other court physicians were baffled and afraid, for none of them knew how to handle a monarch who appeared to have lost his reason. Then Queen Charlotte, his devoted wife, was told by Lady Harcourt of a Lincolnshire clergyman "mad-doctor" who enjoyed an impressive success record with his patients. This was the Reverend Dr Francis Willis of Brasenose College, Oxford. After years of successful practice as a clergyman–doctor, he had taken an Oxford DM degree and set up a private psychiatric care institution at Greatford, near Stamford, Lincolnshire, while also playing a leading role in the founding of the Lincoln General Hospital.

Francis Willis (no relation of Thomas Willis, whom we met in

Chapter 13) was the son of a dignitary of Lincoln Cathedral, and his wife Mary was the daughter of a county clergyman. At Oxford, he had exhibited a passion for medicine and science, attending lectures in anatomy and chemistry. Willis belonged to that parson–doctor tradition which we saw in figures like John Ward, Edward Stone, and Sydney Smith. Where Willis differed, however, was in the nature of his line of interest, for while wholly knowledgeable about anatomy and physiology, his fascination lay with disturbed minds: psychiatry. He possessed a curious gift: a strange charisma by which he could pacify the deranged, sometimes referred to as his "eye" or glance. Francis Willis enjoyed a "discharged cured" rate from his rural Lincolnshire private asylum which might be the envy of a modern psychiatrist. Not only did the patients enjoy rural peace, prayers, and the doctor's reassuring presence, but Willis was also a pioneer of occupational therapy. The gentlefolk under his care worked on the estate farm and pursued other hearty outdoor occupations; much more conducive to restoring a balanced mind, of course, than being locked up, purged, and exhibited like an animal in the zoo.

Behind it all was the constant, reassuring presence of the doctor and his calm, positive, yet quite unchallengeable personality. Willis's technique was known as "moral management", in so far as he related to the characters and needs of his patients, and worked through their own moral and value structures. This was, admittedly, an easier thing to do with disturbed Christian ladies and gentlemen than with the heterodox mixture of "lunatics", syphilitics, deranged scholars, and prostitutes who populated Bedlam and the growing number of county charity asylums. In addition to the obvious power of his personality, the venerable 70-year-old Willis, as he was in 1788, possessed a simple and direct manner: "open, honest, dauntless, light-hearted, innocent, and high-minded", the writer Fanny Burney styled him. Likewise, the popular writer Hannah More described him as "the very image of simplicity, quite a good, plain, old-fashioned country parson".[7]

This was the man whom Queen Charlotte called in to see her poor husband George. Over the winter of 1788–89, just before the

316

horrors of the French Revolution broke in July 1789, the king and his doctor developed a strange relationship. Having made it clear that court protocols would have of necessity to be abandoned, and the king treated like a normal patient, Willis set about "managing" him – making him see that the royal behaviour was not only unreasonable, but also wrong. How could God's anointed sovereign of Britannia so let himself go? Willis's style was a combination of "pull yourself together" – *Sire* – and "remember your sacred duty"; and if you *don't*, Your Majesty will be physically restrained. Willis was to have full control over the royal person.

As the king got gradually better, and winter turned to the spring of 1789, things progressively eased. King and doctor–subject walked around the Windsor estate, conversed, played chess, and read Shakespeare together. Willis was always respectful of His Majesty, yet inflexibly firm, and demanded full obedience. When the king was declared "cured" in March 1789, with a major constitutional crisis averted, Willis was willing to drift away, as a wise doctor should, more than happy with the £1,000 per annum life pension granted to him by Prime Minister William Pitt and a grateful nation.

Yet the king never lost his quick wit, even in the early stage of his illness. George detested doctors, but respected clergymen, and asked Willis why he, dressed as he was in clerical black, was also a doctor? "Sir," replied Willis, "our Saviour himself went about healing the sick", to which His Majesty retorted "Yes, yes,... but had not 700L [£] a year for it"; (a sum later raised to £1,000 by Parliament).[8]

As we know from recent studies, Francis Willis may *not* have cured the king: he may simply have got better naturally. References to the dark colour of the king's urine in October 1788 and to excruciating stomach pains at the onset of his affliction have led to the historical diagnosis that the king had porphyria, a disease passing through the bloodlines of several European royal families. George's illness could have been biochemical in its cause, and not psychiatric. Porphyria may account for later short episodes of mental illness subsequently suffered by the king and may have contributed to the dementia which afflicted his last decade of life up to 1820. Yet to see George III as a "mad" or irresponsible monarch is plainly false.

Francis Willis died in December 1807, a few months before his ninetieth birthday; his son, Dr Robert Darling Willis, continuing as a royal doctor. His other sons became clergymen, and one an admiral. But Willis, largely through his perceived success with his royal patient, had a significant effect upon the future of psychiatric care, leading to the increasing use of humane "moral management" methods.

FROM SCANDAL TO CARE IN YORK, AND THE RISE OF HUMANE TREATMENT

In the late eighteenth century, York was one of those British cities that possessed a "lunatic asylum" for pauper patients, complete with its own physician and care staff, set in pleasant surroundings. In 1790, however, it began to cause concern, after a Quaker named Hannah Mills was admitted, only to die a few weeks later under very suspicious circumstances. Hannah's friends had not been allowed to visit her, and when influential York Quakers eventually obtained access, it was found that the patients were being locked up, entirely neglected, and treated worse than animals. One of the Quakers behind the inquiry was William Tuke, a rich tea-merchant and philanthropist, who in 1796 was to create a new type of asylum in his home town of York.

Yet even after the Hannah Mills scandal, the corruption, misappropriation of charitable funds, and criminal neglect of the York pauper patients continued. Complaints of worried family members of asylum patients were brought to the attention of Godfrey Higgins, a Yorkshire magistrate and active social reformer, who used his legal status as a Justice of the Peace to gain access to the building and see behind otherwise locked doors, and down dark and concealed cellar passages. And what Higgins found revolted him.

Mentally ill pauper patients were found to be starved, beaten, chained, locked away, and living in filth. Fortunately, Higgins had the political contacts and the clout to cause a Parliamentary

318

Inquiry to be set in motion, not just into the York, but into other "madhouses", including Bedlam, in 1815: the year of the Battle of Waterloo. This Inquiry interviewed Dr Thomas Monro, physician to Bedlam, who thought it quite acceptable to chain paupers, but *not* gentlemen![9] We must view Monro's statement in context, for while no military or naval tribunal in 1815 would have dreamt of flogging a disobedient officer, a "taste of the cat" would have been deemed entirely normal should Private Billy or Jack Tar commit the same offence.

It was William Tuke, and then his grandson Samuel, who were to be responsible for advancing the care of the mentally ill, when in 1796 William founded the York Retreat, initially for disturbed Quakers, but later opening its doors to all denominations. The Quakers had no formal clergy, but in his sense of Christian duty, philanthropy, and willingness to innovate therapeutically, William Tuke was very much in the tradition of the English parson–doctor. Like others in that tradition, Tuke was a practical man, financially shrewd, but also compassionate. At first, the Retreat, in its own idyllic rural setting on Lamel Hill, York, had no ideological programme of therapy beyond the simple acknowledgment that most people responded favourably to kindness, and that the "mad" especially just needed somewhere to *be*, to *retreat* awhile. The mentally ill were a bit like misbehaving children, who needed firm yet compassionate handling to slowly nurture them back to sanity. And Tuke's methods were experimental, in so far as he was evolving his techniques as he went along.

On the other hand, Tuke's Retreat was secure, even if its iron window bars were made to look like normal window frames, and its gardens, while relaxing and conducing to peace, were nonetheless impossible to escape from. But there were to be no chains, beatings, or dark cellars, and only *in extremis* was a troublesome and potentially dangerous patient to be physically restrained. As had happened with King George, the firm *threat* of restraint normally produced more compliant behaviour.

Over the years, William and Samuel Tuke evolved a successful treatment regime at the York Retreat. Most of the original clientele were Quakers, with shared religious, ethical, and social values, but the humane methods were found to work equally well with non-Quakers. The impact which the Retreat had upon nineteenth-century attitudes to mental care was enormous, especially after Samuel Tuke published his *Description of the Retreat* (1813) which, backed up by the evidence of his grandfather William, the Retreat's founder, was given to the 1815 Parliamentary Committee set up in the wake of Godfrey Higgins's "whistle-blowing" on the York Asylum.

The "moral management" of the insane was also being explored in France at the time, with Dr Jean-Baptiste Pussin and Dr Philippe Pinel at the male (Bicêtre) and female (Salpêtrière) asylums in Paris. What, if anything, the Tukes, Pussin, and Pinel learned from each other's work is hard to say; however, in spite of Anglo-French wars, scientific and medical information often passed with remarkable ease between the two nations, even during the ideologically driven French Revolutionary and Napoleonic periods. One also wonders how far both English and French "moral managers" had been influenced by the techniques used by Francis Willis both upon his Lincolnshire asylum patients and, most famously, upon King George III himself. One should be careful not to take at face value Tony Robert-Fleury's highly romanticized painting of 1876 depicting the noble Dr Pinel ordering his underlings to unchain the women of the Salpêtrière. For one thing, the unchaining took place sometime after the supposed 1795 event depicted in the painting, while the grateful patients look too young, clean, and attractive to reliably represent the authentic ragged, dirty, and wild-eyed denizens of one of Europe's most notorious psychiatric hell-holes. The girl being unchained more closely resembles a loosely clad seductive and sultry actress than a lunatic. The real Dr Pinel was a pioneer of psychiatry as a branch of scientific medicine and a taxonomist of different types of mental illness, and one of the first clinicians to clearly define schizophrenia, under the name *dementia praecox*.

The Tukes and their York Retreat certainly had an influence on three subsequent English asylums. The first of these was the private one founded in Oxford after 1826 by the Reverend Dr

320

Samuel Warneford: an extremely wealthy and amazingly generous Anglican clergyman and serial philanthropist. Warneford did not practise medicine personally. Instead, in that great tradition of Christian charity, he devoted tens of thousands of pounds from his private fortune, over nearly half a century, to the founding and endowing of hospitals. His Oxford hospital alone received, it has been estimated, a breath-taking £70,000, and that was quite distinct from liberal benefactions to medical charities in Leamington Spa, Birmingham, and elsewhere, along with benefactions to almshouses and religious charities, said to total £200,000.[10]

Dr Robert Gardiner Hill, superintendent to the subscription Lincoln Lunatic Asylum in the 1840s, likewise employed the humane "moral management" method upon his patients. Perhaps the most famous of this new breed of humane "mad-doctor", though, was the English-born Irishman John Conolly. An Edinburgh-trained MD, Conolly took charge of the large Hanwell Asylum in Middlesex, west London, in 1839. Conolly's success at Hanwell soon caught the imagination of the newspaper editors – the mass media of the Victorian age – and the 1848 Twelfth Night celebrations following Christmas at Hanwell even got coverage in the popular *The London Illustrated News*.

Conolly understood the deeply isolating nature of mental illness and did what he could to lighten the atmosphere in his enormous mental hospital. Hanwell, like other asylums, had male and female separation, and the sexes did not generally come together. There were, however, celebratory meals, concerts by local brass bands, and even dances, to try to bring some "normality" and even fun into the lives of Conolly's patients, and Hanwell had an impressive "discharged cured" rate. Conolly attempted to take the personal attention care system of the private asylums into the larger charity "madhouses".

By the mid nineteenth century, however, care standards were getting markedly better across the board. With figures like Pinel in France and the early Victorian Henry Maudsley in England, "mad-doctoring" was turning into the new medical specialism of clinical psychiatry. Yet even by 1890, the essential chemistry of the brain was still a mystery from a clinical, therapeutic point of view, and the first effective "mind-healing" drugs still lay in the future. Others, such as Sigmund Freud and his Viennese and Parisian colleagues, took psychiatry down a different track: that of psychoanalysis. Even so, it was no easy ride to be "mad" in nineteenth-century Europe, as comes over all too graphically in some of the Parisian novels of Émile Zola, set in the 1860s,[11] and England's Wilkie Collins's *The Woman in White* (1859).

Progress was slowly being made, and continues to be made today. Yet mental illness remains one of the most intractable of maladies, and while "Poor Tom" may no longer be "a-cold", his descendants are still very much with us, and demanding our compassionate attention.

CHAPTER 19

Charismatics, Quacks, and Folk Healers into the Early Industrial Age

*D*efining a "quack" before the mid nineteenth century was often as much about legal definitions and language as it was about therapeutic capacity. After 1518 anyone found practising medicine in London who held no qualification from the new Royal College of Physicians was technically a charlatan, or an illicit practitioner, and open to prosecution. As physicians of the "Fellowship" were so small in number, and too expensive to be of service to the vast majority of Londoners, sheer necessity meant that a medley of trained surgeons, apothecaries, and even grocers (who often imported opiates and other "groceries" from the Levant) were involved in healing. Unless they called themselves "doctor" or "physician", they were often left in peace to "physick" ordinary folk, while the gentlemen of the Fellowship concentrated on sick courtiers and rich aldermen.

A *proper* charlatan, rather, was usually seen as an uneducated, untrained, self-advertised, loud wonder-worker, with the gift of the gab, and good at separating people from their hard-earned pennies. *Quack* is thought to derive from "Quicksilver", or "Quack-salver": a practitioner who promised secret cures, often for the embarrassing new disease of syphilis, using mercury. Much of the opprobrium associated with "quacks" derived, therefore, from their boasted cures for sexual diseases. They were also known as "Pox-doctors", not from treating smallpox or chickenpox, but from the sexually transmitted "Great Pox" chancres of syphilis.

As we shall see in this chapter, "quacks" and "charlatans" were a very vigorous and long-lasting breed, ranging from the occupants of Tudor fairground booths to Victorian purveyors of "secret" patent medicines obtained by post, to a diverse medley of "alternative healers" in our own day and age. But what about faith healers?

Valentine Greatrakes: Irish gentleman faith healer

Faith healing in its various forms goes back to prehistoric times, and long pre-dates Greek "scientific medicine". As we saw in Chapter 1, even the Greeks brought their sick to be miraculously cured in the temples of Asclepius. Both the Old and the New Testaments in the Bible are rich in accounts of miraculous cures, as are the sacred texts of other religions. In the late 1970s, the retired Oxford Pentecostal policeman, Fred Smith, held healing sessions in Oxford Town Hall, while I know several people today who claim to have received, or to have witnessed, miraculous cures for a wide variety of maladies.

In the mid seventeenth century, Valentine Greatrakes (or Greatorex), a gentleman of County Waterford, Ireland, who may have received some medical training and had served as an officer in Oliver Cromwell's army, discovered that he had a strange "gift". When he "touched" or "stroked" certain scrofulous persons, they seemed to get better. Scrofula was, by ancient custom, widely believed to be curable by the touch of a king or queen, earning it the name "The King's Evil". Certain other individuals were also believed to be empowered by God to "touch" for the disease.

The nature of scrofula was not understood in the seventeenth century, although we do now know its cause. Scientifically speaking, it is a tubercular-related disease of the lymph glands and nodes, causing the complex lymph system in the neck and lower head to swell into unsightly lumps, often with skin damage. While tuberculosis-related, it does not necessarily mean that the victim

324

also has "phthisis", or coughing tuberculosis. The tuberculosis bacterium can affect various systems of the body, causing, among other things, "Pott's disease", or debilitating spinal nodules, as we saw in Chapter 17.

But scrofula was the disease (along with some others) which the kind and devoutly Protestant Valentine Greatrakes found that he could sometimes cure, *in lieu* of a sovereign, by gentle stroking. As it was a divine gift, he stroked his patients free of charge. Greatrakes came to England in 1662, where he caused something of a sensation, his reputation going before him. Yet when he gave a demonstration of his touching cure before King Charles II at old Whitehall Palace, he had no success, his patients remaining unhealed. The young man who in 1675 would become the first Astronomer Royal, at the new Greenwich Observatory, was also "stroked" by Greatrakes: this was John Flamsteed, who as a youth had what was thought to be scrofula. It seems that he was not cured either.

Many men of high academic standing, such as the "Cambridge Platonist" clergyman–philosopher fellows of the Royal Society, Ralph Cudworth and Henry More, clearly supported Greatrakes and believed his cures to be genuine.

BARTHOLOMEW FAIR AND OTHER FAIRGROUND QUACKS

From the Middle Ages down to 1855, one of the key events of the Londoner's year began on 24 August, when the Smithfield area north of St Paul's Cathedral and hard by St Bartholomew's Hospital became a fairground. Its roots lay in the medieval commercial fairs, where merchants and craftsmen came together to sell their wares, though along with the business, Bartholomew Fair was also a "fun-fair". Tightrope walkers, jugglers, acrobats, conjurers, comedians, and other shows were there, as itinerant entertainers headed for a week of rich pickings in the City of London – along with a legion of pickpockets and thieves.

Bartholomew Fair and its sixteenth- and seventeenth-century equivalents in Nottingham, York, and elsewhere came to showcase what would be thought of as the quintessential "quack doctor". Though changing in dress with the passage of the centuries, the quack doctors, or mountebanks, were a regular feature down to the closure of the Fair in 1855. Generally, they were exotically dressed, sometimes with a fool or stooge, and not infrequently a monkey, as props to help bolster the sale of their nostrums. These invariably included cure-all "secret remedies", pox cures, love philtres, and rejuvenators, to make Jill fall in love with Jack and make the old as frisky as lambs. Not infrequently, the quacks accompanied their spiels with particular medical acts, such as the drawing of teeth, or using "plants" in the audience to come up on to the stage to be wondrously "cured" of their blindness, epilepsy, or lameness.

31. Bartholomew Fair quack with boxes of "medicines", clown, monkey, and persuasive patter, probably seventeenth century. (H. Morley, *Memoirs of Bartholomew Fair* (1859). Chapman collection.)

At Bartholomew, as at other English fairs, the quack doctors would have been competing for attention and patronage with overt entertainers such as jugglers, conjurors, tightrope walkers, dancing bears, and stand-up comedians, whose capers are well recorded. So a colourful show was vital if their stint on the fair was to be profitable. Along with any nostrums that visitors might purchase to heal their aches and pains, they expected fun, frolics, and lots to eat – and drink. It was not without irony that the profits raised from a fair that happily hosted mountebanks were used to help fund the adjacent St Bartholomew's Hospital.

Long after the authorities finally closed down Bartholomew Fair in 1855, however, quack doctors remained a standard feature of fairgrounds. There was, for instance, the famous "Dr Sequah" in the Victorian age: a full-scale commercial concern, and a proper registered company. My late grandfather, Albert Platt, had vivid memories of seeing either the real Sequah or one of his later imitations on Pendlebury and other markets and fairs in Edwardian Lancashire. The Sequah whom he remembered, and of whom he gave me a graphic account (Albert died in 1984, aged 93), incorporated a Wild West show into his act. There were Indian braves and dancing girls and a small brass band as part of the warm-up routine.[1]

Then once the packed crowd was cheering, the "Doctor" made his dramatic entrance in the full garb of an American Indian chief. First, he would offer to "painlessly" remove bad teeth, at a time when, for most working people, dentistry was either "do-it-yourself" or else a visit to the quack. The patient would be taken behind a curtain, and as the deed was being done, Sequah's brass band would strike up a racket to deaden the screams. Then he got down to business: selling his secret remedies, supposedly obtained from the wise tribes-folk of the American Great Plains – Prairie Flower and Sequah's oil which would cure anything from a headache to a fever to a nasty tumour. Very obviously, the Sequah medicine show cashed in on the growing public fascination with cowboys and Indians, and the interest generated by the massive travelling shows of "Colonel" William Cody, Phineas T. Barnum, and others. A fascination with the Wild West was already migrating to the cinema by 1900, and by the 1950s, to television.

Unlike the "good old-fashioned cures", largely ancestral, of John Wesley's *Primitive Physick*, the quack's perennial appeal lay in his supposed unique and privileged access to secret, or arcane, cures, unknown either to scientific medicine or to traditional wisdom. The source of the privileged information could be a Chinese mandarin or an Indian fakir in 1600, or a North American Indian chief or exotic African "witch doctor" by 1895.

LEARNED QUACKERY

It may seem contradictory to speak of "learned quackery". Yet before the mid nineteenth century, and for some time afterwards, the boundary that separated the learned physician or scientist from the Bartholomew Fair quack was often far from clear-cut. Take those two friends, pioneer experimental physiological and chemical researchers and eminent Fellows of the Royal Society, Dr Robert Hooke (MD, Lambeth, 1691) and the Honourable Robert Boyle (DM, Oxford, 1665). In addition to their scientific eminence, both men were long-standing hypochondriacs: victims of headaches, insomnia, eye problems, gout, obstructions, upset stomachs, palpitations, constipation, fluxes, chills, fevers, and other real or supposed maladies. Hooke's "Diary" for 1672–80 in particular, along with his correspondence with Boyle, supplies some fascinating insights into the world of "do-it-yourself" medication. Such an addiction to self-medication seems all the more incongruous because Hooke and Boyle not only possessed essentially honorary doctorates in medicine, in recognition of their scientific researches, but they also mixed socially on a daily basis with many of the leading working physicians of the age, with whom, one supposes, they exchanged medical tips and recipes in coffee houses or at the convivial dinners which often rounded off a Royal Society meeting.[2]

Even admitting that Hooke and Boyle tried out some of these "cures" in the spirit of scientific curiosity, one cannot help but

328

shudder at some of them. On 31 January 1673, Hooke recorded "Mr Boyle told me that *stercus humanum* is good for eye films". The polite Latin *stercus humanum* translates as "human excrement" in plain English. Unfortunately, we are not told exactly how this substance was believed to remove "films" or hazy vision. Then, perhaps more realistically, Hooke went "to Mr Boyle for Spirit of Harts Horn", or liquid ammonia distilled out of animal horn. When sniffed, pure ammonia gives one something of a shock in the head and makes the nose run. According to Galenic physiology, this nasal discharge would release an excess of "animal spirit" in the brain, thereby "purging" the brain, the Eustachian and other tubes, and hopefully relieve a headache. No doubt, Boyle also used it himself.

In the same league as *stercus humanum* was "Usnea *cranii humani*", suggested to Hooke by Sir Robert Moray, FRS, on 12 March 1673. "Usnea *cranii humani*" was also an ingredient in Dr Jonathan Goddard, FRS's famous "drops", along with ammonia, with which Hooke dosed himself. The famous "usnea" was a preparation made from human skull scrapings and skull mould, and may have been related in its therapeutic rationale to the already accepted standby *tinctura mummia*: imported pulverized Egyptian mummy dust.

In common with many genuinely sick or hypochondriac folk, Hooke could not resist a good purge. Some he picked up from doctor or apothecary friends, some from acquaintances he met in coffee houses, some from conversations with passers-by in the street, and one from Mrs Elizabeth Tillotson, wife of his old friend John, FRS, who in 1691 became Archbishop of Canterbury. He often recorded the action of these "pot luck" medical encounters in his diary. On 1 August 1675, he wrote, "Took Hewks Vomit it made me incline to vomit after 1 hour [of] taking. It made me straine after 2 houres but brought up nothing." Another, on 16 February 1673, "wrought… [several] times with me brought up much slime of the guts and made me cheerful".

Then, for 22 December, 1672, there was "a woman in the Tower cured divers [sufferers] of vertigo by stone horse dung" – the Tower of London being not just a state prison, but also a full city parish in its own right. Hooke's friend Mr Wild (26 November

1674) also told him that he believed the blood of a black cat would cure chilblains Then there were the more conventional cures of the age which he tried out, such as "Took Senna. Stayed within [at home] all day", just in case; or the acidic corrosive metallic emetic *aurum potabile*, and the newly imported *Bangue* (bhang, or *Cannabis indica*) which put a man into a "Dream", and might be of value in treating "Distempers of the Head and Stomach".[3]

These "cures" from Hooke's "Diary" and from friends are only a selection of those I have encountered which passed between the learned men of the seventeenth century. So if Hooke and his fellow scientists and scholars and churchmen friends were willing to give credence to "therapeutic" concoctions made up of excrement, skull mould, mineral acids, and cat's blood, who can blame poor Jack the ploughboy and Jill the dairymaid, strolling into London for Bartholomew Fair, for being taken in by the glib patter of the nostrum seller? All that separated these learned gentlemen from "The Wonderful Doctor" on his platform, was a good broadcloth coat – and the absence of a performing monkey!

QUACKS, SHOWMEN, AND DOCTORS

How could men who dissected cadavers, unravelled the complexities of the blood circulation and respiratory physiology, established the science of neurology, and explained the correct functioning of the heart, possibly countenance the use of cat's blood or powdered skulls in medicine, "therapies" more at home with Shakespeare's three witches in *Macbeth* than with a meeting of the Royal Society? The answer lay in the ever-widening gulf that was opening up in medicine: the gulf that separated the academic anatomist and physiologist in his laboratory from the practical business of alleviating human suffering. The disease *process* remained just as mysterious and confusing in 1700, or even 1800, as it had in the days of Hippocrates. To a medical world with no knowledge of bacteria, cellular pathology, trauma-controlling drugs, or modern analytical chemistry and biochemistry, the good old explanations, like the good old cures, still offered the best way forward when faced with typhus fever, misbehaving insides, nasty lumps, or insanity.

Consequently, one reverted to the time-honoured obstructions and blockages theory of illness, where a purge, an emetic, a "clyster" (colonic irrigation, performed with a large syringe and soapy water), or a good bloodletting would at least produce *some* sort of response. Who could say for certain that the blood of a black cat could *not* cure chilblains, or a lump of stone horse dung an attack of vertigo? In this respect, therefore, the MD, FRS often found himself batting on an even field with the Bartholomew Fair quack or his better-dressed equivalent when it came to the lottery of "working a cure".

Eighteenth-century England saw several figures who were derided as "quacks" by the contemporary press, as much for their behaviour as for their medical incompetence. A proper physician was expected to be gravely learned, and in no way loud or showy; and one must assess how far the following figures were ridiculed for not conforming to the accepted norm, even if they did, from our point of view, make genuine contributions to medical understanding which we, with hindsight, might see as laying the foundations for subsequent medical specialities.

Were there any more quacks per head of population in the days of King George I and his Hanoverian successors after 1714 than there had been two centuries previously, in the days of Henry VIII and Elizabeth I? Or could it simply be that a burgeoning popular press has left a more richly documented history? Eighteenth-century England produced an abundance of scandal rags, such as John Dunton's *Whipping Post*, 1706, the *Grub Street Journal*, 1730, and numerous others, that sold by poking fun at the "celebs" of the day. Then even the more respectable London and provincial newspapers, magazines, and periodicals often borrowed news items from each other, to spread stock-market prices, ghost stories, accounts of hangings, scandals, Royal Proclamations, advertisements for "wonder cures", and the antics of "quack" doctors across the nation, while in the burgeoning comic theatre, quack characters and their capers were always guaranteed to raise

331

laughs, whether based on real-life figures (such as "Chevalier" John Taylor) or purely fictitious.

John Taylor, for example, was the son of a Norwich surgeon-apothecary whose family had been producing medical men for five generations before John was born in 1703. As a young man, Taylor had first learned the art of the apothecary, then had studied in London under the eminent William Cheselden, surgeon of St Thomas's Hospital. Travelling to the Continent, he not only obtained MD degrees from the Universities of Basel, Liège, and Cologne, but was even elected to the Fellowship of the Basel College of Physicians. Beforehand, however, the youthful Taylor had published his *Mechanism of the Eye* (1727), a pioneering treatise on the fledgling science of ophthalmology. So how, one might ask, could a man possessing such prestigious formal medical qualifications ever be reckoned a "quack" or charlatan?

Much seems to derive from his manner, for Taylor was, in many ways, a gentleman version of a Bartholomew Fair quack: egotistical, loud, promising the earth, captivating people (especially titled ladies, it was said), calling himself "Chevalier", and playing the part of a grand gentleman all across Europe, from Spain to Russia. On his Continental journeys, Taylor's coach (reputedly painted with pictures of human eyes), flunkeys, and outriders were all orchestrated to cause a sensation as he rattled into Hamburg, Vienna, Paris, or Madrid: Bartholomew Fair in a periwig, brocade coat, and sporting a gold watch, no less!

While most self-consciously "learned" physicians were willing to take pretty well any type of paying case that came along, as a mark of their versatility, quacks were often one-song healers, specializing in just one or two diseases. These often included the pox, the stone, cancers, bad teeth, deafness, blindness, insanity, or "women's ailments". Or else they purveyed panacea "cure-alls". In this respect, "Chevalier" Taylor was immediately marked down as a quack, because he styled himself an *oculist*, or specialist in eye complaints. If his claims could be believed, which is not always easy to prove one way or the other, he was an oculist to many of the crowned heads of Europe, including Great Britain's King George

II. Irrespective of the authenticity of his claims to royal patronage, however, his son and grandson, both named John, really would be recognized as consultant oculists to George III and George IV respectively, down to the 1820s.

"Chevalier" Taylor specialized in the operation of "couching for cataract". This was, as we saw in earlier chapters, an ancient surgical procedure extending back to the Sushruta tradition of ancient India. By the eighteenth century, two techniques were firmly established in the repertoire. Either one could insert a needle into the front of the eye to lever down and break off the clouded lens from its mounting in the ciliary muscle system, so that it would sink down in the vitreous humour; or the operator might make a fine slit in the side of the eye and then remove the broken-off lens with tweezers: a modification of the sucking-out technique discussed by the Arab Albucasis in Chapter 4. By the eighteenth century, however, with the easy availability of spectacles, a lens-less eye, now enjoying a full intake of light, could often "see" once again when fitted with the right glasses, as we saw in the case of the Reverend Patrick Brontë in 1846.

Several eighteenth-century medical men, including Albrecht von Haller, commended Taylor's genuine skill as a cataract surgeon: a brief procedure taking no more than 10–15 minutes if the patient were sufficiently brave and steady not to flinch or blink, but could maintain a fixed gaze as the needles and tweezers were inserted. Taylor was also skilled in the correction of squint and of drooping eyes. His *An Exact Account of 243 different diseases to which the Eye and its Coverings are exposed* (1759) suggests that, for all his circus-like bravado, Taylor possessed a thorough knowledge of the anatomy of the eye, and was an adroit and skilful operator.

Neither Taylor nor any other surgeon before 1865 had any control over bacterial infection, for irrespective of manual dexterity and compassion, unsterile hands, instruments, and dressings made all surgical procedures something of a lottery, especially when performed on sensitive regions such as the eye. One wonders, therefore, how far the accounts of Taylor "blinding" his patients (including, perhaps, his reputed "blinding" of both of the composers Bach and Handel) may have stemmed from this wholly unpreventable source of surgical complication. If Taylor had followed a less outlandish lifestyle, gone about his business more discreetly, and sought a respectable hospital appointment in London or some other major European city, he might have ranked alongside his old surgery teacher William Cheselden in terms of historical reputation.

Another eighteenth-century celebrity quack might also have left a different mark on the historical record had he behaved differently. This was Joshua Ward: nicknamed "Spot" Ward, from a birthmark on his face. Ward, born in either 1684 or 1685, was the son of an industrialist, his father being the owner of an alum works in northeast England. Alum-making was a lucrative trade, as the chemical was used in textile processing.

Ward never trained in medicine, though perhaps, owing to his chemical manufacturing background, he had a flair for pharmacy and practical chemistry. Having picked up some recipes during travels on the Continent (probably necessitated because of his Jacobite sympathies), on his return to England, he did two things. He got himself "elected" Member of Parliament for Marlborough, as the result of a deadlock in the town council; and then he began to manufacture his celebrated nostrums: Ward's Pills and Ward's Drops.

Like so many of the patent medicines of the age, Ward's concoctions contained a mixture of vegetable material, mercury, ammonium chloride (similar to hart's horn), nitric acid, and alcohol. The logic behind them all was that they were to be purgative – inducing profuse sweating to shake off fevers, and violent vomiting to cleanse one's insides – the golden road to health, in eighteenth-century medical thinking. Being a genius at self-promotion, Ward not only became the darling of high society, but even came to move in court circles – the court of the very Hanoverian kings to whom one presumes that, as a Jacobite, he had been politically opposed – ending his exile in France, and receiving a pardon from King George II. The shrewd Ward seems to have been on good terms with George and possessed sufficient anatomical knowledge to realize that the king's painful thumb was caused *not* by gout, as was supposed, but by dislocation. He successfully clicked it back into place.

Joshua Ward seems to have been a cunning networker, a not entirely honest businessman, a possibly shady politician, and a highly successful quack, who, when he died at his home in Whitehall, London, left some £16,000 – a very handsome fortune. Yet what could have redeemed Ward's reputation, both contemporary and historical, had it been better known, was his generosity to the poor. Rather like the legendary outlaw Robin Hood, Ward fleeced the rich and gave his services free to the poor. He opened, and paid for, a small hospital for the poor in Westminster, and another in the City of London. His clinics and consultations were entirely free, and it has been estimated that he gave away a staggering £3,000 to charity, even throwing handfuls of silver coins to groups of street beggars. It was said that he was plain-speaking to the rich, and even to royalty, yet treated the poor with special respect.

Joshua Ward was a fascinating combination of paradoxes, and while not an academically trained quack, he was certainly a highly successful medical entrepreneur, whose purgative nostrums were no more outlandish than many prescribed by conventional MDs of the age. A resident of Westminster, the bachelor Ward had a right to be buried within the abbey, and was duly interred, with much ceremonial, on 26 December 1761, beneath the flagstones of Poet's Corner.

In 1736 William Hogarth published one of his comic caricatures (or cartoons) depicting a "Company of Undertakers" – *Et Plurima Mortis Imago* ("The Many Faces of Death") – or a group of contemporary high-profile quacks.[4] Ward, with his disfigured face, is there, while Chevalier John Taylor is recognizable from the eye engraved into the head of his golden physician's cane. Between them, however, is an extremely ugly, almost androgynous, figure, done up in the chequered suit of Harlequin. The figure is that of the ungainly Sarah Wallin, who took the name Mapp from her brief marriage to a husband who quickly absconded with her savings. This Wiltshire lass was also known as Sally Mapp, and even as "Crazy Sally", from her manifold eccentricities. But she became

335

renowned, over her short 31-year life, as the "Bonesetter", who had a genius for dealing with breaks, dislocations, and bad bruises of the body. And in Hogarth's "caricature", she, like Taylor and Ward, holds up the symbol of her art: a large bone.

Sarah may have learned the art of bone-setting from her father, who also seems to have had a knack with injuries, though this was pure folk empiricism. It used to be popularly believed that a seventh child of a seventh child (same sex as the gifted parent usually) inherited a sort of divine touch when it came to injuries. Sarah Mapp was certainly no trained anatomist, yet she could do wonders for the injured and not only appears to have cured, but also to have inspired patient confidence: always a valuable trait in any healer. Her combination of sensitive fingers, strong grip, and massive strength of arm enabled her to locate a subcutaneous injury. Then, by instinct and experience, she would twist and pull as was necessary to realign the jagged ends of a broken bone, or to snap a dislocated bone back into its socket.

Sarah took up residence at Epsom, famous in the 1720s, as it is today, for horse racing, and won a considerable reputation for treating what we might now call sports injuries. She even acquired a carriage in which she rode up to London to see patients and, like Chevalier Taylor and Joshua Ward, was patronized by high society and even by royalty.

Yet just as Taylor has subsequently been recognized – his circus-act persona notwithstanding – as a genuine pioneer of ophthalmology, and Ward as a medical entrepreneur and a charitable benefactor, so one might say that Sarah Mapp was something of a pioneer of physiotherapy and maybe even osteology. Unlike the learned generalists of the formal medical profession, Taylor, Ward, and Mapp pioneered specialist skills and approaches to healing.

One formally trained learned physician of the eighteenth century, who was sometimes accused of being not so much a quack as a breaker of professional etiquette in pursuit of wealth, was Dr Robert James. Around 1746, he invented a powder compounded, as subsequent analysis showed, of antimony, calcium phosphate, and other additives. It was intended to treat, or break, fevers, working in accordance with the purgative rationale of inducing copious sweating on the time-honoured assumption that this would drive out the toxins causing the fever.

The main criticism of James, however, was that, as a doctor, he should not have patented a supposedly life-improving treatment for personal gain. When he died in 1776, the legal right to manufacture the medicine was passed on to a commercial partner. In spite of the toxicity of antimony, the medicine enjoyed enormous and continuing popularity, and was still available into the twentieth century. It even got into English literature, being mentioned in Oliver Goldsmith's story *Goody Two-Shoes* (1765) and subsequent Christmas pantomime versions.

DOVER'S POWDERS AND NOSTRUMS GALORE

One eighteenth-century patent medicine which enjoyed a very long life was "Dover's Powder". Concocted by the academically trained Dr Thomas Dover (MA Oxford, BM Cambridge) in the early eighteenth century, this treatment for colds and fevers endured into the 1960s, its well-known ingredients being easily available to any dispensing apothecary via the standard pharmacopoeia.

The main ingredients were powdered ipecacuanha and opium. Ipecacuanha is a vegetable alkaloid drug from South America, deriving from the plant *Cephoelis ipecacuanha*, and essentially has an expectorant action. In small doses, it was generally believed to be good for congested chests – to "get up the phlegm" – while large doses could render one as sick as a dog, thus moving "heaven and earth". If the ipecacuanha provided the action, then the opium provided the counter-action, by acting as a calmative and sleep-inducer, and pacifying the cough reflex.

Much traditional pharmacy contained this action and counter-action rationale: first shock and then calm. In spite of its wholly out of date clinical rationale, Dover's Powders, in various forms, remained available into the mid twentieth century. Whether it was Dover's I cannot be sure, but my dear late grandmother, Lily Platt, born in 1892, fancied herself a bit of a doctor and believed

337

that a feverish person with a bad cold, cough, or the 'flu needed a "powder" to "sweat it out of 'em". The "good old cures" were slow to die, and she was promoting this one in the late 1950s.

Three late eighteenth-century medical men, possessing impeccable professional qualifications, were inspired by the medical potential of new scientific discoveries.

Franz Anton Mesmer was a Viennese society physician who in 1774 began to experiment with the human body's susceptibility to magnetism. He claimed that there were "magnetic tides" of "animal magnetism" in the human body that could be influenced for therapeutic and psychological purposes. *Mesmerism* became a sensation, and while it was discredited in itself, Mesmer's influence was later to be felt in the new nineteenth-century medical discipline of psychiatry. For what *did* power the brain and central nervous system, and why could people be seemingly helped by mesmeric trances, or what would after 1843 come to be termed *hypnotism*? Relating to the new passion for magnetism, there was the American Elisha Perkins, who claimed to have devised magnetic instruments, or "tractors", which could draw off or disperse "electrical fluids" causing gout, inflammation, and other diseases. With a sincere faith in his methods, Perkins went to New York to use his "tractors" to disperse the yellow fever epidemic raging in 1799. There he sadly died of yellow fever.

Yet "animal magnetic" and electrical ideas would become the rage, as early doctors and writers of the Romantic Age came to equate magnetism and electricity with the *life force* itself. Mary Shelley's *Frankenstein* (1818) is the enduring literary achievement in this line of thinking.

Around 1793, on the basis of his numerous dissections and clinical studies of skulls, the Viennese physician Franz Joseph Gall proposed that the human brain had 27 "organs", or "faculties" which influenced our mental and emotional lives. These brain regions could be delineated by the careful examination of the bumps on a patient's skull, which were believed to accord with the faculties. Gall's ideas came to be popularized, and to some degree plagiarized, by his disciple Johann Spurzheim, who brought phrenology to Britain. All controversy and contemporary name-calling apart, however, Gall's work was significant in further developing neurology and craniology, or the scientific study of the human skull.

Many factors contributed to the "Golden Age of Quackery". These included overt charlatans, theatrically inclined yet well-meaning medical men in the process of evolving new specialisms, and their colourful brethren fascinated by new scientific discoveries and their therapeutic possibilities. Quackery was never a simple divide between "doctors" and "charlatans". A combination of growing national prosperity and an expanding free press carrying patent medicine advertisements fuelled the phenomenon. For as the late Professor Roy Porter showed, one single newspaper, *The Bath Journal* in the 1780s, regularly carried ads for over 40 "Genuine Medicines". It truly was an age of "nostrums galore".[5]

CHAPTER 20

Sewers, Soap, and Salvation: The Origins of Public Health

*I*n 1800, the population of London approached one million persons, while the first proper census in 1801 would indicate that some 8.9 million people lived in England and Wales and 1.6 million in Scotland (the first Irish census was 1821). This was a massive increase since Gregory King's 1688 figure of 5.5 million for England and Wales. Evidence also suggested that populations were rising across Europe – in spite of the French Revolution and related military slaughters – while the new Republic of America was growing fast. Why was the population of the Western world increasing at such an unprecedented rate? As we have seen in Great Britain in particular, that growth of "scientific" agriculture sometimes called the "Agricultural Revolution" was producing larger supplies of food, while a spreading network of canals was creating the first wholly national distribution system. More people had a chance of being fed, and hence, of going on to breed new generations.

Likewise, that transformation of industrial potential made possible by first Thomas Newcomen's (1712), then James Watt's (1775) steam engines was generating wealth at an unprecedented rate by 1800, as machine power replaced human and animal power, and the production of coal, textiles, and everyday artefacts was escalating. By no means everyone prospered, and evicted landless labourers and industrial hands were often acutely short of money and sometimes homeless. Yet even these social tragedies would give rise to movements to improve the lot of the poor, establish charities, and provide food and shelter for those multitudes who had missed out in the prosperity boom.

Death from disease was ever present, though, and affected the entire population: smallpox, typhus, typhoid, scarlet fever, tuberculosis, and a whole medley of other maladies banged on the doors of rich and poor alike. And in 1831, a new fever would arrive from the East: Asiatic cholera – a terrifying by-product of faster and increasingly global commerce.

One eighteenth-century clergyman and FRS, however, was to take a less rosy view of endless progress and beneficent population growth.

THE REVEREND THOMAS ROBERT MALTHUS, FRS

Daniel and Henrietta Catherine Malthus, Thomas Robert's parents, were independently well-off Surrey county gentry, with Daniel having an admiration for radical thinkers including David Hume, William Godwin, and Jean-Jacques Rousseau. Schooled locally at first, then studying at the famous Warrington Academy – a university-standard college for non-Anglican "dissenters" – Thomas entered the Anglican Jesus College, Cambridge, in 1784, when he was 18. There he won academic prizes and was elected a fellow of Jesus, ordained in 1797, and went on to become an independent academic gentleman and fellow of the Royal Society.

Malthus's interests were not especially medical, but lay rather in the infant science of demography – population studies – and economics. In *An Essay on the Principle of Population* (1798) and subsequent works, he examined the relationship between food and economic resources on the one hand and population sizes on the other. Far from being sanguine about a rising population, Malthus saw disaster looming.

What was necessary for a stable and prosperous population, in Malthus's analysis, was a parity between mouths needing to be fed and adequate food supplies with which to fill them. Yet this was rarely the case in reality, as it was the tendency of populations

341

to increase faster than society's capacity to produce food and other necessary resources. Consequently, famine and want set in as resources became inadequate, and people died until a natural equilibrium was established once again.

Malthus saw this as a divinely ordained mechanism, for three factors providentially clicked in to keep escalating populations in check. These factors were *famine*, *disease*, and *war*. He even came to give this demographic mechanism a mathematical expression: for while food and other resources increased in an *arithmetic* ratio (or 2, 4, 6, 8, 10, and so on), population tended to rise in a *geometric* or multiplication ratio (or 2, 4, 8, 16, 32, and so on). In short, we were doomed to fight a losing battle for what Malthus termed the "means of subsistence"; for as food and other necessaries of life increased, population would expand in an enhanced ratio. This scarcity would lead in turn to "misery and vice", as those who could not prosper turned to theft, crime, and prostitution as a way of surviving. It is hardly surprising, therefore, that the emerging discipline of political economy, or economics, came to be termed a "dismal science".

The only way to exert any kind of control upon the upward population spiral was by limiting the birth rate. *Not* by contraceptive techniques, although several methods, including the sheep-gut condom and unreliable post-coital douches and primitive spermicides, were already in use, as were abortions. These were firmly associated with prostitutes, brothel-keepers, political radicals advocating free love, and godless libertines doomed to die young from the pox or drink, rather than with respectable, God-fearing British couples. What *respectable* people should do, rather, was postpone marriage until the husband was financially secure and then, within marriage, exercise *restraint*.[1]

It is hard to overestimate the impact of Malthus's writings on post-1798 discussions about economic resource management, politics, social medicine, and religious thinking. Malthusian ideas about the "surplus population" run through Charles Dickens's *A Christmas Carol* (1843) and *Hard Times* (1854), Victorian Parliamentary debates about industrial poverty, and the Irish Potato Famine. They inspired the young Charles Darwin to frame an idea of a competitive mechanism acting in *all* living populations, and his medical cousin Sir Francis Galton's ideas of "eugenics". As populations exploded in the twentieth century, a host of social engineers and dictators, including Stalin, Hitler, and Mao, used Malthusian ideas as a way of justifying their wholesale extermination of "degenerates".

On a much more positive note, however, the growing surplus population and the widening social divide between the successful and the unsuccessful inspired great acts of humanity and charity. Irrespective of the causes behind the Victorian population explosion, one faced a stark choice: Did a very self-consciously Christian society help its destitute, or did it simply let them die? A choice which it was possible for fiercely Protestant officialdom to overlook in its tragic neglect of the Potato Famine in Catholic Ireland. Yet even here, private charities did whatever their meagre resources allowed, given the magnitude of the Irish disaster. Generally speaking, however, it was a foregone conclusion, as early prime ministers, enlightened noblemen, trades unionists, "friendly" society organizers, Methodist preachers, and the bishops in the House of Lords all recognized a common duty towards the "surplus population". One global engine of that duty after 1865 would be the Salvation Army.

In no way can the wholesale annihilation of anti-communist Russian communities or the death camps of the Third Reich be remotely laid at Malthus's door, nor at that of Charles Darwin. The Reverend Mr Malthus, FRS, and the reluctantly agnostic Charles Darwin, FRS, were both gentlemen of compassion and Christian charity, who would have been *horrified* to realize where their world-changing ideas had led. But it is a fact of life that when one releases a great idea into the world, one cannot predict where it might go.

From cow to human: Dr Edward Jenner and the impact of vaccination

One major factor in the nineteenth-century British population explosion from 10.5 million in England, Wales, and Scotland in 1801 to over 38 million by 1901 was the discovery announced by a Gloucestershire country gentleman physician between 1798 and 1801. Dr Edward Jenner, FRS, had been born in 1749, the son of the Reverend Stephen Jenner, vicar of Berkeley, Gloucestershire. A countryman to his fingertips, Jenner would, in addition to his medical work, contribute learned papers on ornithology – he was an expert on cuckoos – to the *Philosophical Transactions* of the Royal Society.

As a young man, Jenner had studied surgery and medicine with various Gloucestershire medical men, before going up to St George's Hospital, London, to study anatomy and surgery under the illustrious John Hunter FRS. Then in 1792, he obtained an MD degree from the University of St Andrews, Scotland. Most of Jenner's life, however, was spent in his beloved rural Gloucestershire, taking patients, marrying Catherine Kingscote, the daughter of a local landowner, residing in a large and elegant house, and living the life of a country gentleman. He and some friends also formed what became the Gloucestershire Medical Society.

It was his closeness to the daily life and gossip of rural Gloucestershire that helped Jenner to come upon his momentous discovery. It seemed to be a rural truism that milkmaids did not contract smallpox. Early in her working life as a milkmaid, a young girl would develop pustules on her hands and perhaps feel a bit feverish. But this "cowpox" would soon clear up naturally, leaving her healthy – and with lifelong immunity to smallpox. Jenner had picked up the cowpox–smallpox connection from colleagues, for Dr John Fewster had discussed cowpox as a possible smallpox preventative with Gloucester medical colleagues in the 1760s, while other doctors had commented upon its apparent immunological powers.[2] No one had taken the matter further until Jenner decided to subject the truism to experimental testing, in May 1796. He then did something that might shock our modern sensibilities: he performed his *experimentum crucis* on an eight-year-old boy, James Phipps, the son of his gardener.

His reason for choosing a child was obvious, for smallpox was so widespread in eighteenth-century society that it would have been hard to be sure that an adult had never encountered a mild dose of the disease and obtained immunity. Jenner needed a guaranteed smallpox-free person, with no prior immunity, which made his gardener's little son a natural choice.

A local milkmaid, Sarah Nelmes, currently had cowpox, picked up from a cow named Blossom. Jenner took some pus from her hand and rubbed it into a small cut on each of little James Phipps's arms. After the usual passing mild fever, James recovered. But the acid test came when, some time later, Jenner twice injected James with active smallpox matter, in the same manner that he might have injected a patient using the older and much riskier "inoculation" method. To everyone's delight, the boy displayed a complete immunity to the active smallpox.

Edward Jenner would publish the details of the James Phipps experiment, along with many other cases, in his *An Inquiry into the Causes and Effects of the Variolae Vaccinae* (1798), and additional cases and observations in 1799, 1800, and 1801. Vaccination (deriving from the Latin *vacca*, "cow") caused a sensation. Here was a simple, cheap, quick method, using easily accessible ingredients, that had the power of conferring lifelong immunity against one of the most feared scourges of the age. Smallpox, even if it did not kill, could destroy eyesight, leave faces horribly pock-marked, and cause infertility. Yet a mere scratch from a lancet, a dab of active cowpox pus, and a brief mild fever rendered one immune for life.

Jenner's vaccination technique rapidly spread to Europe (where Napoleon Bonaparte had his soldiers vaccinated), America, Australasia, and elsewhere in the world where European medicine was practised. The cartoonists and comedians had a laugh at it, with comic pictures of vaccinated people suddenly sprouting horns, developing cow-like faces, and mooing.

Vaccination immediately rendered obsolete the immunization method discussed in Chapter 17, where one had to take the risk that the (hopefully) mild active smallpox with which one was injected might turn virulent and kill or maim. Vaccination was almost child's play in its simplicity, gentleness, and reliability. The great drama of the discovery is captured in Giulio Monteverde's 1878 statue "Jenner", in Genoa, depicted in the very act of injecting Phipps.

345

Other memorial statues to Jenner are to be seen in Kensington Gardens, London, and Gloucester Cathedral. Jenner, the real founder of immunology, was said to have saved more lives than any other individual in history, and more than Bonaparte's wars had taken. Yet while Dr Jenner, the parson's son, understood why people were grateful to him for his discovery, and was thankful for the public rewards he had received, he wondered why people were not even more grateful to God, for making him the mere "instrument" whereby vaccination came into the world.

Vaccination, however, was not a cure for active cases of smallpox, but a preventative, or "blocker", which took away the patient's susceptibility to the disease. Living as he did in that pre-bacterial and pre-viral age, Jenner had no idea of the physiological mechanism behind the disease: the virus *Variola major*. Mercifully, since 16 October 1975, not a single smallpox case has been reported worldwide; a true triumph of medical science, and humanitarian zeal.

CHOLERA

If folk empiricism first stimulated Edward Jenner's research into smallpox immunity, it was civil engineering that furnished the route by which a new epidemic disease would come to be managed, over 20 years before medical science discovered that disease's true cause.

Asiatic cholera had raged with especial virulence in India and the East over 1817–23, though it was not a new disease. From the late 1820s, however, it began to move north – fanned by human migrations and growing trade with India. The spearhead seemed to come through Afghanistan and across the Asiatic Steppe. Astrakhan, on the Volga River, got it in 1830, as did Moscow, and soon after St Petersburg, while by 1832 cholera was in Paris, and then Lisbon. It is believed that a merchant ship from Danzig, Poland, to Sunderland brought the disease into England on 19 October 1831, and by 1832 it was pretty well everywhere in the British Isles: sped on by stagecoaches, canals, and the first few railways.

In 1883, that great German pioneer of bacteriology, Dr Robert Koch (unknowingly in the wake of the Italian Filippo Pacini's badly publicized prior discovery in 1854) and his team would discover that cholera was caused by the bacillus *Vibrio cholerae* or *Vibrio comma* (from its curved "comma" shape), which makes its home in the human intestinal tract and then proliferates, with dire consequences. In 1830, the classical explanation of a miasma or a foul poisonous air was reckoned to lie at the heart of most mysterious, supposedly wind-blown, infections. Exactly which miasma caused cholera, or smallpox, or "gaol fever" (typhus) was a mystery, but most doctors of the time still worked within the miasmic theory. Dr Thomas Shapter in his detailed account of the cholera epidemic in Exeter, Devon, in 1832, reported various avenues whereby the *materies morbi* was thought to pass from person to person: "put a body in a coffin", "smelt effluvia at a funeral", "from using a bedstead", among many.[3]

Mode of transmission apart, the symptoms of the new disease were clear enough. First, a patient would feel dizzy, and this would be followed by violent and sustained vomiting, diarrhoea, excruciating intestinal cramps, a burning sensation in the guts, and the body expelling a seemingly endless involuntary flow of grey slime – the "rice water" stage of the disease. The pulse would fade, and life would ebb out of the sunken, cadaverous body. Cholera was not the "romantic" death beloved by poets and novelists!

While some well-off people succumbed to the cholera, one fact emerged from the mass of medical statistics accumulated about the disease: cholera was at its deadliest in the slums and in places where poor people lived in congested conditions. Yet what could be done to treat it? Dr William Braithwaite's *The Retrospect of Medicine* journal, which he edited, offers some fascinating insights submitted by contemporary physicians and surgeons, some of them Indian Army doctors, and others practising in England. The 1848–49 volumes are particularly insightful, being published at a time when medical men were already fully experienced in managing cholera cases. Opium was widely used, not only for its soporific and analgesic properties, but also as a bowel calmative. Calomel, a popular mercury chloride preparation, was also employed, where, oddly enough, its bowel-

347

irritant action was thought to clear away the "poisons" that lay at the heart of the disease; while that eternal favourite – bleeding – was also recommended. Copious cold drinks, too, were found to be very effective and relief-giving. But Dr Stevens's "Saline Treatment" was proving astonishingly effective by 1849, when it was reported that, out of 1,000 cholera cases, only 6 per cent died.[4] Stevens recommended administering Seidlitz powders (tartrate of sodium and potassium and sodium bicarbonate) and magnesium sulphate, in the form of an effervescent, fizzy drink "to expel the poison from the blood and the body". The patient was also given beef tea "well-seasoned with common salt".[5]

From what is known today about body chemistry and metabolism (*not* known in 1848) one might hazard a suggestion why Dr Stevens's treatment might have been successful. Cholera is a rapidly dehydrating and desalinating disease, which would have seriously damaged the body's basic chemistry, pitching it into a descent to death. Yet by giving the patient plenty of water, laced with sodium, potassium, and magnesium salts (and beef extract), one might just help to repair the damage and cast a lifeline.

Yet if cholera hit the great unwashed in their slums more than it hit the clean and better-off in their spacious villas, could the disease be related in some way to dirt, which might act as a vehicle for the *materies morbi*? With the creation of the national Registry of Births, Marriages, and Deaths in 1837, an accurate national statistical instrument for measuring and monitoring cholera came into being. The Astronomer Royal, [Sir] George Biddell Airy, was even invited to begin detailed wind, weather, and even river Thames temperature studies at the Royal Observatory, Greenwich, to see if there was a correlation between environmental factors and cholera outbreaks. And in 1851, James Glaisher, FRS, the superintendent of the Greenwich Meteorological Department, told a reporter from Charles Dickens's popular paper *Household Words* that "cholera was found greatest in those places where the air was stagnant", and that improving ventilation and even lighting fires to create air movement, "are thus proved to be of the utmost consequence".[6] But, as it would come to be discovered during the 1850s, the bad air or miasmic approach to cholera was found to be misguided, yet without anyone quite knowing why.

SANITATION, STATISTICS, AND THE BROAD STREET PUMP

By the 1840s, London was reaching crisis point. With a population now approaching two million and a rapidly growing sprawl of congested, overcrowded, poverty-stricken districts on both sides of the Thames, this was the city of Charles Dickens's *Oliver Twist*. The Thames had become an open sewer, and in the summer of 1858 the "Great Stink" would become a national scandal that finally prompted action. Yet while London was the biggest public health disaster area in Britain, the rapidly expanding industrial towns, such as Manchester, Leeds, and Glasgow, were not much better, given their respective population sizes.

Once it had taken up residence, cholera made regular return appearances in Continental Europe, Great Britain, and the USA: 1841, 1847–48, 1853–54, 1866, and into the 1890s. Two factors contributed to these Asiatic epidemics having such a fertile human terrain: poor sewage disposal and sewage-contaminated drinking water supplies. And this is where civil engineering and medicine came together.

As things stood in 1840, London's growing population received its drinking water from a medley of private companies, most of which drew their water directly from the Thames. While the more gross contents of the water may have been filtered out, there was no way of preventing a host of bacterial and chemical pollutants coming out of the taps and pumps of the communities they served. Similarly, there was no shortage of flushing water-closets in London by that date, especially serving better-off homes and public buildings, yet all these closets did was keep one local environment clean, while discharging the waste into the Thames. The densely crowded poorer homes, tenements, and cellars were probably served by simple privies: mere brick-lined holes in the ground with a seat over the top, which might get emptied by shovel and bucket once

349

or twice a year, if the residents were lucky. A household might deal with its own refuse, or a paid provider might do so, the excavated material sometimes being sold to market gardeners – or else tipped into the river. (Elderly relatives of mine who grew up with privies in late Victorian Manchester told me that the men who emptied them were known as "muck-misers".)

One real health hazard occurred when a household or tenement drew its water from a private well sunk downstream from the privy. Nor were things made any better by the appalling over-burial of fresh corpses of cholera or typhoid victims in the old city churchyards. Yet if cholera was blown in the wind, what did it matter if the water you mixed with your gin, or used to rinse out the baby's drinking-cup, just happened to be yellowish, and smell?

DR JOHN SNOW AND BREAKTHROUGH AT LAST

Dr John Snow was not the first to doubt the truth of the miasmic theory, but he, as with Edward Jenner and smallpox, was the first doctor to take things further. John Snow was remarkable in many respects, being the son of a City of York labouring man, who got his formal medical training first in the Newcastle upon Tyne area, then in London. In addition to his work on cholera, he became the first specialist anaesthesiologist, who administered chloroform to Her Majesty Queen Victoria at the birth of Prince Leopold and Princess Beatrice, as we shall see in Chapter 23.

Snow had first announced his doubts about choleric miasmas in 1849, but it was the 1854 outbreak that really got things moving. The outbreak around Broad Street and Golden Square, in London's Soho district, especially caught his attention, as 500 deaths had occurred in this one small district over the first 10 days of September 1854. A pioneer of medical statistics, Snow plotted the deaths on a map. Assisted by a local clergyman, the Reverend Henry Whitehead, and other residents, both men came to the conclusion that what all the deceased shared was that they drew their water from the common pump on Broad Street (now Broadwick Street).

Snow then got an engineer to examine the well beneath the pump, along with the water coming from the Southwark and Vauxhall Waterworks Company, which drew its supply from the Thames. Both were contaminated with faecal matter. He then advised the local Board of Guardians authority simply to remove the pump handle, which caused the epidemic to quickly tail off.

Snow was by no means the first Victorian doctor to use maps and statistics as a way of identifying the source of infection – Robert Baker had done the same in the Leeds epidemic of 1833 – but he was the first to discover the crucial link. Cholera was a *waterborne* disease, and not the product of a vague miasma. Many of his "miasmatist" colleagues disagreed, and it would not be until the 1866 epidemic that the weight of evidence would really shift the argument in favour of Snow's position.

It would not, however, be until Robert Koch's well-publicized microscopic identification of *Vibrio cholerae* in 1883 that the precise pathogenic mechanism of cholera would be demonstrated; for cholera, like smallpox 60 years before, would first be contained empirically before it was understood scientifically. And in that containment, the civil engineer would be crucial.

As a historical observation, one wonders to what extent the English habit of tea-drinking, already firmly established across much of English society by 1830, might have helped to lessen the impact of cholera in Britain. For irrespective of the colour of the water poured into the teapot, its sustained boiling in a kettle beforehand could well have destroyed the cholera bacteria that it contained.

SANITATION, CIVIL ENGINEERS, AND SALVATION

By 1830, Great Britain was by far the most technologically sophisticated country in world history. Its ironworks, steam-powered plate-iron rolling mills, and threading, turning, and milling machines would provide the first "front" to help defeat cholera: long before germs were thought of. The mass production of rolled-iron water and sewer pipes, pumps, taps, and ancillary fittings may,

351

before 1854, have accidentally helped to get infected Thames water into countless London homes, but once the infection mechanism of the disease was recognized, miles of iron pipes could protect against it, and bring in safe, pure water.

By the late 1850s and 1860s, iron sewer pipes, combined with the mass production of *billions* of canal- and railway-transported bricks, combined with the new building material, concrete, were ringing the death-knell of the leaky brick privy, although much of the matter they conveyed still ended up in the river. But cleanliness was becoming the operative concept, even if new high hospital ward ceilings, prison barracks, and ship ventilators were still aimed at generating air currents that would waft miasmas away from the inmates. For before John Snow's 1854 discovery, it was the civil and sanitary engineer who was at the forefront of the fight against infection.

Alongside ran those Victorian passions for good works, statistics, and Parliamentary legislation: all aimed at making the world a better place. The statistician and reformer Edwin Chadwick was the driving force behind Great Britain's first major piece of health legislation. This was the Public Health Act of 1848 which, six years ahead of John Snow's discovery, provided the first legislative and administrative levers for cleaning up the cholera-haunted towns of Britain. The Act created a Civil Service Board of Health, while empowering towns and cities across the land to curb dirt and infection – even while still unaware of what exactly infection was in clinical terms. If the municipality had a cholera death rate of more than 48 in 1,000 persons, then special administrative powers could be invoked. The 1848 Act also created municipal Medical Officers of Health, or public doctors, whose duty it was to compile health-related statistics that might be used to frame public policy. These published Reports are nowadays a mine of information for medical historians.

But it was the "Great Stink" during the summer of 1858, which even occasioned Parliament to postpone some of its sittings, that really got the ball rolling to transform London's sewage disposal system. Sir Joseph Bazalgette (he was subsequently knighted) was engaged as chief engineer by the exasperated Metropolitan Board of Works. Bazalgette's plan remains one of the most visionary of all pieces of civil engineering, which would first transform the health of Londoners and then go on to inspire municipalities worldwide.

Bazalgette designed two vast sewer systems, one on each side of the Thames, driven by gravity-feeders, subterranean water-courses, rainwater, and the lie of the land. Two huge brick and Portland cement pipes ran alongside the Thames, oval-shaped to enhance the flow, and each big enough for a modern-day double-decker bus to drive down. They still lie beneath the Thames Embankment, the Embankment itself being part of Bazalgette's system to narrow and speed up the river's rate of flow.

Between 1859 and 1865, Bazalgette dug some 450 miles of main sewers, feeding into several miles of giant interceptor sewers, designed to receive the contents of several thousand miles of smaller domestic district sewers. When the sewage came downhill from high ground to the north and south of the Thames, it eventually flowed *not* into the river, but into the giant sewers *below* river level, following the course of the river. Then, having flowed eastwards to the Thames estuary, still deep underground, all the effluent was sucked up by gigantic steam-driven pumping engines and discharged into the outgoing tide, and into the North Sea. By the 1870s, the seeming miracle had been achieved: the Thames was sweet once more, could sustain fish, and even be safely drunk. If a man ever deserved his knighthood, it was Sir Joseph Bazalgette!

By the time of the completion of Bazalgette's sewer system in 1874, which effectively served the eviction order on cholera in London, other aspects of cleanliness and social care were starting to make their impact. The development of chlorine-based bleaches and bleaching powders – the first commercial washing powders – was making clothes easier to wash, while better-quality yet cheaper soaps did the same for human bodies, as Victorian soap advertisements testify. The proliferation of gas lighting and supply technology spawned other if sometimes dangerous aids to cleanliness and domestic comfort – gas boilers for water, and gas fires and even central heating for wealthy homes. Awful as the great slums of London and other major cities were, local charities and municipal councils were being empowered by public health and other legislation to provide better facilities for the slum-dwellers. These included public wash-houses and laundries, bath-houses, and even swimming baths.

The Victorian maxim about cleanliness being next to godliness had become a major weapon in the fight against grime and disease for many people by the 1860s. And irrespective of religious belief, the increasing mass production of iron bathtubs and plumbing fittings began to make cleanliness a matter of good behaviour, especially if one had a public profile. Figures as diverse as the Duke of Wellington, Queen Victoria, Charles Darwin, and Charles Dickens were all "into hygiene", and Dickens had a shower of such force installed in his bathroom that he called it "The Demon".

So in spite of the coal smoke, the thick winter fogs, and the resulting bronchial coughs, a cleaner, better world was slowly emerging, and major social problems, such as hunger, poverty, homelessness, ignorance, and political and social estrangement, came to be both officially recognized and *acted upon*, as reform, public education, and the extension of the franchise to the working man became the hot topics of the day. When the inspired and enterprising General William Booth entered upon his Salvation Army mission to the poor of Whitechapel between 1865 and 1867, he recognized that, before starting to talk religion, one had first to address the practical needs of the body: soup, then soap, and only then talk about salvation.[7]

CHAPTER 21

"Them Damn'd Murderin'
Anatomists"

*M*édical men and their students had been stealing corpses since at least the days of Vesalius in early sixteenth-century Padua. There had been a perfectly legitimate supply of human corpses, namely, the bodies of criminals who had been executed for especially nasty crimes; but it was an increasingly inadequate supply. Besides, these legitimate cadavers were designated for the formal, almost theatrical anatomies, conducted by a professor before an audience of colleagues and students. If one wanted to conduct private, or small-group research dissections, or fathom the complex operation of the brain or cardiovascular system, one was obliged to look elsewhere.

If one were fortunate enough to hold a major hospital appointment, in the London or the provincial hospitals of Edinburgh, Glasgow, and Dublin, then one might be able to procure a legitimate supply, such as by taking the bodies of unclaimed paupers, beggars, and people who had died naturally in prison — anonymous persons with no families to bother about them. Yet even this supply was being outstripped by demand by 1800, as burgeoning hordes of medical and surgical students wanted to get hold of what were euphemistically referred to as "subjects", or "things" for the surgeons.

So pressing had this demand become by the early decades of the nineteenth century that in somewhere like London, with its teaching hospitals, Royal College of Surgeons, and private medical schools, a "nice fresh thing" might fetch £7, £8, £10, or more during term time, and almost as much in the Scottish universities. A labourer would have to work for perhaps six to eight weeks to earn this amount legitimately, but it might be got in an evening if one visited the churchyard after dark with a shovel following a funeral. A "subject" could be obtained even more easily if he or she were a conveniently deceased penniless lodger who still owed rent, or an unloved relative, eased into heaven with the application of a wet towel over the face!

THE TRADE IN "THINGS"

Before the legitimate supply of corpses became more plentiful after the passing of the Anatomy Act of 1832, body-snatching became something of a national hysteria, as newspapers, magazines, poems, and songs highlighted the vulnerability of the "loved one" during that week or so between death and serious decomposition setting in. This was when "them damn'd murderin' anatomists", or more likely, their hired "resurrection men" might be on the prowl. The term "resurrection men" was a macabre allusion to those shadowy, lantern-carrying figures who would "resurrect" you well before Judgment Day.

There was abundant mileage in macabre and comic verse to be extracted from "resurrecting". Take Robert Southey's grimly admonitory "The Surgeon's Warning" (1796).[1] It narrates the last hours of a surgeon, now terrified that his own students will "carve me bone from bone" as soon as he has drawn his last breath – while still beautifully fresh! Even after a respectable funeral the danger would not diminish: for "... I who have rifled the dead man's grave / Shall never have rest in my own." The old doctor orders that paid blunderbuss-wielding guards should be posted at his graveside at night until "... I think that I may stink / Enough to rest in my grave".

Thirty years later, there was the mischievous Thomas Hood's "Mary's Ghost" (pre-1830).[2] It tells how "'Twas in the middle of the night" when the ghost of the recently deceased young Mary

356

appeared at the bedside of William, her grieving sweetheart. She tells him how "The body snatchers... made a snatch at me", "But tho' I went to my long home, / I didn't stay long in it". This had been in Marylebone Burial Ground. "But from her grave in Mary-bone/ They've come and boned your Mary". Poor Mary then supplied William with a grisly catalogue of which bits of her went to which particular London medical schools, not to mention parts sent elsewhere. "As for my trunk, it's all packed up / To go by Pickford's van": perhaps to Scotland, Leeds, or Manchester.

What is so informative about Hood's macabre comic masterpiece is his knowledge of who got which parts of Mary within the London medical world. Her arm has gone to Dr Vyse, while her legs "are gone to walk / The hospital at Guy's". And as for her pretty little feet, "There's one, I know, in Bedford Row / The t'other's in the City". The eminent anatomist and surgeon John Abernethy, founder of St Bartholomew's Hospital Medical School, lived at No.14 Bedford Row, and both the surgeon's house and his hospital were in the City of London.

The hand that poor Mary had pledged to William in marriage is "at Dr Bell's / In spirits in a phial": Sir Charles Bell, the neurologist who in 1833 would write his *Bridgewater Treatise* vol. 4 on *The Hand. Its Mechanism* & *Vital Endowments Evincing Design*. She then tells William that Dr Carpue could probably tell him of the whereabouts of her head. Joseph Constantine Carpue would certainly have been a grateful recipient of Mary's head had he received it, as he was a pioneer facial repair and transplant surgeon, who after 1814 developed an operation to restore lost noses – rhinoplasty – based on accounts of an Indian folk-operator. Then the catalogue of London's contemporary medical "great and good" is concluded with:

The cock it crows – I must be gone!
My William, we must part.
But I'll be yours in death, altho'
Sir Astley has my heart.

357

This was Sir Astley Cooper, surgeon to the king, the most famous and probably richest anatomist in London in 1830, who quite likely would also have encountered Mary's severed legs, as he was the chief surgeon at Guy's and St Thomas's Hospitals, immediately south of London Bridge in the Borough of Southwark.

The only surgeon named in *Mary's Ghost* with whom I am unfamiliar as a real, historical person with a known special interest is Dr Vyse. He does not appear in the London or Edinburgh Royal College of Surgeons lists, and I am still trying to trace him. Thomas Hood's tragic-comic poem was more than just a pot-boiler, for he had clearly done his medical homework beforehand.

How to snatch a "thing": The practicalities of "resurrecting"

While some students, and even the occasional young surgeon, such as the daring and controversial Edinburgh anatomist Robert Liston, sometimes did their own "resurrecting", the trade in illicit medical cadavers came to be overwhelmingly dominated by skilled professionals: characters also known as the "sack-'em-up men", from the bags used to conceal their wares.

How did known and documented characters such as Ben Crouch and his brother Joe, Jack Harnett, Bill Hollis, and other body snatchers go about their grisly yet lucrative business? The law, perversely enough, was on the snatchers' side, for a naked corpse did not constitute property under English law. So as long as you left the shroud, cap, and other grave-clothes in the empty coffin, you were technically within the law. Disturbing a graveyard was an offence, but nowhere near as serious as stealing something of even token economic value, such as a winding-sheet. The biggest danger to which the body-snatcher was exposed, however, were the spring-gun, blunderbuss, and angry mob if caught in the act. Anyone who escaped such an armed mob need not expect much sympathy from the authorities.

The first thing was to clearly identify the grave one intended to rob. As Sunday was the only day of the week that most working people had free, their dead would often be interred on the Lord's Day. Therefore, the wife, or girlfriend, of a "snatcher" might visit a likely churchyard on a funeral day, dressed in mourning and looking sad and respectable, perhaps on the pretence of putting flowers on a loved one's grave. But she would carefully note and mark, maybe with a few arranged pebbles, a newly filled-in grave with a nice fresh "thing" inside. It was even easier if the body was in an old grave with two or three people already decomposing beneath the latest resident, for graves were not generally deemed full until there were three or four corpses in them, the latest only a couple of feet underground; an easy night's work for her men friends.

Skilled body snatchers did not dig out the entire grave they were robbing. Rather, a shaft would be sunk, maybe a couple of feet square, down to the coffin lid. Armed perhaps with a crow-bar and strong ropes, a snatcher would be let down into the hole head first, his mate holding his feet and the shaded lantern. The upper part of the coffin lid would be prised up, forced, and hopefully snapped off. A pair of hooks on ropes could then be placed beneath the arms of the "thing", and the deceased person pulled out of the grave.

Then he or she would be stripped naked of all valuable grave-clothes, doubled up, and "sacked". The grave would then be carefully filled in, and the earth patted down to make it look undisturbed, so as not to attract the attention of the constable and start a hue and cry next day. The "thing" would now be placed on a hand-cart or barrow, covered with some old carpets, and promptly delivered to whichever surgeon or medical school the snatchers were working for. Joe, Bill, or Ben would now be more flush with cash from their brief escapade in the churchyard than they would otherwise have been after a couple of months of coal-heaving on London docks.

32. Bodysnatchers at work. These look like novices because they have exhumed
the entire coffin. They are also stealing the shroud and grave clothes:
technically a property theft, with potentially dangerous legal consequences.
Accomplished "snatchers" would only have excavated sufficient earth to
allow the coffin lid to be forced up and broken off, to facilitate a speedy
extraction of the corpse, and would have stripped it of grave clothes. Our
snatchers are clearly so delighted at the prospect of a quick ten guineas that
they have not noticed Death holding their lantern. Nor have they spotted
the vigilant loved one of the deceased about to discharge his blunderbuss
at them. (Original drawing by A. Chapman, based on a cartoon by Thomas
Rowlandson, *c.* 1810.)

Body-snatching also involved serious physical dangers. Better-
off grieving relatives might, for example, set up a swivelling "spring-
gun" on a spike near the grave, with trip wires extending around it.

Should a "resurrection man" walk accidentally into a trip wire in the dark, then the large-barrelled gun would swivel round and blast a hail of lead shot, scrap iron, or old nails along the wire. Even if this did not instantly transform a bodysnatcher into a "thing", his shrieking would probably attract an angry crowd, who might finish him off anyway.

Relatives might also, as did the dying surgeon in Southey's "The Surgeon's Warning", set armed guards to watch over the grave until serious decomposition had set in. These trigger-happy fellows might be got merry on "a keg of Holland's Gin", and be paid a swingeing five guineas "if he shoot / a resurrection man". Nor would most magistrates enquire too closely if the next morning Bill the bodysnatcher were found dead a few yards away from a discharged spring-gun. For in 1825, Great Britain was still an armed society, where many a householder legally possessed a pair of pistols or a blunderbuss to defend his property, in the absence of a proper police force. Besides, were not body snatchers the scum of the earth?

There were, however, more peaceful ways of protecting a recently deceased loved one. There were patent iron coffins with spring-catch lids. Once the body was inside and the lid hammered down, it was practically un-openable. The only problem was that clergy and churchwardens refused to inter them, for they would never rot away as would a wooden coffin. Then there were "mort safes", or cages of heavy iron bars anchored firmly over a new grave.

It was in Edinburgh, however, with its large and illustrious medical school, that body-snatching – and "murderin" – reached their apotheosis.

EDINBURGH: THE MEDICAL LION OF THE NORTH

If you were an aspiring young medical man in 1810, the magic letters to have engraved upon your name-plate were "MD Edin.". You might already have degrees from Oxford, Cambridge, Dublin, or a Continental university, but Edinburgh was the clinical crème de la crème by that date. It was an exhilarating place to be, full of intellectual adventure, where dissecting a corpse stolen from Greyfriars churchyard could be combined with chemical experiments and discussions in philosophy, literature, and theology. Edinburgh was wonderfully *cheap*, economically speaking, as well; much cheaper than Oxbridge or even London. When the French Revolutionary and Napoleonic Wars made Paris or Leiden inaccessible to Britons between 1794 and 1815, Edinburgh, and to a lesser degree Glasgow and St Andrews, were *the* places to go to do medicine.

Edinburgh must have been an inspiring and exciting place in which to study, for wealth and dire poverty collided head-on in a relatively small city. Ill-lit at night, ill-drained, largely unpoliced, it was a place where scintillating genius lived cheek by jowl with cut-throats and ragged prostitutes. And it was packed with students. Because Edinburgh did not have any religious requirements, a Scottish Presbyterian, a Birmingham Congregationalist, a London Jew, an Irish or Lancashire Roman Catholic, a Welsh or Cornish Methodist, and a Gloucestershire Anglican were all obliged to jog along together.

Though Edinburgh University was founded in 1582, its medical school only really began after 1726, by which time Dr Alexander Monro *primus* had returned home from his studies at Leiden. Monro *primus* was a brilliant physician, clinician, and teacher, who by 1726 was already teaching new anatomical techniques in Edinburgh, and attracting students. He was succeeded to the anatomy chair by his son, Alexander *secundus*, and he in turn by Alexander *tertius*.

The Monro dynasty dominated Edinburgh anatomy for 128 years, getting progressively worse: from genius, to good, to hopeless! The young Charles Darwin attended Monro *tertius*'s lectures in the mid 1820s, by which time they were virtually a joke. Yet this descent in quality, oddly enough, worked to Edinburgh's advantage, because it gave opportunities for independent teachers, who often made their livings not from a salary, but from direct pay at the door student fees, to establish themselves.

These private anatomy teachers included men like John Barclay, the notorious Robert Knox (whom we will meet shortly), and

Robert Liston. Some of them would stay in the lucrative private sector, while others went on to occupy chairs, either in Edinburgh, other Scottish universities, or London. Anatomy, experimental physiology, dentistry, chemistry, and surgery flourished cheek by jowl in Edinburgh. What is more, the increasingly overcrowded Royal Infirmary provided a fascinating museum of human misery, where students who had obtained the right tickets might walk the wards with their teachers and learn how to study symptoms and compare cases.

Perhaps one factor that caused Edinburgh to catch fire and become the inspirational place it was for young doctors throughout the nineteenth century and beyond was its democratic or egalitarian nature, made possible largely by its cheapness. The bright young Highland crofter's son might tramp south to the Athens of the North with a hoarded £25 in his pockets. This might see him through a year – fees, bed, food, and drink. Then he would return home, help get in the harvest, look after the sheep, and tend the family smallholding, and when he had scraped enough money together, back he would tramp to Auld Reekie for more studies. Eventually, and with luck, he would win the coveted "MD Edin.", and the world became his oyster.

James Young Simpson, the discoverer of chloroform anaesthesia in 1847, was the seventh son of an impoverished West Lothian baker. His parents, brothers, and sisters scrimped and saved to send bright "wee Jamie" to Edinburgh – and the rest is history. We will meet wee Jamie again, as Professor Sir James Young Simpson, FRS, FRSE, MD Edin., in Chapter 23. In an aspirational open society, all manner of things are possible, and one finds similar stories of London and even Oxbridge students. Contrary to popular belief, British and often Continental education was not restricted to a social elite, for the intellectual gene pool needs constant enrichment. In many ways, it was the Scottish model that would be influential in shaping the study and research practices of many US universities in the nineteenth century.

"True murderin' anatomists": The Burke and Hare scandal, Edinburgh, 1828

While wet towels over faces may have hastened some fresh "things" to the dissecting tables in the past, what transpired in Edinburgh over 1827–28 truly became the stuff of legend. Instead of digging up their bodies, William Burke and William Hare resorted to plain murder to obtain their "subjects". Although corpses sold for slightly less in Scotland than in London, the handful of guineas that might be paid for auld Wullie or drunken Jenny in the slums of Edinburgh often meant that, quite literally, many people were worth more dead than alive: an interesting idea to the proprietors of cheap lodging-houses and impoverished relatives lumbered with a useless family member.

William Hare and William Burke had come to Glasgow from Ireland to work as labourers and ended up running a slum tenement doss-house in Edinburgh, where for three pence per night a medley of street beggars and destitute people could huddle down, wrapped in filthy, unwashed blankets. At the end of November 1827, one Donald, a drunken old soldier and one of the lodgers, died naturally on the premises, and Burke and Hare had the idea of selling him to the surgeons. They received an amazing £7 10s. for Donald's corpse: more than a month's joint earnings for the pair, for simply wheeling Donald's concealed body around to Surgeons' Square. Burke and Hare realized that they were on to a good thing.

A medical student of whom they chanced to enquire had advised them to take Donald around to the private medical school of Dr Robert Knox. Knox was probably the most successful and popular of all the private teachers, with a class of some 500 students, to whom he lectured in three separate sittings. An arrogant, argumentative, showy man with one eye (he had lost the other to childhood smallpox), Knox cultivated professional enemies and jealousies, but he was a brilliant anatomical and surgical teacher, adored by his students. As Knox prided himself on maintaining a constant supply of cadavers for his students to work on, and had apparently invited them to fetch more bodies, Burke and Hare probably realized that they had struck a gold mine. Besides, heavy whisky-drinking was common, and many poor folk died of drink anyway, so why should anyone feel especially suspicious if a nice fresh "subject" stinking like a distillery were wheeled around? Who would care if some inconsequential folks were given a push?

In total, Burke and Hare murdered around 16 people. The usual formula would be to get someone tipsy, invite them back to their doss-house in Tanners' Close when the residents were out begging, ply them with cheap spirits, hold the poor wretches down, and smother them. Dr Knox so liked their human wares that he was willing to pay £8 or £10 for a fresh "subject": usually via David Paterson, his dissecting-room porter. Questions were not asked.

But the cadavers of two well-known young Edinburgh characters in particular did cause flutters in 1828. The first was Mary Paterson, alias Mitchell, a strikingly beautiful golden-haired 18-year-old prostitute. It is said that when the sheet was pulled off her body on the dissecting table, many young men gasped: they had been more familiar with the fair Mary in less academic surroundings!

The other was James Wilson, a well-known simpleton with the mind of a child, universally known as "Daft Jamie". Jamie was a strapping fellow, and when Burke and Hare tried to hold down and smother him, all the whisky notwithstanding, he put up a fearsome fight, as a result of which his corpse was suspiciously battered and bruised. "Daft Jamie" had deformed feet, and as he never wore any shoes, they supplied a key to his identity. Neither Mary's nor Jamie's disappearance provoked formal police investigations, but one wonders how far Knox, his staff, and his students knew why the newly supplied corpses were so fresh?

Only after the body of a recently murdered woman was accidentally uncovered during a drunken party at their lodgings did Burke's and Hare's crimes come before the law. They were arrested, and Hare turned king's evidence and ratted the whole story in return for indemnity from prosecution. Then all hell broke loose in Edinburgh, but while the medical fraternity was generally in a state of civil war within itself, on this occasion it closed ranks. Knox was never prosecuted, though a mob smashed all the windows of his house at 4 Newington Place. Among the crowds, in the popular press, and at all levels of Edinburgh and then British society, "Burking" was the hot topic of conversation, as were violent anti-anatomist outbursts.

The rascally Hare was later released, but Burke was hanged in Edinburgh on 27 January 1829. Burke, it was said, displayed no remorse for his crimes, but a blazing hatred towards Hare for his treachery. He also hated Dr Knox for a different reason. It seems that Knox had failed to pay him £5 still outstanding on a cadaver, which Burke had intended to use to purchase a new waistcoat for his execution!

DR ANDREW URE OF GLASGOW TRIES TO RAISE THE DEAD, 1818

In Bologna in the 1780s, Professor Luigi Galvani discovered a most curious phenomenon. When he applied a pair of metallic electrodes to the dissected leg nerves of a dead frog, they induced spasms, making the legs jump. "Animal magnetism" caused a sensation. For if the natural "Galvanic current" induced muscular spasms, was electricity the "spark of life"? Doctors across Europe began experimenting, especially after Alessandro Volta of Pavia devised the first metallic and acid chemical action "Voltaic pile" battery in 1799. As we saw in Chapter 19, Anton Mesmer's and Elisha Perkins's magnetic therapies seemed connected to electricity, for both produced sparks and shocks, though it would not be until Michael Faraday's experiments in 1831 that the physical association between magnetism and electricity was conclusively demonstrated. But chemical battery-generated current electricity seemed to hold much greater experimental potential than either frictional static, or animal magnetism.

During the French Revolutionary genocides of the 1790s, Xavier Bichat in Paris electrified guillotined heads and produced frightening muscular contortions. Even John Wesley in *Primitive Physick* (1747) had realized that a static electricity friction machine could induce tingling and shock reactions, but current electricity seemed much more precise. For if Bichat could make a severed head blink or grin by electrifying the correct muscles, then could the natural, living force of the body be *electricity*? In 1802–03, Giovanni Aldini, working in London, used a powerful "Voltaic pile" to electrify the head of the executed murderer Thomas Forster, producing grotesque facial distortions. But it was Dr Andrew Ure of the University of Glasgow Medical School who became the real-life Dr Frankenstein in 1818.

33. Gothic horror medical science, 4 November 1818. Dr Andrew Ure of Glasgow University attempts – unsuccessfully – to revive the freshly-hanged murderer Matthew Clydesdale. The anonymous artist in Louis Figuier's *Merveilles de la Science*, 1867, has played up the shock horror of the experiment, as Clydesdale's corpse is made to twitch and jerk in a grotesque manner from the current of a powerful "Voltaic" battery. (Copy of a rare engraving, drawn by A. Chapman.)

Andrew Ure did something unique in the annals of body-snatching, for he *bought* the body of a condemned yet still living man, Clydesdale the murderer, the day before his execution. Clydesdale seems to have been keen on the transaction, for not only was he able to buy new clothes in which to be turned off the gallows, but he had the chance of a fresh start in life, if Dr Ure could resuscitate his body. Clydesdale was hanged in Glasgow on the morning of 4 November 1818. He hung for the required one hour, and on being cut down was immediately conveyed to Ure's laboratory in the university. Here, before an audience, powerful electric shocks were applied to Clydesdale's body from a chemical battery. The cadaver so convulsed, grimaced, and jerked his face as to horrify some watchers. But he remained dead, and it proved impossible to restore Clydesdale's life force. This was true Gothic horror medicine for the Romantic Age, as a copy of a contemporary engraving of the incident in my possession amply demonstrates.[3]

BISHOP AND HEAD, THE LONDON "BURKERS", AND THE *ANATOMY ACT*

It seems that Burke's activities inspired a pair of London resurrection men. John Bishop was an old hand in the trade, claiming to have raised between 500 and 1,000 "subjects" over the years, while his mate, Thomas Head (alias Williams) was no better. They hung out at The Fortune of War tavern, Smithfield, hard by both London's great edible meat market and its other, less edible "meat market" of St Bartholomew's Hospital. They were strong, strapping fellows, equally able to carry half an ox to the butcher's or a corpse to an anatomist – although the corpse could earn £10, but the ox only a couple of shillings. Bishop might even have carried Thomas Hood's poor Mary's head to the facial repair surgeon Dr Carpue, whose lecture rooms in Dean Street would have been familiar to him as a destination for "things".

Finding murder an easier way of acquiring nice fresh "things" than digging them up, Bishop and Head (or Williams) began to walk in the footsteps of Burke and Hare, becoming the "London Burkers". But they likewise were detected, arrested, and both hanged before a vast crowd of 30,000 people outside Newgate Gaol, Old Bailey, London.

The Burke and Hare murders, followed by those of Bishop and Head, really brought what by 1831 was the sheer *scandal* of body-snatching to its grand finale. The rapidly expanding medical schools of Scotland, London, the English provinces, and Dublin meant that no one was safe, either living or dead. Bodies, or body parts, of exhumed or stolen "subjects" were sent across Great Britain to anatomy teachers by stagecoach, canal, or sea, and it was said that any suspicious or malodorous package could trigger fear and suspicion. The whole business was especially upsetting because many Protestants saw the spiritual resurrection at Judgment Day as involving the revivification of the deceased physical body. Either way, body-snatching, let alone murder, had by 1830 become an intensely hot topic of public concern.

This led to the agitation for and in July 1832 the final passage into law of the *Anatomy Act*. The Act sought to provide legal medical cadavers from the unclaimed remains of indigent people in workhouses and hospitals, as well as regulating where anatomies could be performed. The impact of the Act was salutary and did much to allay public fears for deceased loved ones. But the demand for cadavers was such that the problem was by no means wholly solved, and incidents of both grave-robbing and unburied corpse stealing continued to be talked of in the press, though to a much lesser extent. One man who provided some fascinating mid-century insights was "Aesculapius Scalpel", the pen name of Dr Edward Berdoe.

St Bernard's, The Romance of a Medical Student, 1888

The real-life Edward Berdoe was born in 1836, some four years after the passing of the Anatomy Act, and would probably have been at "St Bernard's" (St Bartholomew's) medical school during the middle and late 1850s. *St Bernard's* is a piece of fiction, but it bears the hallmarks of autobiographical fiction, written by a 52-year-old literary doctor about his student days 30-odd years before, and is packed with incidents which have the ring of truth about them, as we know from other independent sources. He even includes the elderly Dr Robert Day in his story, whose career had been effectively ruined, so we are told, by the old Burke and Hare scandal, and who was now residing in London, and a familiar figure on the wards and in the dissecting rooms of St Bernard's. "Dr Day", I suspect, is a linguistic play on the name "Knox" (Latin *nox* = "night"), for Robert Knox really *was* living and working in London during the 1840s up to his death in December 1862, aged 71.[4]

While he was not legally implicated in the Burke and Hare scandal of 1828, Knox's public reputation had been effectively ruined by the mob. He had attempted, unsuccessfully, to return to his earlier career as an army surgeon, but ended up living as quiet a life as possible in London, working as a doctor and medical journalist, and holding an appointment at the London Cancer Hospital. But everything that Aesculapius Scalpel says about "Dr Day" accords with what the historical record tells us about the 60-odd-year-old Robert Knox.

In this post-*Anatomy-Act* age of medicine, it seems, from what Aesculapius Scalpel tells us, that some shady practices may have survived when it came to obtaining and dealing with anatomical "subjects". There are the usual student pranks with body parts: pranks of a kind that have survived the centuries and which, I know, are still played by senior upon "fresher" medical students today. In "St Bernard's" there were smelly, ill-lit dissecting rooms – perhaps not the best place for sensitive souls to be alone on a winter's night. (I have read in several other accounts of how it was a common practice in many dissecting theatres for students to throw scraps of cadaver onto the floor to discourage hungry rats from climbing up on to the "subject" on which one was currently working.[5])

Aesculapius Scalpel, like many of his chums, lived in the lodging-houses adjacent to "St Bernard's": medical student landladies, apparently, being a breed all of their own, some married to hospital porters and beadles, and some with sisters who were nurses in this pre-Florence-Nightingale age. After 1832, all anatomies should have been performed on authorized premises, yet Aesculapius tells us that a young hospital "gent" might smuggle an arm or a head back to his lodgings for private study. And though a 1850s' landlady might be furious if she chanced to find a half-dissected head on Mr Tom's table when servicing his room, this would most likely have been tolerated in the past. It seems that the back gardens of many of the medical student lodging-houses around "St Bernard's" would have had odd thigh-bones, mandibles, finger bones, and other fragments, remnants of former private dissections, buried just a few inches down – not to mention what would have been interred in the hospital's official burial ground. The same would have occurred in the other London, Edinburgh, Glasgow, and provincial hospitals – wherever cheroot-smoking (to kill the stench) medical students were to be found.

Aesculapius Scalpel also recounts some stories from the "olden Time of procuring bodies" some 50 years earlier. As with most academic institutions, old alumni and friends called at the "St Bernard's" medical school to say "hello" to the rising generation. Aesculapius recalls how "One old fellow who had been a demonstrator [or postgraduate assistant] in that room... some fifty years before" told him a story, probably dating back to 1803–10, where the beadle, or senior servant of the medical school, promised to supply a cadaver. "Can't get you a fresh 'un before this day week, sir... You see, sir, it is my mother-in-law. She only died last night. She will be buried on Thursday. We'll have her up the same night, and she'll be ready next morning for you."[6] The poor lady was duly "resurrected" from Bow cemetery, stripped of her shroud, and taken off in a cart. But then, fearing pursuit and capture, the beadle and his cronies tipped the dead woman into a ditch and ran off. Next morning, after the body had been found, it was assumed that she had suffered "horrible murder".

One presumes that the young demonstrator did not receive his promised "thing". The story puts into perspective what was meant by a "fresh 'un" in the pre-refrigeration age, if the "subject" had already been dead for a week before being expected to appear on the dissecting table. It also brings over the obvious appeal of true

371

"fresh 'uns" such as "Daft Jamie" and Mary Paterson alias Mitchell who had passed from drunken frolic to dissecting table without the formality of a funeral in between.[7]

FINDING BONES: A POSTSCRIPT

Aesculapius Scalpel tells us that by his time, in the 1850s, body parts dissected in "St Bernard's" medical school were eventually put into coffins, often in a higgledy-piggledy fashion, and buried as departed "brethren and sisters" – presumably under the auspices of the hospital chaplain in the official burial ground, as the law now required. But around the year 2000, two sites in Oxford being excavated to provide new building space yielded a grisly harvest. When the old ditch behind the Museum of the History of Science, Broad Street, was dug out, it yielded a medley of disarticulated human and animal bones: the remnants of dissections performed in the museum's back semi-basement chamber after 1683. No formal interment was suggested, as the skeleton parts were all jumbled together, human and animal – relics of comparative anatomy classes, perhaps. Some of the most interesting pieces are now on display in glass cases only a few yards from where they would have been dismembered.

Around the same time, when Oxford Castle Gaol was being converted into an up-market hotel, more bones were unearthed. This had probably been the old prison burial ground. In most cases, the bodies appear to have received the dignity of some kind of funeral, although knife-marks on some of the bones suggest that post-mortem dissections could have taken place. It is worth remembering that, had not Drs Petty, Willis, and their colleagues succeeded in resuscitating the hanged Anne Greene in 1650 (whom we met in Chapter 13), her post-dissection remains would probably have joined the anonymous Oxford Castle skeletons.

Let us now move on from the grisly circus of the "resurrection men" to that body of escalating discoveries from which would emerge efficient, curative, modern medicine.

CHAPTER 22

The Miracle of the Microscope

*W*hat has always driven progress in all of the sciences has been new evidence: new facts about the structure of the natural world. It was increasingly accurate geometrical measuring instruments that drove astronomy; then, after 1609, the telescope. Optics, similarly, really came into being with the invention of lenses in the Middle Ages; and the physics of motion by a progressive ability to make exact line- and time-measuring devices. Yet as we have seen in previous chapters, medicine was the "Cinderella" of the sciences, for in spite of anatomical and physiological advances, one could not see or physically measure the cause of pain, fever, or madness. The doctor might identify an illness on a descriptive level, yet even in 1820, the causal mechanisms behind most diseases were just as obscure as they had been in the days of Hippocrates.

Then things began to change quickly and profoundly. During the nineteenth century, the whole physical and conceptual landscape of medicine would be transformed. In many ways, this would begin with the development of the high-powered "achromatic" microscope after 1826.

The microscope, as an instrument containing a pair of simple matched lenses in a tube, was almost as old as the telescope: probably being devised by the Dutchman Zacharias Janssen in 1611. Galileo himself had discussed the microscopic beauty of insects, and in 1661, Marcello Malpighi had used an early microscope to observe small capillary vessels that helped to clinch the physiological case for Harvey's theory of blood circulation. But it had been Robert Hooke's monumental *Micrographia, or some Physiological Descriptions of*

Minute Bodies made by Magnifying Glasses with Observations and Inquiries thereupon (1665), with its beautiful plates and descriptions of minute structures, that had captivated the world. Hooke had even coined the term "cell" to describe the structural globules that made up cork, and which he speculated conveyed the *succus nutritius* ("nourishing juice") that kept the plant alive. Yet Hooke had no coherent idea of what cells actually *did*; he coined the term because of their enclosed "box"-like nature.

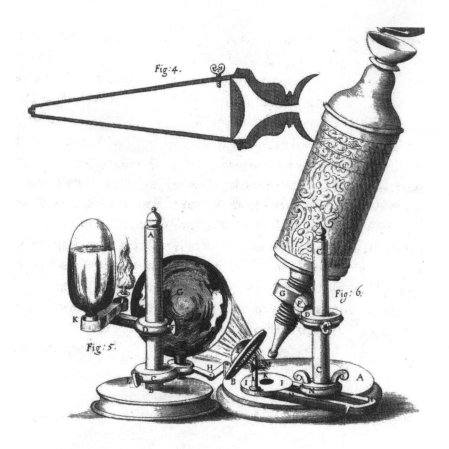

34. Robert Hooke's simple-lens non-achromatic compound microscope, 1665. Notice the brine-filled glass ball used to focus lamp-light on to the specimen. This instrument has a tube and fittings of cardboard, wood, and small metal parts, with a magnification of around ×100. (Hooke, *Micrographia* (1665), frontispiece. Christ Church Library, © Governing Body of Christ Church, Oxford.)

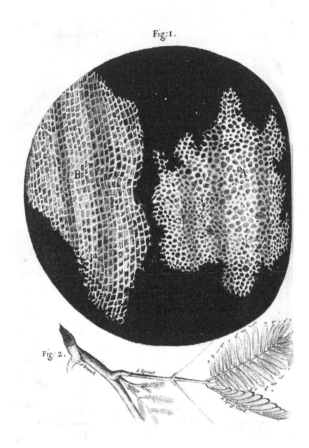

35. The cells of cork. Although Robert Hooke had no idea of the physiological function of cells, he was the first scientist to coin the term – for an organic fabric made up of enclosed, "box"-like structures – and to depict large cork cells as seen under his microscope. (Hooke, *Micrographia* (1665), Observation 18, Scheme 11. Christ Church Library, © Governing Body of Christ Church, Oxford.)

All these early microscopes were of a limited magnification and image clarity, however, with their two or three simple lenses: a short-focus object glass, a magnifying eyepiece, and sometimes an intermediary field lens. I am very familiar with these "Hooke-type" *Micrographia* microscopes, having built one myself and used it to examine a variety of objects. The low practicable magnification was determined by the simple glass lens, which, as a fact of physics, could not bring all the rays of light passing through it to an exact point of focus. Consequently, the resulting images might have false colour fringes of ill-focused light (chromatic aberration), along with an inevitable tendency to blur and break down above a certain level of magnification, usually around 70–100 times. But things changed in the 1820s.

JOSEPH JACKSON LISTER FRS: QUAKER, MICROSCOPIST, AND GENTLEMAN OF SCIENCE

Joseph Jackson Lister belonged to that class of men who, more than any other, drove British scientific innovation between *c.* 1550 and *c.* 1900: namely, private gentlemen of ample independent means, often devout, with a passion for rational knowledge. If the name "Lister" rings a bell, then let me mention, ahead of Chapter 24, that Joseph Jackson's son, Joseph Lister, would be the founder of antiseptic surgery after 1867.

The Listers were London wine-merchants who had done extremely well. While unable, as a Quaker, to attend Oxford or Cambridge, Joseph Jackson, like so many of his denomination, had received an excellent schooling and independent education, and had been imbued with a passion for all knowledge. He delighted in natural history, optics, and the microscopic realm, and in 1832, on the strength of his work on improving microscope lenses, he would be invited to join the Fellowship of the Royal Society.

Though exquisite brass microscopes and fittings, complete with finely crafted mahogany cases, were commercially available in 1820, their magnifying powers were still relatively limited, enabling one to see little more than Robert Hooke had done in 1665. Lister, however, with his passion for mathematical optics, and knowing several fine London instrument-makers, approached the problem from a new angle. He set about redesigning the microscope's optical parts, incorporating principles already established in telescopic optics.

Around 1758, the London optician John Dollond had found that when he made a telescope *not* with the usual simple objective (or light-gathering) lens, but with a specially designed *compound* lens, the image clarity and potential magnification could be greatly improved. His compound lenses were made from two different types of glass: clear "crown" and greenish "flint" glass, their optical curves figured to match each other perfectly. Because the two lenses had different refractive, or light-bending, properties, each lens could be made to cancel out the aberration, or faults and false colour fringes, of the other. All of a sudden, telescopic magnifications could be increased; and as they minimized colour distortions, Dollond's lenses were known as "achromatic", or with no false colours. The new lens patent would make the Dollonds' fortune, and establish a firm that would endure over the centuries.

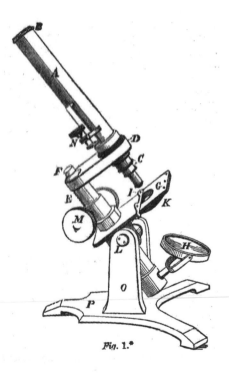

Fig. 1.*

36. Achromatic compound microscope, *c.* 1860, made in heavy brass with fine lenses and precision adjustments. (Edwin Lankester, *Half-hours with the Microscope* (London, 1876), p. 5. A. Chapman collection.)

Telescopic object-glasses are large, being 3, 5, or 6 inches in diameter, and it is much easier for a skilled optician to grind and figure a 4-inch-diameter lens than a lens half the size of a shirt button, such as a microscope objective lens. This was only part of the problem facing J. J. Lister and other opticians. For as well as being tiny, microscope objective lenses had to be of a very short focal length; and here one encountered additional problems when it came to mathematical lens design. The higher the intended magnification, the shorter the focal length had to be, so that for very high magnifications, the lens was only a fraction of an inch above the specimen being examined.

Lister was not the first optician to address the matter of *achromatic* microscope lenses, for Dutchmen, Italians, and Frenchmen as well as Englishmen had been working on the problem, and manufacturing achromatic microscope objective lenses. Yet while better than simple lenses, these crown and flint glass "doublet" and "triple" component lenses could not magnify effectively beyond about 120 times. It was at this point that J. J. Lister began to get involved.

There were two quite distinct optical problems. Firstly, there was "chromatic aberration", or the problem of designing a lens which neutralized its own colour-fringe distortions. Secondly, one had to overcome "spherical aberration", caused by the spherical curves of the lenses then in use, which led to image-blurring when one tried to increase the magnification. The ideal lens would be one where all the light coming up from the object under examination came to a perfect focus-point.

Combining practical knowledge with mathematical analysis, J. J. Lister began to experiment, beginning initially by analysing an achromatic microscope objective lens by Charles Louis Chevalier (and perhaps his collaborator Camille Sebastien Nachet) of Paris. Lister came to the conclusion that the solution lay in the exact proportional spacing of the three crown and flint glass lenses, in accordance with a geometrical "tracing" of the path of the light rays through the lens.

Guided by Lister, the London optician Charles Tulley began to make triple-component lenses, such as his celebrated "9/10" objective, giving ×300 or more magnification: an unprecedented magnification for 1825, yielding all kinds of wonders. Lister then tried his hand at making his own lenses, in accordance with his pre-calculated designs and proportions, and found that he could now get even higher magnifications.

Quaker philanthropist that he was, J. J. Lister, far from trying to obtain a patent for his crucial improvements to the microscope, simply gave them, gratis, to the world. In 1830, he read a paper to the Royal Society which laid out the road ahead for anyone wishing to try: "On some Properties of Achromatic Object-Glasses Applicable to the Improvement of the Microscope", in 1830.[1] Nor did Lister's work stop here. He continued to improve upon his original principles and worked closely with microscopic manufacturers, thereby helping to make the reputations of Victorian microscope manufacturers such as Beck, Ross, and Powell in London, and Continental firms. Lister would provide the basic optical design of research microscopes until Ernst Abbe of Jena University made the next improvements around 1880, leading to the emergence of the renowned German optical firm Karl Zeiss of Jena. By 1883, microscopes giving magnifications of ×2000 or more were commercially available, as a published "photomicrograph" of that date in my own collection substantiates.[2]

The medical research potential of Lister's achromatic lenses was immediately recognized. One of Lister's close friends, a fellow-member of the Royal Society and a Quaker, was Dr Thomas Hodgkin of Guy's Hospital, London. As early as 1827, Lister and Hodgkin had examined and published an account of blood cells as seen under high magnification: they had measured the sizes of the cells with a microscope micrometer, and noticed how they came together and stacked when clotting took place. This was a foundational study in haematology, entitled "Notice of Some Microscopic Observations of the Blood and Animal Tissues", and published in *The Philosophical Magazine*.[3] Lister would publish a microscopic study of tubular and cellular polyps in the *Philosophical Transactions* in 1834.[4]

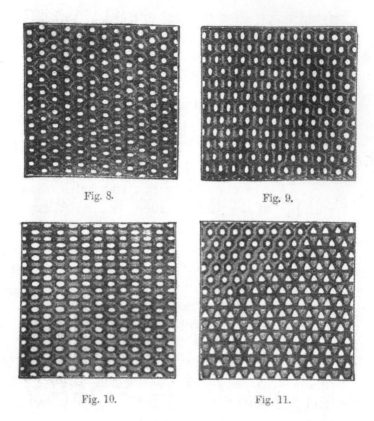

Fig. 8.   Fig. 9.

Fig. 10.   Fig. 11.

37. High magnification. Part of the siliceous valve of *pleurosigma angulatum*: a popular lens test among Victorian microscopists. *Pleurosigma* is a water-borne diatom revealing beautiful cellular structural detail at ×2000 magnification. (Article "Microscope", *Encyclopaedia Britannica*, 9th edn. (Edinburgh, 1883), p. 264. A. Chapman collection.)

While its impact was not recognized at the time, Thomas Hodgkin and the high-powered achromatic microscope would help establish scientific oncology, or cancer study. His work included a study of carcinomas in 1829, and, portentously, a paper delivered to the London Medico-Chirurgical Society (now the Royal Society of Medicine) in 1832, entitled "On Some Morbid Appearances of the Absorbent Glands and Spleen". Over 30 years later, Sir Samuel Wilks would make the same discovery, which he generously named in honour of his predecessor: Hodgkin's disease.

CELLS: PROFESSOR VIRCHOW IDENTIFIES LIFE'S
BUILDING BLOCKS

By the late 1830s, achromatic microscopes incorporating Lister's optical principles were coming to be manufactured and used across Europe and the expanding USA, and one thing that their enhanced magnifications made possible was an answer to the ancient question "What is the stuff that we are made of?" While, as we have seen in the language employed by Lister and others, the word cell was in use to describe discrete physiological entities, such as magnified blood, no one had any coherent idea of what a cell was, and what was its biological function.

At the time, there were a variety of explanations on offer as to what living material actually was: "tissue", which was somehow believed to form out of blood and lymphatic secretions, to make skin, bone, nerves, and hair. In his *Traité des membranes* (1799), Xavier Bichat classified "tissue" into 21 types. Or perhaps things mysteriously *grew*, impelled by some life force, such as that which made infant bones grow to adulthood? Or there might be *blastemas* or seed "principles" or "origins" in living things which possessed a tendency to proliferate.

German scholars were especially fascinated by this question, particularly when viewed from a metaphysical perspective. *Naturphilosophie* set about exploring the differences between living and dead, organic and inorganic things. Were the substances being investigated by the young Giessen-Darmstadt chemistry professor Justus von Liebig and his colleagues – such as milk, alcohol, and fat – unique *organic* creations, fundamentally different from minerals like clay, stone, or tin, or were they simply more complex chemically? Germany in the 1820s was the birthplace of that new discipline, organic chemistry: a discipline that would exert an immeasurable influence upon the ensuing development of medicine, both conceptually and in terms of laboratory techniques. And it was Germans, such as the physiological and evolutionary thinker Lorenz Oken, who asked what impelled things to grow up. Oken, like many others, was fascinated by the parallel stages through which all foetuses seemed to pass, be they frogs, cats, or humans. First there was a blob, then it developed a heart, a spinal cord, limb projections, and some sort of brain. Why? What was the "stuff" that all living things shared? All of this detail, and much else besides, was visible through a powerful microscope.

The university philosophers and the laboratory scientists pursued their own disciplines, but in the fermenting intellectual culture of Berlin or Paris, it was inevitable that strict boundaries became confused, and a physiologist might validly ask questions about the metaphysics of living matter. This is where the new high-powered microscopes came in.

Several scientists, including Matthias Schleiden, Theodor Schwann, and Johannes Mueller, all have a valid claim to contributing towards the development of the cell theory and that of the growth of living matter by the early 1840s. But it was Rudolf Virchow of the Charité Hospital, Berlin, who was to appropriate much of the credit. In particular, Virchow's propensity to seize priority led to a rupture with his colleague Robert Remak, who had demonstrated that new cells were always formed from the division of existing cells. But Rudolf Virchow was in no way a metaphysical or philosophical doctor. Rather, he was an instinctive experimentalist, microscopist, and evidence-based researcher. The 1840s were the breakthrough decade, however, when it came to recognizing that all living things were made of cells, and the achievements of that period laid the foundation upon which an experimentally verifiable understanding and eventual cure of a whole range of diseases could be built. Perhaps most significantly, they provided the first flash of light at the end of the tunnel for humanity's many-millennia-long struggle with cancer.

Central to all of this research was the new, high-powered achromatic microscope. As a young man, at the Charité Hospital, Virchow became something of a virtuoso in the use of that instrument. It was the identification of that disease styled *leucocytosis*, and soon after *leukaemia* (from Greek *leukos*, "white", and *haema*, "blood"), that would be the first decisive triumph of the new cell pathology. Exactly who was the first person to identify the disease, whether John Hughes Bennett in Edinburgh or Rudolf Virchow in

382

Berlin, is a matter of dispute. Suffice to say that the identification fell around 1845 or slightly after. Hughes Bennett's defining paper "A Case of Hypertrophy of the Spleen and Liver…"[5] and the 24-year-old Virchow's microscopical identification were published at about the same time. This is yet another example of that not infrequent nineteenth-century phenomenon of two wholly independent researchers, working hundreds of miles apart but using a similar technology and thinking along the same lines, making a simultaneous discovery.

Virchow came eventually, partly as a result of Karl Thiersch's suggestion, to see leukaemia as caused by a metastasis of malignant cells, which destroyed the correct physiological function of the blood and killed the patient. In the wake of Robert Remak's realization, Virchow would come to understand cell replication as the primary process of body growth, decline, and malignancy. In an extension of William Harvey's Aristotelian doctrine *Omne vivum ex ovo*, or "All life comes from an egg", so Virchow formulated *Omnis cellula e cellula*, or "Every cell comes from a cell." Virchow would hammer home this doctrine to students in his lectures as director of the Pathological Institute, Berlin, and in his monumental *Die Cellular-pathologie* (1858).

According to Virchow's new pathology, a disease was not so much an entity in itself as a *response* to abnormal cell replication. As one might put it, defective, "rogue" cells proliferated unchecked and ran riot through the body, disrupting its normal healthy function. One great impetus which this approach to pathology gave to medicine was to study all manner of cells: for now cells needed to be collected, identified, measured, and classified as an essential preliminary to any scientifically coherent therapies being devised. One vital tool towards accomplishing that end was the high-powered microscope.

Ironically, however, his realization of the primacy of cell function in every aspect of health and disease led Virchow to be sceptical about the pathological importance of that other great microscope-related discovery in the 1860s, namely, bacteria, to which we will turn shortly.

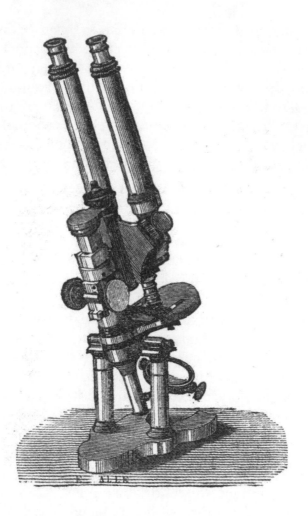

38. Binocular compound microscope, capable of very high magnification. (Revd. W. Houghton, *The Microscope and Some of the Wonders it Reveals* (3rd edn., London, undated, *c.* 1880), p.14. A. Chapman collection.)

UNDERSTANDING CANCER

Being able to pin down leukaemia pathologically, however, was not the same as being able to do anything for sufferers. Arsenic compounds were tried, but the grim reality in 1850, and for well over a century thereafter, was that a diagnosis of leukaemia was a death sentence. So it was for all other types of cancer.

Breast cancer was what most people meant by "cancer" down the centuries, because it was by far the most obvious manifestation of the disease. Ancient Egyptian literature mentions the condition, while it was the Greek word for "crab", or *karkinos*, which gave us our modern term "carcinoma", and the Latin for "crab", *cancer*, which gave us "cancer". The term derives from the nasty, crab-like lesions of certain cancers. Rhazes, the *c.* AD 890 Arabian physician, believed only the application of the red-hot cautery iron was of any use against cancer.

Generally speaking, cancer was believed to be connected with obstructions in the movement of black bile, though by the eighteenth century John Hunter and others had come to recognize its relation with the lymphatic system. Cancers, however, were seen as obstructive lumps or blockages until the work of Rudolf Virchow and then Wilhelm Waldeyer in 1867 indicated that, being based upon faulty cell replication, they were much more complex and multifaceted, and capable of occurring *anywhere* in the body. In particular, this was the case with systemic cancers, such as leukaemia.

But what actually *caused* the defective cell replication of cancer? One theory was that a blow or an infection might somehow cause cell replication to go awry. Another theory, particularly favoured, was that it was caused by toxins. Did people develop cancers because of lifestyle, food, or ingested chemicals? In 1894, the eminent Glasgow surgeon Dr Robert Bell, senior physician to the Glasgow Hospital for Women, publicly renounced surgery for breast cancer, arguing that 50 per cent of breast lumps were benign anyway, and that cancerous ones operated on invariably returned within three years, and killed the patient: a circumstance suffered by Joseph Jackson Lister's daughter and sister to the great Joseph Lister, Isabella Pim, who in 1867 was operated on using the latest safeguards of antiseptic surgery. After three years, the cancer returned and Isabella died.[6]

Dr Robert Bell came increasingly to devote his attention to finding the *cause* of cancer, which he laid at the door of increased meat-eating. Bell used statistics to attempt to correlate the rise in meat importation between 1868 and *c.* 1910 with a rise in cases of cancer. The real problem, in his view, lay with the sluggish bowel action occasioned by digesting large amounts of red meat, resulting in constipation, and the absorption of toxins into the bloodstream. Robert Bell was a pioneer histological researcher and one of the first scientists to use photomicrographs, or photographs taken through a high-powered microscope, of diseased and healthy body tissue. He came to the conclusion that cancer cells formed in the blood, which then led to metastasis.

Robert Bell was in no way alone in his concern with finding a toxin cause for cancer. Could cancer even be prevented by speeding up the action of the gut, so that food did not remain in it for so long? He became an advocate of what we today might regard as a healthy, balanced diet: vegetables, nuts, cheese, cereals, eggs, and milk, to reduce toxins and speed up bowel action. In this respect, he was thinking along the same lines as Maximillian Oskar Bircher-Benner in Switzerland, inventor of muesli breakfast cereal, and John Harvey Kellogg in the USA, who gave us cornflakes; much healthier than the piles of sausages, bacon, liver, kidneys, and fish which constituted a middle-class Victorian breakfast.

It was partly in response to a perceived increase in and diversity of cancer types that the modern health food movement was born. By 1900 the word that was circulating in certain oncological circles was *autotoxaemia*, or the blood being poisoned by food remaining in the gut longer than was healthy: chronic constipation, in fact.

In addition to healthy diets, some doctors devised new surgical procedures to deal with the problem. As anaesthesia and antiseptics, as we shall see in later chapter, had rendered surgery both painless and increasingly safe by the 1890s, surgeons on both sides of the Atlantic were performing "heroic" colonectomies: opening up the patient and removing a few feet of intestine, thereby shortening and speeding up the journey through the body. The new "epidemic" of appendicitis, also ascribed to constipation and too much meat-eating, initiated a further surgical procedure, and Sir William Arbuthnot Lane, who in his time probably removed several hundred feet of constipated and potentially cancerous gut, incidentally performed the world's first heart resuscitation procedure in 1902, as reported in *The Lancet*.[7] A patient's heart suddenly stopped in mid operation. But as his abdominal cavity was already opened up, the quick-thinking Arbuthnot Lane simply stuck his hand and arm up inside the patient's diaphragm, grabbed the heart, gave it a few squeezes, and restarted it. The patient survived.

Was cancer really on the increase, and was it specifically a disease of industrial civilization, as many believed? Two medical statisticians, George King and Arthur Newsholme, set to work on the problem, the Royal Society publishing their results in 1893.[8] Cancer, they concluded, appeared to be on the increase only in countries where modern scientific medical services *and* efficient mortality registration were firmly in place.

If a person died in Manchester, Berlin, or New York, their death would be certified and registered by a qualified physician. If they died in a remote hamlet in rural Ireland, Poland, or Spain, the family might simply tell the parish clerk that they had succumbed to "a great pain", "a running sore", "an old lump", "spitting blood", or had just "wasted away". As better nutrition and science-based medical care led to increasing longevity over the nineteenth century, the population became statistically more susceptible to degenerative diseases such as cancer. Cancer is more likely to affect a person over 50 than it is a 25-year-old. But it is unlikely that any of these breakthroughs would have been made had not the new post-1825 achromatic microscope opened up a vast new realm of diagnostic and pathological potential. In addition to revealing the true nature of cancer, the microscope would hold the key to first identifying, and then helping to prevent, bacterial infection.

THE FRENCH CHEMIST AND THE GERMAN
PHYSICIAN

As we have seen in previous chapters, infections in their diverse forms had been the perennial plague of the human race, from leprosy and the plague of the Philistines in the Old Testament to the medieval Black Death, from Victorian cholera to surgical gangrene in the

American Civil War, down to Ebola in our own time. Was infection caused by miasmas and wind-blown foul air? Was it occasioned in some way by dirty drinking water and unemptied cesspits? Or was it generated from fermenting matter, or overheated or bad blood? Or was it simply generated spontaneously? All of these possibilities were still on offer in 1840.

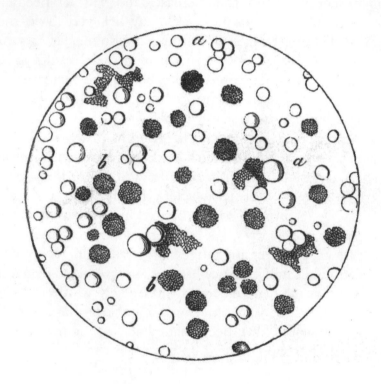

39. Dark "pus corpuscles" (bacteria) seen in milk, which can lead to souring. (Revd. W. Houghton, *The Microscope* (*c.* 1880), p. 102. A. Chapman collection.)

One thing came to be revealed by the new high-powered microscopes, however, and that was an even greater abundance of "animalcules", or wriggling things, than had been visible in the old simple microscopes of Robert Hooke and Antonie van Leeuwenhoek. Yet how could those beasties be the cause of infection, for we *all* had them, even when in the prime of health?

They were there in teeth and tongue scrapings, in blood and other body fluids, in faecal matter, in milk and numerous everyday foodstuffs. Between 1840 and 1890 the whole scientific landscape surrounding infection would change beyond recognition as the new microscopes – in connection with rapid advances in organic chemistry and biochemistry – gave rise to the new pathological specialism of microbiology.

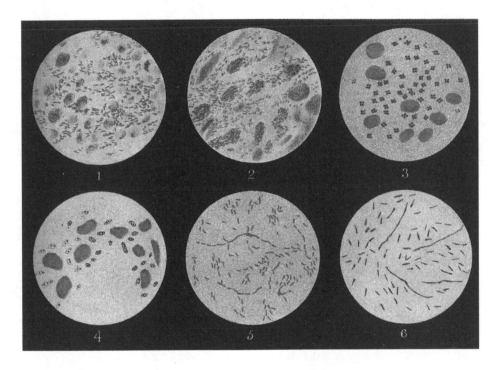

40. Bacteria. Left to right: (1) tuberculosis, (2) leprosy, (3) micrococcus tetragenus, (4) pneumonia diplococcus, (5) cholera, (6) typhoid. (*The Family Physician* (*c.* 1905), Vol. 1, colour plate. A. Chapman collection.)

As with all new scientific disciplines, things began with rigorous taxonomy and classification. Ferdinand Julius Cohn was in many ways the father of cell-type classification, working with his microscope first on botanical and animal cell structures, and then, after 1870, on bacteria. Bacterial classification, however, was not always straightforward, for until the development of efficient chemical staining agents by the 1870s, which enhanced and gave visual contrast to different germ types, it was often hard to *see* them in detail, irrespective of the power or quality of the microscope optics. And quite independently of the microscope itself, the rapidly progressing science of organic chemistry was elucidating the structure of proteins, carbohydrates, and the complex hydrocarbons that compose all living things.

Both a skilled analytical chemist and a microscope user, Louis Pasteur addressed himself to one of the age-old questions: the nature of fermentation. Born in Dole, eastern France, in 1821, the young Louis Pasteur had been a very average student and not especially good at chemistry. Yet he would be the scientist who proved beyond doubt that fermentation (and by experimental association, infection) was caused not by spontaneous generation or miasmas, but by bacteria, or microbes.

Pasteur's studies on fermentation began in 1856, when he was approached by a Lille wine manufacturer to investigate the souring that sometimes destroyed vintages. Working also with lactic acid, Pasteur found that fermentation took place only when yeast, an organic, microbial culture, caused sugars to decompose into alcohol and release carbon dioxide. (This could also explain the old phenomenon of why a lighted candle held over a fermenting beer or wine vat was suddenly extinguished.) Pasteur's paper, published by the Société des Sciences de Lille in the autumn of 1857, sounded the death-knell for the ancient doctrine of "spontaneous generation", or living matter being formed by some kind of occult or mysterious process related to the air. Pasteur demonstrated from his experiments that fermentation could only take place when microbial spores were already present to begin a reaction, the air playing no biological or chemical role whatsoever. Organic spores might be introduced deliberately to begin a fermentation process, or unwanted microbes might be blown in on the wind (the wind being no more than a mechanical agent of conveyance) and get into, and ruin, a vintage.

Recognizing that it was the skin of the grapes that provided the vital yeast, Pasteur sterilized some grapes, and found that they would not ferment: common sense to us, but an astonishing realization in 1860. Taking this line of reasoning further, Pasteur then devised his *experimentum crucis*: his swan-necked flask experiment.

Soup was placed in a laboratory flask and boiled, thus killing any airborne microbes which it contained. Now if the flask were fitted with a long, graciously curved swan-necked glass tube with a trap to prevent external air from getting inside, the soup would stay clear and fresh. This demonstrated beyond doubt that fermentation, souring, or decomposition were caused *not* by mysterious spontaneous generation, but by an external, microbial agent getting in from outside. This boiling and sealing process could now be applied to food preservation – such as milk or wine – to pasteurize it and increase its edible lifespan.

In the 1870s, Pasteur began to work towards preventing diseases such as chicken cholera in birds, anthrax in farm animals, and rabies in dogs and humans, by developing the immunological principles set out by Edward Jenner and others with regard to smallpox. But as there was no naturally occurring weakened strain of these diseases, as there was with cowpox for smallpox, Pasteur had to experiment with developing a laboratory-weakened strain with which to immunize his experimental animals, so that an immune reaction would (hopefully) be triggered when they encountered an active, deadly strain of the same disease. This he did by heating and desiccating cultures taken from creatures that had succumbed to the killer strain and died.

He discovered that the technique did work, and immunity from the killer disease could be conferred by injecting an animal with a laboratory-weakened strain. But what about humans? Pasteur had, by the early 1880s, a vaccine for rabies, but he was a chemist and not a licensed physician, and potentially liable if he injured or killed a human being. An ideal case presented itself in early July 1885.

Joseph Meister, a nine-year-old boy, had been badly mauled and bitten by a rabid dog, and could be in danger of losing his life to rabies. Pasteur injected young Meister with his rabies vaccine: the boy did not develop rabies and recovered fully from his injuries. Pasteur became a hero, and the Parisian Pasteur Institute which came to be named in his honour, and of which he was first director, became the global prototype bacteriological and immunological research institute. And Pasteur named his new weakened-strain drugs *vaccines*, in honour of Edward Jenner's vaccination method of 1796.

Louis Pasteur had demonstrated beyond doubt that many diseases were transmitted by bacteria and could be prevented from becoming active by *pasteurization* techniques, or by *vaccines*, but it was his younger German contemporary Robert Koch who supplied the crucial link between specific strains of bacteria as seen through the microscope and specific diseases, such as anthrax, cholera, and tuberculosis.

Born in 1843, near Hanover, Robert Koch's early academic life was different from Pasteur's. Koch was an academic high flyer from childhood, and qualified as a doctor at Göttingen in 1866. He then became an army surgeon, and following his marriage, a physician at Wöllstein, and later a professor in Berlin.

Koch had been fascinated by the power of the microscope since his student days, and became a skilled microscopist and the founding father of microbiology. Developing sophisticated techniques for cultivating bacteria in the laboratory, he was able to identify the very *Bacillus anthracis* that caused anthrax. Then, travelling to Egypt and India to study the cholera outbreak, Koch isolated the bacterial causal agent of the disease: *Vibrio cholerae*. Yet as we saw in Chapter 20, this was really a rediscovery of Pacini's original, yet largely unknown, microscopic identification of *Vibrio cholerae* of 1854. When Koch was born, most people attributed cholera to vague miasmas; when he was a schoolboy, John Snow in London had demonstrated that whatever caused cholera, it was conveyed in drinking water; and by the time that he was a Göttingen medical student, Sir Joseph Bazalgette's sanitary engineering was protecting Londoners from its ravages. Then in Alexandria in 1883, and confirmed in Calcutta in 1884, Koch and his team identified the very bacterium that caused cholera.

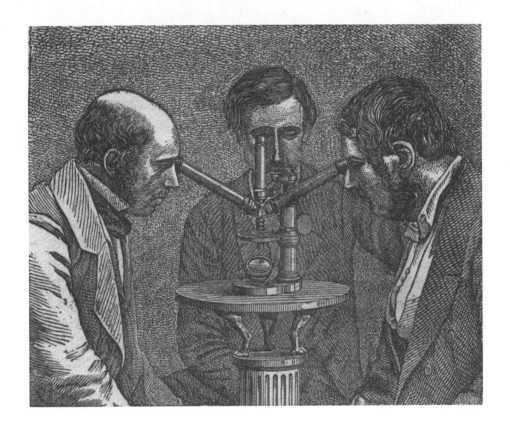

41. "Triple-bodied microscope", enabling three people to observe the same
specimen simultaneously. (Revd. W. Houghton, *The Microscope* (*c.* 1880),
p. 18. A. Chapman collection.)

The year before, in 1882, Koch had identified the cause of
that great scourge, tuberculosis, the slow-growing *Mycobacterium
tuberculosis*. This piece of research would win him an early Nobel
Prize for physiology and medicine in 1905, though an effective *cure*
would have to await the development of the first antibiotics in the
late 1940s.

All of this amply demonstrates the astonishing rate of medical
advance in the nineteenth century. Lying at the heart of it all were
the bacteriological revelations made possible with the new high-
power microscopes, working in tandem with rapidly advancing
laboratory techniques in organic chemistry and biochemistry.

393

Yet one of the greatest medical breakthroughs ever made was taking place at the same time: anaesthesia and the chemical control of pain.

CHAPTER 23

Chemistry and the Control of Pain: Anaesthesia and Beyond

*M*uch attention has been paid in previous chapters to the development of surgery and its relation to anatomical and physiological understanding. Yet whether the surgeon treated the wounded in Pharaoh's army, as in the Edwin Smith papyrus of *c.* 1500 BC, whether he were Galen with his gladiators, John of Arderne with his medieval knights, John Hunter or Sir Astley Cooper at Guy's and St Thomas's Hospitals in 1830, they all had one thing in common: the necessary infliction of appalling suffering upon their patients.

There were stories of opiates, soporific sponges, and getting the patient thoroughly drunk beforehand, yet these techniques could only, at best, confer a transient numbness upon the victim. Conversely, one might try to deaden sensation in an arm or leg due to be amputated by attaching a screw-clamp, which when tightened up could help compress the nerves running down the limb. But one suspects that such devices were no less painful to endure than the amputation itself. Or perhaps, like Alexander Monro in Edinburgh, one might bleed the patient to the point of exhaustion prior to operating, or try hypnotism. But apply a saw to a thighbone, begin to amputate a cancerous breast, or open up a bladder to remove stones in a fully conscious patient, and screams of terror and agony would follow. Speed was the only possible mercy.

By the eighteenth century, and into the nineteenth, surgeons would plan an operation just like a military campaign, with well-trained assistants, juniors, and dressers playing their well-rehearsed parts at each stage of the procedure – all with the aim of doing an excellent piece of surgery, from preliminary incision through closing severed arteries to tying the last stitches, in the shortest possible time. Between two and four minutes could see a leg scientifically amputated, while Robert Liston's 33-second arm amputation is still the shortest *scientifically* executed operation, on a conscious patient, on record.[1] Many of the great master surgeons in London, Edinburgh, Paris, or Boston would have had a sharp-eyed assistant with a stop-watch at the ready. This was the kind of virtuosic surgery, refined and perfected upon cadavers, that made professional reputations.

CHEMICAL ANAESTHESIA: 16 OCTOBER 1846, BOSTON, USA

After *c.* 1770 several discoveries in chemistry began to suggest a way forward in medicine, especially after it came to be realized that air was in reality a collection of chemically specific *gases*, such as oxygen, carbon dioxide, and nitrogen, and not a homogeneous elastic medium. In the 1790s, Dr Thomas Beddoes of Bristol, opened a "Pneumatic Institution", or hospital, where attempts were made to treat various illnesses, such as tuberculosis, with volumes of laboratory-generated gases. Beddoes's brilliant young apothecary assistant was one Humphry Davy.

Davy was a gifted experimental chemist, who would go on to become a knight and president of the Royal Society before his death at the age of 51 in 1829. This early death probably resulted from inadvertently self-inflicted lung damage, as he had breathed and observed the effects of gases such as chlorine. Perhaps as early as 1795, while still a 17-year-old apprentice apothecary in Penzance, Davy had prepared and inhaled nitrous oxide, and noticed its laughing gas effects. He continued this line of investigation while working for Dr Beddoes, and at 22 published his *Researches, Chemical and Philosophical; Chiefly concerning Nitrous Oxide* (1800). Davy noticed that in addition to inducing light-headedness and silly behaviour, nitrous oxide tended to dull pain and cause drowsiness – with no nasty side effects. He even suggested that it might be of use in surgery. But sadly, he took the matter no further, and as his career as a science expositor escalated following his move to London's Royal Institution after 1801, nitrous oxide got superseded by other concerns.

By 1824, however, Henry Hill Hickman of Shropshire had found that inhaled doses of carbon dioxide could induce unconsciousness in animals, enabling him to operate on them without pain, although as a respiratory suppressant, CO_2 was scarcely an ideal way of producing a surgical slumber. What was emerging was that some gases and chemical vapours could induce stupor states in humans and animals.

By the early 1840s, nitrous oxide and sulphuric ether "frolics", parties, and commercial shows had become a part of the entertainment scene on both sides of the Atlantic. They were a bit like getting merrily drunk, but without the ensuing hangover! Also in the 1840s, several American dentists began to experiment with chemical pain-deadeners for their own professional purposes, for by that time the USA led the world in dental innovation: gold tooth fillings and early "vulcanite" suction-action dentures in particular. To get the best suction fit, however, it was often best for the patient to have *all* teeth removed, upper or lower, and if a safe, reliable dental anaesthetic could be found to render extractions painless, huge profits stood to be made.

The Boston, USA dentist Horace Wells had one of his own teeth painlessly extracted under nitrous oxide on 11 December 1844 and claimed to have done likewise for several of his patients by late January 1845. He then approached the great Professor John Collins Warren of Harvard College and the Massachusetts General Hospital, to make a demonstration before the medical school. Wells applied the nitrous oxide to the patient, a medical student with toothache, and went ahead with his usual skill. But the young man, who was rather fat, uttered a cry in the middle of the battle, which led the onlookers to bellow "humbug" and dismiss the whole thing as a joke. We now know that overweight people are sometimes harder to anaesthetize than slimmer folk, and it is possible that the student guinea-pig might have felt something. However, it was also possible that he might have given an unconscious, involuntary groan, and it was later claimed that he had actually felt no pain at all.

Nitrous oxide would have been prepared beforehand by gently heating ammonium nitrate and collecting the colourless gas, with its faintly sweet smell, in a rubber bag or bladder, similar to a modern balloon. The bag would then be unsealed and the stopper tube placed in the patient's mouth, ready for inhalation. This was how it was generally administered at "laughing gas frolics", and one cartoonist of the period shows a hen-pecked husband contentedly sucking at his bag of laughing gas while his wife gives him a good telling-off!

Yet while nitrous oxide would come to be widely used for the milder anaesthesia required for dentistry, it was sulphuric ether, a colourless, somewhat volatile liquid, that was destined to become the world's first effective *surgical* anaesthetic. Dr Crawford Long would perform the first successful surgical operation under ether, when in Jefferson, Georgia, in March 1842, he removed a tumour from a friend's neck, although Long would not report this case until 1849. In the meantime, however, the Bostonian dentist William T. G. Morton would win the laurel for being the first recorded medical man to successfully apply an anaesthetic for a major surgical procedure performed before an academic audience.

The date upon which surgery became painless was Friday, 16 October 1846, and the place, the Massachusetts General Hospital, Boston: a date that should be etched into every school curriculum on the planet, and an achievement of which every American should be proud. This was the birth-date of truly modern surgery, and the USA's first great gift to medical science.

The principal actor in this brief, 10-minute-long drama was the above-mentioned Dr John Collins Warren, Hershey Professor of Anatomy and Surgery at Harvard, dean of the Harvard Medical School, and chief surgeon to the Hospital. A tall, strongly built New England Yankee aged 68 at the time, Warren came from the aristocracy of Old Boston and had a distinguished Harvard doctor father. (I am told by an old Harvard graduate surgeon friend that Warrens continue to be active in Harvard and its Medical School to this day.)

Aged 21, J. C. Warren had been involved, almost inevitably, with body-snatching in London in 1799, when a medical student in England. After beginning his studies at Harvard, he had crossed the Atlantic to undertake further studies at the then "United Hospitals" of Guy's and St Thomas's, Southwark. In a letter home, in December 1799, he tells of the antics of the "resurrection men" and relates what was in itself a tragic incident. It seems that a starving local beggar had acquired some food and proceeded to eat it so avidly that he suffocated himself, dying in the street. One of the "resurrection men", passing, immediately claimed the man as his brother, took him to the dissecting room of St Thomas's, and secured a good price.[2] By 1846, J. C. Warren was America's foremost surgeon, and famed for his skill and operative speed.

Following the failure of the previous tooth extraction experiment under "laughing gas", Warren was sceptical about the whole business of serious surgical anaesthesia. Nonetheless, he was willing to give William Morton's drug a try. The patient, Gilbert Abbott, had a large neck tumour – requiring a procedure potentially more problematic than a simple amputation – and Morton applied the anaesthetic. Once Abbott was under, Warren set to with his usual expertise, dissecting-out the tumour and finally stitching up. Not only did Gilbert Abbott fail to utter the slightest cry during surgery, but he reported, after coming round, that he was only aware of a "scratch" in his neck. The world had truly changed, and aware of the operation's significance, Warren addressed the audience with "Gentlemen, this is no Humbug!"[3]

Morton, it was said, had tried to disguise the liquid he was using by adding various aromatics, with the intention of earning a public reward or securing a patent for his "discovery". But the very well-practised nose of J. C. Warren sensed the active ingredient, as he recorded in his diary: it was sulphuric ether. Sulphuric ether was an already familiar chemical in the 1840s, being a complex hydrocarbon, built up from hydrogen, oxygen, and carbon molecular bonds, and manufactured from alcohol and sulphuric acid. It even had some medical uses before its discovery as a general surgical anaesthetic. Being highly volatile, as well as inflammable, it could cause short-lived local numbness of the skin, as well as being a useful laboratory solvent. Surgically, it was generally administered from a spherical or round glass vessel through a flexible "gutta-percha" (early rubber) tube which was placed in the patient's mouth, the warmth of the anaesthetist's hands around the glass causing the liquid to vaporize. A spring clip would often be attached to the patient's nose, thereby obliging them to exhale back down the tube and re-inhale their own CO_2 as well as ether fumes. Sets of ether administration equipment survive in many medical museums.

42. Ether apparatus, 1847. The bell-shaped jar contains sulphuric ether, the vapour of which is conveyed to the patient via a gutta-percha or early rubber tube. A spring clip would often be put on the nose to ensure correct respiration. (*Illustrated London News*, 9 January 1847. A. Chapman collection.)

News of the discovery spread like wildfire, down the American seaboard and expanding interior, then on to Europe as fast as wind and steam could carry it. Warren, aware of its significance, had the idea of creating a photographic record of the operation. The

Boston Daguerreotype photographer, Josiah Johnson Hawes, was later invited to bring his camera to the operating theatre, where Warren and his colleagues posed for a re-enactment of the great event. (Like anaesthesia, the Daguerreotype photographic process was only seven years old in 1846.) Hawes would do further work with Warren, and, in spite of his proclaimed dislike of the sight of blood, he became the first medical photographer.

Yet surprisingly, it was two months before the first ether operation was performed in Europe: in Paris, on 15 December 1846. Surprising, because by that date steamers, including the first Cunard liner SS *Britannia* and Isambard Kingdom Brunel's *Great Western* and *Great Britain*, were routinely crossing the Atlantic in two weeks. (Charles Dickens had a rough 18-day crossing, Liverpool to Boston, in SS *Britannia* in January in 1842.[4])

The first scientifically performed and recorded public ether operation in Great Britain was at the new University College Hospital, London, on 21 December 1846. The surgeon was the exotic, quarrelsome, sometimes body-snatching Scotsman, Robert Liston, who, as we saw above, had performed a classic 33-second unanaesthetized arm amputation some years before. Quite simply, the reputation-driven medical world of Edinburgh had become too hot to hold him, so he moved south to the new University of London Hospital. In his audience that day was Joseph Lister, the 19-year old medical student son of Joseph Jackson Lister the pioneer microscopist, who 20 years down the line would become the pioneer of antiseptic surgery, as we shall see in the next chapter.

As we have seen, surgeons as a breed have never been the wilting violets of the medical profession, and as far as Liston was concerned, combat and acerbic wit came more easily than shyness. He addressed his audience of students and colleagues with the usual clinical lecture, prior to the patient being brought into the medical amphitheatre to start the operation. Liston, like Warren and most experienced surgeons, was sceptical about putting a patient into a deep and painless sleep, but told the audience that he was going to try out a new "Yankee dodge". I suspect that Liston's "Yankee dodge" quip carried other possible meanings that would have rung a bell with most of his audience. Firstly, in 1835, the American journalist Richard Adams Locke had caused a sensation by publishing a series of bogus articles claiming that the British astronomer Sir John Herschel, currently in remote South Africa, had discovered intelligent beings on the moon. This scoop was no more than journalistic hype, aimed at boosting the *New York Sun's* circulation. Secondly, Liston may have been alluding to the sensation caused in London by the visit of Phineas T. Barnum's freak and illusions show. The "Lunar Hoax" in particular damaged the reputation of serious American science for some time afterwards, especially when outlandish claims appeared to be made. And to most surgeons, a painless operation, where the patient slept like a baby through all the cutting and sawing, seemed no less fanciful in 1846 than the purported discovery of intelligent moon-folk had been a decade earlier.

Liston's amputation operation that day, on one Frederick Churchill, a butler, was a complete success. So much so, it was said, that when the patient came round after surgery, he could only be convinced that the operation was all over when Liston showed him his severed limb. Like John Collins Warren, Liston purportedly recanted his previous opinion, for ether was no humbug: "This Yankee dodge, gentlemen, beats mesmerism hollow."[5]

CHLOROFORM: THE SCOTTISH WONDER DRUG

Science-based technology, including medicine, is a curious thing. Something can be an accepted truism since time immemorial, such as the inevitability of pain during surgery; yet as soon as a transformative breakthrough is made, ingenious men, and later women, immediately begin to search for a way of improving upon it. In October 1846, there was the miracle of ether; and by November 1847, there was another: chloroform. Ether was one of the greatest humanitarian discoveries of all time, but it was not a perfect anaesthetic. It was soon found that some people reacted badly to ether. Some were hard to put under, and the sleep could be unpredictable in its duration; and ether often had respiratory and nauseous effects, making people cough or vomit afterwards. Nor was it ideal for childbirth cases.

None of these things would have mattered a jot before 16 October 1846, when tying the patient down, enduring the screams, and working like fury were the ancestral lot of the surgeon. But once blissful slumber was suddenly available, then how best to eliminate the side effects became the pressing concern. Improving upon a discovery or invention is the next thing that takes place in an ingenuity-driven society.

In 1847, Dr James Young Simpson was professor of surgery at Edinburgh University, and at the age of 36, was not only one of Scotland's most renowned surgeons, but also a leading authority on obstetrics. Could an anaesthetic with fewer side effects be found, to use as an alternative to ether? Medical tradition tells us that Simpson, and his friends Drs Duncan and Keith, began – to our eyes – the hazardous task of sniffing a variety of substances to see what happened. When either David Waldie, a Scottish-Liverpool pharmacist, or Messrs Duncan Flockhart, pharmacists of Edinburgh, sent Simpson a bottle of the new chemical solvent chloroform, Simpson and friends, seated around the dining-table, poured a quantity into table glasses, and began to inhale. And when the servant later came into the room, Simpson and his friends were found sprawled in their chairs in deep sleep. This was on 4 November 1847.

Chemically speaking, chloroform was another complex hydrocarbon compound, produced by distilling alcohol, slaked lime, and chlorine bleaching powder. Three chemists – Justus von Liebig in Germany, Samuel Guthrie in the USA, and Eugène Soubeiran in France – had discovered it more or less simultaneously in 1831–32.

Drs Simpson, Duncan, and Keith were very lucky, because the colourless sweet-smelling liquid is very dangerous, and had they inhaled more, they could have died from heart arrhythmia and seizure. Simpson then tried it on a relative, Miss Petrie, who – before she passed out – thought that she was an angel. People would die on the operating table over the next few years, but the value of the drug was obvious. It was also, at least in these early days, much simpler to apply than ether, requiring no more than one or two teaspoons of chloroform on a piece of soft cloth held over the nose to produce an anaesthetized slumber. One major aspect of his practice in which Simpson used chloroform was obstetrics.

ANAESTHESIA, CHILDBIRTH, AND THE BIBLE

One myth that grew up in the late nineteenth and early twentieth centuries is that the Christian church condemned anaesthesia, especially when used in childbirth, on the ground that women in particular were condemned to suffer. For does not Genesis 3:16 say "In sorrow thou shalt bring forth children" as a penalty for Eve's eating the fruit of the tree of knowledge? *Some* clergy and devout laity did have their qualms; yet historical scholarship has shown that the supposed fulminations of the church against ether and chloroform in childbed owe more to latter-day secularist agendas than to contemporary fears. Right from the start, Simpson argued for the divine sanction of anaesthesia. When God created Eve, by opening up Adam's side and removing a rib, in Genesis 2:21, He "caused a deep sleep to fall upon Adam". Was not God the first to use anaesthesia in the delivery of another human being?

Within weeks of his first use of chloroform, in December 1847, Simpson published a pamphlet entitled *Answer to the Religious Objections Advanced against the Employment of Anaesthetic Agents in Midwifery and Surgery*, as though expecting an attack. Yet the most significant qualms about obstetrical anaesthesia came from *within* the ranks of the medical profession. Was not childbirth a *natural* and not a *pathological* process, best left to nature? Did not a woman in labour need to be conscious to push and facilitate a natural birth? Only after anaesthesia had caught on did it become clear that the mother's muscular contractions necessary to a successful birth were *involuntary* and would take place naturally even if the woman were asleep.

In the USA, George Rapall Noyes, professor of Hebrew and oriental languages at Harvard, made an original textual study of the relevant Genesis passages and could find no problem. How did

404

the suffering in childbirth differ from God's parallel curse upon the disobedient Adam and his descendants, who were condemned to back-breaking toil? Were steam engines and other labour-saving devices equally wicked, because they made life easier? For more detail on this matter, however, may I refer the reader to Rennie B. Schoepflin's excellent chapter in Ronald Numbers (ed.), *Galileo Goes to Jail, and Other Myths about Science and Religion.*[6]

It was Queen Victoria's inhaling of chloroform vapour at the hands of Dr John Snow (of cholera fame) for the birth of Prince Leopold in 1853 and Princess Beatrice in 1857, however, that settled the matter for most people.

DR JOHN SNOW: FOUNDER OF SCIENTIFIC ANAESTHESIOLOGY

It is hard to overestimate John Snow's importance in the history of the new medical specialism of anaesthesiology. In the same way that, in his cholera researches in 1854, he would start by collecting data about the distribution of sufferers in the affected area, so he would begin to collect data about patient's responses to ether and chloroform. He made a meticulous study of the effects of ether and chloroform upon patients, and came to quantify doses for each anaesthetic, depending on a variety of individual patient and wider physiological criteria. Snow also devised the first proper – as opposed to essentially experimental – sets of apparatus for the administration of each drug, as well as a proper face-mask for chloroform, which was a significant improvement upon the original technique of simply placing a piece of chloroform-dampened cloth over the patient's mouth and nose.

In 1847, Snow had published a study on ether inhalation, though sadly his major work *On Chloroform and other Anaesthetics and Their Action and Administration* was published shortly after his death in 1858. Anaesthesiology, however, would still take some time to become a fully established academic specialism within medicine, although Lord Nuffield's creation of a chair in that discipline within the Oxford Medical School in 1937 would become an international landmark. Anyone who undergoes an operation using a general anaesthetic should thank John Snow for setting anaesthesiology onto that firm scientific foundation upon which it now stands.

MORPHINE, COCAINE, AND THE HYPODERMIC

Once it had been demonstrated in 1846–47 that even the most severe surgical pain could be controlled chemically, a whole new raft of possibilities came to be explored by doctors and laboratory chemists. Could opium be made safe and deaden pain without the need for unconsciousness? Could other alkaloids, such as the juice of the South American coca plant, which both stimulated the brain and had the power to deaden local pain, be brought into proper clinical usage? Then what about Indian "bhang" (*Cannabis indica*), tried by Robert Hooke in the early Royal Society, which seemed to fire the imagination in strange ways and reduce stress? In their natural or indigenous doses, all these drugs could be dangerously erratic, for one never knew the exact strength of a given sample. But the onward march of organic chemical analysis would soon isolate the key ingredients, and make exact and hopefully predictable doses available for medical usage.

The first of these new, isolated active ingredients had come as early as 1805, when Friedrich Wilhelm Sertürner in Germany succeeded in isolating the most powerful of the several natural alkaloids present in the juice of *Papaver somniferum*, the plant from which raw opium is derived. It was initially christened *morphium*, from *Morpheus,* the god of dreams in the Roman poet Ovid's *Metamorphoses.* Administered orally, *morphine* became a much more effective and predictable way of alleviating pain than did laudanum, or poppy juice in alcohol. The main problem, however, was that morphine is extremely addictive and led to people who used it for a painful medical condition becoming dependent. It was not suitable as a surgical anaesthetic, but its chemical isolation opened up possibilities of chemical pain control for the future.

Many people who became opium or morphine addicts were genuine medical cases, though others began to take cheap apothecary's *laudanum* for recreational kicks, or to dull the pain of depression or despair. At around sixpence a bottle, laudanum could give you a score of days or nights of oblivion, make pink elephants appear, or fly you to the moon, whereas a similar expenditure on gin would give you little more than a bad hangover. The opium eater became an early Victorian type, as one sees in Thomas De Quincy's *Confessions of an English Opium Eater* (1822). De Quincy became addicted to the drug in an attempt to alleviate a painful gastric ailment and succeeded, heroically, in weaning himself off it. The starving unemployed Manchester mill workers in Elizabeth Gaskell's *Mary Barton* (1848) took the drug as a cheap appetite suppressant.

Various explanations for opium addiction circulated within the medical profession. Was it caused by an oral craving for the bitter-tasting drug? It was this, along with other factors, that led Dr Alexander Wood of Edinburgh, around 1855, to administer medical morphine by a route that bypassed the mouth – by means of the new hypodermic syringe, invented a decade before by the French surgeon Charles Pravaz. Physicians soon discovered, however, that addiction was not related to oral administration, which would eventually lead to the recognition that addiction was physiological. Whole new lines of research, involving physiology, neurology, and psychology would then begin to open up, to explain, and hopefully treat, drug addiction. But that is another story.

It was Sir Christopher Wren and his proto-Royal Society friends in Oxford in 1656 who first used a primitive syringe to inject a dog with a mixture of sack wine and unrefined poppy extract (see Chapter 13). But this was done in pursuit of deepening our understanding of how the circulating blood affected the brain, rather than for medical treatment purposes. The hypodermic syringe offered a new instrument by which not only morphine but many other drugs as well could be delivered into the human body. As an intravenously injected drug was conveyed straight to the brain by the bloodstream, rather than having to mix with the digestive juices and pass through the stomach linings as when administered orally, its action was not only much faster, but could be controlled with far greater precision.

There was one particular branch of surgery that could not benefit from either general anaesthesia or morphine injection: eye surgery, where it was often necessary for the patient to retain conscious control over their eye during the operation. It was here that the next major anaesthetic agent came into its own: cocaine. Dr Albert Niemann of Göttingen had first isolated this biochemically active ingredient from the South American coca plant in 1860, though its earliest use, like that of chewing the natural coca leaf itself, was as a psychological stimulant.

In Vienna, Sigmund Freud was to play a connecting role between cocaine as a stimulant and as a surgical anaesthetic. By 1884, he was experimenting with cocaine as a way of lifting himself out of his depressions.[7] (The fictional Sherlock Holmes would do the same in several stories, such as *The Sign of Four*, 1890.) Freud had the idea, at this dangerously innocent infant phase of psychological drug chemistry, that cocaine was not addictive in the way that morphine was. So could a morphine addict be weaned off the drug by administering cocaine? Tragically, what happened, so doctors began to find, was that people become addicted to *both* drugs.

Yet the 29-year-old Freud found that when he took cocaine orally, it made his tongue numb. He mentioned the fact to his ophthalmologist colleague Dr Carl Koller, who realized that cocaine could be invaluable for eye surgery, as it would render the eye temporarily impervious to pain, so that a patient could be operated on while fully conscious and able to follow the surgeon's instructions. Cocaine thereby became the world's first effective *local* anaesthetic, opening up all manner of new surgical – and dental – possibilities; and, as with general anaesthetics, researchers began to hunt for new local anaesthetics. The Merck drug company began to synthesize and manufacture pharmaceutical cocaine in 1885.

PEACEFUL SLUMBERS: NEW DRUGS TO COMFORT AND CALM

In the same way that the microscope brought about a pathological revolution, first in the study of the healthy and the abnormal cell, followed by the advancement of the new sciences of microbiology and immunology, so anaesthesia lit a parallel 'fast-burning' fuse of discovery. Between 1846 and 1885, a whole fresh and wonderful range of possibilities in pain management had come into being, starting with William Morton and his ether, to create the new specialism of scientific quantitative anaesthesiology by 1855, and thence to intravenously administered general pain-deadeners such as morphine, and on to local anaesthetics.

As scientific pharmacology, the intellectual child of organic chemistry and biochemistry and the grandchild of old-fashioned apothecaries' pharmacy, began to be driven forward not only by both clinical and commercial laboratories, new lines of fruitful research began to open up. There was an enormous demand for mild, effective, non-addictive analgesics: drugs that might deaden the pain of headache or arthritis, help one to sleep, or enable one to keep going against a stinking cold. On a purely commercial level, vast fortunes could be made from the easy availability of such drugs. For the mentally distressed, there were calmatives, which offered peace, if not a cure, and then, by the mid twentieth century, psychotropic drugs such as lithium (1949) and Largactil (early 1950s).

The first safe calmative was chloral hydrate, introduced by Matthias E. O. Liebreich in 1869. Taken in a drink, such as orange juice, chloral might give an overwrought and mentally disturbed person, or even a normal insomniac, something like a good night's sleep. Unlike an opiate, chloral was not physiologically addictive, nor did it necessarily produce bad hangover symptoms. In many respects, chloral transformed asylum management, by giving a good night's sleep to the patients. Yet chloral was a dangerous drug if one happened to be allergic to it, or if one accidentally took too much. The eminent Irish physicist and mountaineer John Tyndall died accidentally in 1893, when, in the middle of a sleepless night, his sleepy wife inadvertently got his chloral and solvent mix wrong, and he passed out. Hot coffee poured down his throat, shouting, shaking, and banging all failed to revive him, and the accidental extra-strong dose of chloral carried him off.

What probably qualifies as the most widely used proprietary drug of the twentieth century came out of the Bayer laboratories and on to the market in 1899: aspirin. In Chapter 16 we saw how the Reverend Edward Stone announced the fever-breaking powers of willow bark. Then in that escalation of organic chemical discoveries that began in the early nineteenth century, Italian and French chemists discovered that the active ingredient in willow bark was salicin, and the same organic salicylic acid was then extracted from another popular folk-remedy plant, meadowsweet, by Carl Jacob Löwig. Research now turned to designing an organic compound that was easier for the stomach to digest, leading Felix Hoffman to develop acetylsalicylic acid. He was inspired to undertake this line of research partly in an attempt to manage his father's arthritic swelling and pain. The new compound could reduce inflammation and bring down temperature, and in 1900 the German pharmaceutical company patented the formula, and began to market the drug as aspirin (from the Latin botanical name for meadowsweet, *Spiraea*).

Once the scientific possibility of a mild, soothing, analgesic drug became a demonstrated reality, the pharmacological hunt was on to find alternatives or rivals. Today our pharmacies, let alone hospital dispensaries, are replete with prescription and proprietary drugs to control or to reduce almost every kind of pain and trauma. But it was the laboratory chemists who pioneered ether, chloroform, morphine, cocaine, chloral hydrate, and aspirin, who first showed the way.

CHAPTER 24

Glasgow, 1865: Young Jimmy Greenlees Meets Professor Lister

*A*s we saw in recent chapters, once key discoveries were made in medicine, such as smallpox vaccination, the connection between cholera and contaminated drinking water, anaesthesia, and microscopic pathology, further discoveries followed. It was as though, by *c.* 1850, European and American civilization had reached a flash point where a discovery in one field of scientific endeavour led to unexpected ones elsewhere. In addition to the more obvious scientific components of this process, such as microscopes and organized chemical laboratories, other, collateral factors spurred it on. Rapid rail and steamship travel, photography, electric telegraphy, steam-press printing, and the mass distribution of books, newspapers, and periodicals all contributed to the cascade of ingenious discoveries. Ideas were now beginning to fly between London, Paris, Vienna, Berlin, New York, and Boston in a way that would have been unthinkable in 1800.

We saw with smallpox and cholera how it was becoming possible to control the impact of a disease by technology, even before doctors discovered their exact pathological mechanisms. The same would apply to those other great scourges: infection and gangrene. Why did wounds become infected, even when a learned and conscientious hospital surgeon applied his knife to a patient's anaesthetized body? John Hunter had come to the partly correct realization that poking around in an open wound somehow let in some agent which caused the blood to putrefy. Why did a simple bone fracture that did not break the skin invariably heal without septic complications, once the bones had been adjusted into place and the limb splinted up, while a compound fracture that broke the skin, even ever so slightly, often led to septicaemia? Why did a woman who gave birth in a clean, well-appointed bedroom at home, attended by a midwife, normally make an uncomplicated recovery, whereas a poor woman giving birth in a crowded public maternity hospital, attended by medical students, often developed "childbed fever" and died? Was it caused by the thick miasmas believed to lurk in crowded public places? This question would first be pursued by a Hungarian doctor working in Vienna.

PRELUDE: VIENNA, 1847

Dr Ignaz Semmelweis was working in the great Vienna General Hospital at a time when many of his countrymen were agitating for independence from Austro-Hungarian imperial rule. Semmelweis could not understand why the two adjacent maternity wards for which he was responsible had such glaringly different mortality rates. One ward, where the women were attended by doctors and medical students, had a horrendous 29 per cent death rate among the new mothers. In a similar ward, only a few yards away, where the women were attended by nuns and midwives, the death rate was only 3 per cent. This was not so much about women dying *in* childbirth, from complications or blood loss, but rather dying some days *after* a normal healthy birth, when the dreaded puerperal fever suddenly set in.

Semmelweis, who was clearly a very astute clinician, examined the working procedures of both wards. When a woman went into labour in the ward attended by the medical men, a bell was rung in the dissecting rooms nearby, and one or more young men would come, with their sleeves rolled up, and their hands having received only a quick cold rinse at best, to attend her. All of this was a good 40 years before Goodyear in the USA manufactured the first surgical rubber gloves, so that it was normal for all doctors, students, and nurses to handle patients with their uncovered, unsterile hands. But the nuns and midwives in the adjacent ward did *not* dissect cadavers.

What is glaringly obvious to us today – and would appear so to every doctor in Europe or the USA by 1890 – was simply not obvious in "pre-bacterial" 1847. But as Semmelweis realized, one could hardly talk of miasmas affecting one ward but not another only a few yards away. Could the root cause be related to hygiene? Just as John Snow in London would realize that cholera was spread via sewage-contaminated water, so Semmelweis came to realize that students with bits of cadaver clinging to their cold-rinsed hands or under their nails might be conveying something lethal to the birthing women.

This realization was powerfully impressed upon Semmelweis following the puzzling death of his friend and hospital colleague Jakob Kolletschka in 1847. Kolletschka had accidentally nicked his own finger with the scalpel while dissecting a woman's corpse, only to go down a few days later with a generalized infection which killed him. Both prior to his death, then at his own post-mortem, the same pathological characteristics were found to be present as in the woman, who had died of puerperal fever. Yet how could a *man* die of childbed fever?

Recognizing that some connection existed between cadavers and puerperal fever, Semmelweis instituted a cheap and simple procedure. Whenever a medical man heard the bell ring and left his cadaver to attend a living woman in the ward, he must first give his hands a thorough washing in chlorinated water. Chlorine (discovered by Carl Wilhelm Scheele in 1774) was already widely used as a disinfectant and washing agent. Once this disinfecting regime was established, Semmelweis was delighted to see his mortality statistics collapse to the same low level as in the nuns' and midwives' ward.

Semmelweis's hygienic procedures were not popular, and in 1851 he left Vienna to become physician to a maternity hospital in Budapest. In Budapest, he introduced his chlorinated hand-washing regime, and mortalities fell to below 1 per cent. In 1861, he published his results in *Die Aetiologie, der Begriff und der*

413

Prophylaxis der Kinderbettfiebers ("On the Cause, Understanding, and Prevention of Childbed Fever"). Sadly, Semmelweis began to lose his reason, through, it was said, his obsessive concern with puerperal fever and its prevention, and died in a Viennese mental hospital in 1865.

GLASGOW, AUGUST 1865

Even a couple of decades after the introduction of anaesthesia, in the mid 1860s, surgery was still fraught with a major problem: pyaemia, septicaemia, and other gangrenous complications. They could sometimes account for over 40 per cent of deaths in surgery, and in fact the infection problem had even got *worse* after anaesthesia, as surgeons attempted bolder procedures on their now-sleeping patients. More complicated operations took longer to perform, and hence incurred a greater risk of infection.

In 1865, most non-military surgery was orthopaedic, involving the bones and joints of the limbs, as it had been for centuries. With the exception of bladder stones, complicated hernias, cataracts, mastectomies, and skull trephinations, few surgeons cared to open up body cavities, especially major ones such as the abdomen or thorax, even after anaesthesia. Infection would invariably set in and the patient would die.

Especially mystifying and distressing, were compound fractures of the arm or leg bones. A broken tibia, for instance, might just break through the skin to form a small open wound, and a skilled surgeon would reset it and splint up the limb. All should have been well, but about three days later, the wound would often start emitting a nasty smell. Gangrene had set in, from – as we now know – bacteria in the air, on the patient's clothes, or even on the surgeon's hands, getting into the open cut. The limb would have to be amputated from above the break, and everyone would hope and pray that the amputation wound itself did not also become infected, and the patient die. Even in 1860, surgery was fraught with terrible hazards, irrespective of the individual surgeon's skill and compassion.

414

But it would be orthopaedics, and compound fractures in particular, that would lead to a transformation in surgery. On 12 August 1865, an accident patient, a boy named James Greenlees, was admitted to Glasgow Royal Infirmary. He had been run over by a wagon in the street and had sustained a compound fracture in his leg. This was just the type of case that Dr Joseph Lister, Regius Professor of Surgery at Glasgow University and son of Joseph Jackson Lister the microscopist, had been looking for.

43. Joseph Lister, founder of antiseptic surgery, 1902. (*The Family Physician* (*c.* 1905), Vol. IV, frontispiece. A. Chapman collection.)

Lister, like most of his colleagues, was fully aware of medical statistics and was distressed at the 45 per cent surgical fatalities in his hospital. But he had recently read, and been inspired by, Louis

Pasteur's new published researches into fermentation, and by the concept of airborne microbes causing wine or soup to putrefy. Could these microbes also be the agents that caused wounds to putrefy and become gangrenous? How could these bodies be kept out of a surgical wound, which by definition could not be subjected to Pasteur's intense heat?

In that fascinating cross-fertilization of disciplines going on in Victorian science, Lister found medical inspiration in both chemistry and sanitary engineering. He had recently read how the City of Carlisle had begun to treat sewage with a new industrial chemical, carbolic acid, or phenol: a complex organic compound obtained from distilling coal tar, which was a by-product of the gasworks. If carbolic acid could neutralize sewage, then might it not neutralize bacteria? Skilled distillation techniques enabled Lister to extract a refined fluid from the otherwise sludgy chemical.

He then used this fluid in the treatment of James Greenlees's leg. The wound was thoroughly washed with phenol, the bones reset and splinted, and phenol-soaked dressings applied. (Lister later came to wrap the dressings inside a tinfoil overcoat, to prevent evaporation.) Six weeks later, James Greenlees walked out of the hospital, his leg mending beautifully by infection-free "first intention".

Over the next few months, Lister used a similar technique on more compound fractures before announcing 11 cases in *The Lancet* journal for 16 March 1867.[1] It seemed like a miracle that 11 such cases should *all* recover without sepsis. What made Lister's work so uniquely successful was not his use of a chemical to treat wounds: since the distant past, and into the recent Crimean and American Civil Wars, various antiseptic substances had been tried. These included alcohol, wine, vinegar, lemon juice, salt water, and various other substances, usually on the basis of their astringency. None had been especially efficacious. Lister's carbolic (phenol), however, was not simply splashed around, but used in accordance with an exact method: initial cleansing, followed by the creation of a post-operative anti-microbial shield that would be maintained and replaced until healing was safely under way.

Lister's next challenge was to find how this method could be applied to general surgery in the operating theatre. Working on Pasteur's assumption that microbes were essentially airborne, Lister devised a spray machine whereby refined phenol could be pumped to produce an antiseptic aerosol cloud around the operating table. Once surgery was over, the same regimen of carbolic bandages and dressings, as used for compound fractures, would be observed. His results were decisive. Mortalities fell from 45.7 per cent for a group of 35 cases before or without antisepsis, to 11 per cent for 40 cases between 1867 and 1870 with antisepsis.

Not all medical men were convinced. Nineteenth-century British science, including medicine, was famously experimental, based on demonstration of cause and effect, and regarded unsubstantiated theory with suspicion. As Lister's professional opposite number in Edinburgh, John Hughes Bennett (of leukaemia diagnosis fame) pointed out, Lister could not scientifically demonstrate which "little beasties" actually caused surgical sepsis. As we saw above, even the microscopes of Hooke and van Leeuwenhoek two centuries before showed that the human body hosted legions of "animalcules" which did no harm. A major drop in surgical mortalities was all well and good, but Pasteur's and Lister's "beasties" need not be connected with them. Even so, Lister's method and spray came to be increasingly used, with dramatic results, at home and abroad, including in the field hospitals of the Franco-Prussian War of 1870, yet just missing the American Civil War. A demonstrated connection between specific microbes and specific diseases, however, would only be forthcoming after 1879, when Robert Koch began to publish his research into specific strains of bacteria and the clearly defined diseases with which each was associated.

FROM ANTISEPTIC TO ASEPTIC SURGERY

One serious critic of Lister's methods was the young Birmingham gynaecologist, Robert Lawson Tait. Eighteen years younger than Lister, Lawson Tait had developed a remarkable operative procedure which enjoyed astonishingly low infection rates. It was not based upon any theoretical premises, but it did involve a level of general cleanliness that was much in advance of that of most hospital operating rooms. It was a generally accepted truism among surgeons, however, that operations performed in private houses had much lower infection rates than those performed in crowded public hospitals.

What Lawson Tait seemed, by chance, to be practising in the 1870s was what by the late 1880s would become the proper technique of surgery: not *anti-septic*, but *a-septic*. In this method, there were no sprays to irritate the eyes and hands of operating theatre staff. Instead, there was a surgical environment that was so scrubbed clean and disinfected as to be *sterile*. Flat polished surfaces with no places to harbour germs and even new instruments without the bone or ebony handles of traditional surgical knives and saws. Everything was designed to be heated or autoclaved prior to use. Even the surgeon, instead of wearing an old frock coat to protect his good street clothes from blood and pus, would now roll up his shirt-sleeves, don a clean, hopefully sterile, gown, and even put rubber gloves over his well-scrubbed hands – as would his assistants and nurses. By the 1890s, a radical new approach to surgery and a new breed of surgeons were coming into being: men who did not see surgery as confined to the operating theatre, but as working in conjunction with the research laboratory, and with a growing array of technological aids.

THE NEW SURGERY

These new surgeons could be just as much at home looking down a microscope or working in a laboratory as they were at the operating table, as a whole host of factors were coming into play and setting surgery itself within a wider medical context. These "medical scientist surgeons" were coming to work within an international culture of interconnected scientific disciplines, which included physics, chemistry, optics, precision engineering, instrument design, and photography. After its invention in 1880, even the new electric light bulb entered surgery, not merely for flood-lighting the operating table, but also for making visible those parts of the body to which it was impossible to routinely direct sunlight.

Among their ranks was the Berliner Christian Albert Theodor Billroth, the Englishman Sir Frederick Treeves, the Americans William Stewart Halstead and the doctor–surgeon brothers William and Charles Mayo, the Canadian Sir William Osler, and many more, on both sides of the Atlantic. This was the age of the new integrated *clinic*: where operating theatre, hospital, and pathological laboratory occupied a single, invariably expanding, site. Halstead's new Johns Hopkins clinic in Baltimore and the Mayo brothers' clinic in Rochester, Minnesota, soon to become world famous, were but two in the USA alone.

As the nineteenth century progressed, surgery, just like the rest of medicine, would become increasingly technology based, as new devices bestowed a precision upon diagnosis and technique that transcended, yet complemented, old-fashioned experience.

Histological and biopsy studies involving sophisticated biochemical tests and microscopy, increasingly in conjunction with photographic recording, plus Claude Bernard's discovery of the importance of good body hydration and electrolytic levels, were fast becoming normal parts of scientific surgery.

What were the new instruments? They included René Laennec's post-1816 stethoscope, enabling a doctor to "auscultate" or sound out healthy and diseased body organs and cavities; Carl Wunderlich's medical thermometry, making fever and crisis points easier to diagnose, predict, and manage; Adolf Kussmaul's "Gastroscope", 1868; and the American Howard Kelly's "Rectoscope" of 1895. It was said that Kussmaul got the idea of the "Gastroscope" for stomach interior inspections from watching a circus sword-swallower, for if a man could let a sword down his oesophagus into his stomach, then could not a doctor do the same with a scientific instrument containing lenses, mirrors, and a light source? Likewise, Scipione Riva-Rocci's sphygmomanometer of 1896 not only enabled blood pressure to be monitored non-invasively, but much could be learned thereby about the action of the patient's heart and overall cardiovascular system. Hermann von Helmholtz's "Ophthalmoscope" of 1851–54

419

revolutionized eye surgery (especially after cocaine local anaesthesia in 1885) by enabling a doctor to safely examine the interior structure of the living eye, while E. Siegel's post-1864 "Otoscope" did the same for the living ear.

Surgical safety before, during, and after the operation was also greatly increased after 1900, following the Viennese immunologist Karl Landsteiner's work on the chemistry of blood antigens, and why blood transfusions sometimes worked and sometimes did not. Landsteiner came to identify specific antigen groups within blood, which he named "A", "B", "AB", and "O", which, from a prior laboratory test, made it possible to establish which group of blood could be given to a patient. Without transfusion, many modern-day surgical operations, which can take many hours to perform, with inevitable high levels of blood loss, would be impossible. But it was Wilhelm Conrad Röntgen's discovery of "X" rays in 1895 that would complete the great diagnostic and surgical transformation of the nineteenth century, and lay the foundations for what would follow in the twentieth.

What all of these discoveries did in the wake of anaesthesia and antiseptics was to make surgery safer and more predictable in its outcome, for these and subsequent discoveries would greatly reduce the hit-and-miss component of more traditional surgery, by providing the surgeon with something of a diagnostic road map of the patient's condition even before he took up his scalpel.

THE NEW OPERATIONS

The first hitherto inoperable region of the human body upon which the 1880s' surgeons began to work with a growing certainty was the abdomen, especially in the wake of Billroth. We saw in Chapter 22 how Sir William Arbuthnot Lane (of heart resuscitation fame) and his colleagues were by the 1890s performing "heroic" colonectomies in the hope of curing abdominal cancer or severe constipation, but one of the first conditions for which a surgical cure came to be found was typhlitis. Like most afflictions, it had been around for centuries under different names, such as the "iliac passion", and

John Woodall had left a good account of its symptoms in his *The Surgeon's Mate* (1617). But it would be Reginald Heber Fitz, of the Harvard Medical School, who would provide an accurate pathology and an enduring name for the illness in 1886: appendicitis.

At first, in the early 1880s, surgeons such as Robert Lawson Tait preferred to open up the patient, drain the inflamed and perforated *vermiform appendix* of pus, then stitch up, letting nature take its course. By the 1890s excision of the inflamed appendix was coming to be practised, yet when Sir Frederick Treeves operated on King Edward VII in 1902 – occasioning the postponement of the Coronation – Treeves very wisely stayed on the side of safety when operating on his elderly and distinctly corpulent patient, simply draining the royal appendix rather than removing it. Appendectomies were still dangerous, even by the early twentieth century, especially if post-operative infections set in, for antibiotics still lay some 40 years in the future. In 1907 the future prime minister David Lloyd George's beloved daughter Mair Eluned died from appendectomy complications, and it was touch and go when Sir Winston Churchill had appendicitis in 1922.

The brain had been a subject of fascination since antiquity, though it was Thomas Willis's *Cerebri Anatome* (1664) that really began scientific neuroanatomy. Yet with the exception of trepanning the skull to relieve pressure, usually after a severe head injury, the brain itself was out of bounds surgically. But in the wake of the "new surgery", things started to happen.

Using "Listerian" antiseptic techniques, the Glasgow surgeon Sir William MacEwen successfully drained a brain abscess in July 1879, though when John Rickman Godlee attempted to excise a tumour in 1884, the patient died following complications. MacEwen was later to turn down the new Johns Hopkins surgical chair, but it would be that same medical school which would become in part the training-ground for Harvey Cushing before his return to his alma mater, Harvard. In 1905, the 36-year-old Cushing began his work on pituitary brain tumours, and on that gland's connection to "gigantism", or excessive growth. Cushing was to pioneer what was really a *system* of brain surgery that would be fundamental for

421

the new specialism of neurosurgery. It involved a silent, disciplined operating theatre, the use of electrical instruments to precision-cauterize, specially designed silver arterial clips, constant blood-pressure monitoring, and other control factors.

Cardiology and cardiac surgery as exact medical disciplines would also trace their origins to the late nineteenth and early twentieth centuries. The endocardiograph would build upon earlier cardiological research with the stethoscope, by employing new electrical technology to measure and record heart arrhythmias. First developed by Augustus D. Waller at St Mary's Hospital, London, in 1887, the endocardiograph would come to be used alongside the sphygmomanometer, making it possible to establish the concept of essential hypertension, or high blood pressure. It was becoming possible by 1900 to define normal and abnormal cardiac states in precise, measured terms: all of which would be essential before any serious surgery on the heart was possible.

Although early twentieth-century surgeons had operated on the pericardium and elsewhere around the heart, it was, in many respects, World War I that opened up the possibilities for cardiac surgery, simply as a result of the necessity of treating battle injuries. In 1920, the British surgeon Sir Charles Ballance reported in *The Lancet* a case from 1917, when his French colleague Paul Delbet removed a 1.5 cm piece of a German grenade from the right ventricle of a French infantryman. The soldier was later discharged in good health. Ballance himself opened up the chests of wounded men and held their beating hearts in his rubber-gloved hand.[2]

In 1919, Sir John Bland-Sutton, stimulated by recent injuries in the trenches, published a fascinating historical paper in *The Lancet* entitled "Missiles as Emboli"[3] (from Greek *embolus*, "stopper"), where he reviewed a series of historical and modern cases in which accidental wooden splinters, Napoleonic War musket balls, and contemporary German rifle bullets had entered the arteries and been carried by the blood flow, and even entered the heart, including a soldier who not only suffered a leg amputation, but also survived 12 days with a rifle bullet lodged in his right ventricle.

Then in May 1925, Sir Henry Sessions Souttar operated on a

19-year-old woman, designated "L.H.", at the London Hospital. She suffered from a long-standing stenosis, or hardening, of her mitral valve, which controls the blood flow from the left auricle into the left ventricle of the heart – a condition extremely rare in one so young. Souttar opened up her chest, cutting through ribs and causing a five-inch-square section of her left rib-cage to fold back – a bit like opening the cover of a book – to give full access to her beating heart and thoracic cavity.

In a minutely detailed account of his procedure in *The British Medical Journal* (1925), Souttar tells us that he next made an incision enabling him to insert his gloved finger into the left auricular chamber of her heart. Then, having explored the interior of the auricle by touch, he pushed his finger downwards through her mitral valve, in the direction of the blood flow, and was able to remove some of the obstructing material by finger-pressure, and thereby improve the valve's action. "L.H.", the daughter of a labourer, was next sent to the country, where after six weeks in bed, followed by three months of convalescence "she declared that she felt perfectly well, although she still became somewhat breathless on exertion".[4]

From the work of colleagues in Great Britain, Europe, and America cited by Souttar, one gets a good impression of how rapidly cardiac surgery was moving even by 1925, though its achievements then were extremely modest compared with what would be viable 50 or 90 years ahead. The same would apply to eye, ear, nose, throat, and other rapidly advancing branches of surgery at that time.

It seems incredible that when the elderly John Collins Warren applied his scalpel to the neck of the sleeping Gilbert Abbott in October 1846, the purely curative aspects of many branches of medicine and surgery had not significantly changed since the days of Galen. So one wonders what that great man would have thought had he been able to revisit earth in 1925, when not only anaesthesia, but also bacteriology, anti- and a-septic techniques, cellular pathology, pharmacology, and the industrial manufacture of pure drugs and medicines had become "normal" wherever Western civilization had gone. What would J. C. Warren have made of the healing arts today?

CHAPTER 25

The New Professional Healer

*E*ven by the time that anaesthesia, vaccination, public hygiene, and early microbiology were beginning the transformation of medicine across all its branches, the British medical profession was a curiously heterogeneous body. There were the Doctors and Bachelors of Medicine trained in the universities and the great teaching hospitals, and scientifically trained surgeons. The old Worshipful Company of Surgeons, chartered in 1800 to become the Royal College of Surgeons, taught, examined, and legally licensed highly trained practitioners.

The Company of Apothecaries did the same for *proper* druggists and pharmacists, as well as venturing into wider aspects of general medical practice. Many of the men who served the nation as general practitioners by 1850 were not university-trained, but had qualified under the Apothecaries, starting from a good grammar school education. Indentures (or being apprenticed) to a surgeon-apothecary would follow, then a spell of listening to lectures and ward-walking in a recognized London or provincial teaching hospital. The appropriate examinations would then be sat, and after undergoing an oral or viva voce examination by a group of established practitioners, the candidate, if successful, would be licensed to practise. Though the training was not academic in its focus, such men could go on to become excellent *practical* GPs, rich in common sense and experience. Many who qualified under the Apothecaries in 1850 could still be in practice in 1900: deeply respected pillars of their communities.

Yet it was still not uncommon for a charitable and scientifically minded country clergyman to "physic" his rural parishioners if no one else were available, as did a sprawling diaspora of more modern Bartholomew-Fair-type quacks. These could include the shabby frock-coated itinerant dentist, who would – if you were lucky – give you a whiff of chloroform and grab hold of the offending tooth with pincers, or sell you a wonder cure for sixpence. Or he might be a fashionable homeopath, a mesmerist (or hypnotist), the owner of an up-market health spa promising water cures, or a blatant fraud. The respected Glasgow physician "Dr" Edward William Pritchard was only found to have *bought* his MD, from a foreign university, after he had poisoned his wife and mother-in-law in 1865.

THE *MEDICAL ACT* OF 1858

That heterogeneous landscape was to change, however, as the Act of 1858 brought in much tighter regulation. The General Medical Council, with its registrar, was now responsible for compiling a register of those persons deemed qualified to take patients and practise medicine and surgery. It was necessary for them to have undergone a formal medical training, which could be a university MD, British or foreign. Even if a surgeon of the older generation had received his training via indentures or apprenticeship, it needed to be formally recognized by the Royal College of Surgeons or other professional body. By 1858, however, most surgeons would have received a university-level education prior to being registered. But now it was possible, from a glance at the "Register", to tell who was a doctor and who was a quack.

Of great historical importance, Dr Elizabeth Blackwell was entered on the Register: the first academically trained and qualified female doctor. Born in Bristol in 1821, making her a year younger than Florence Nightingale, she and her family emigrated to the USA in 1832. Fascinated by medicine, and a devourer of medical books, she finally succeeded in being admitted to the Medical School of Geneva, New York, qualifying in 1849. She came back to Europe to begin practice, but inevitably met with much resistance, so in 1851 she returned to New York, where she became a successful physician. Then in 1868, Elizabeth once again returned to England, where she would reside for the remaining 42 years of her life. Relatively few women would become formally trained physicians in the nineteenth century, in spite of centuries of unofficial practice. But the Medical Register from 1858 onwards would define members of the "medical profession", be they male or female.

One physician who declared war on the now legally identifiable quacks was Dr Walter Rivington, in *The Medical Profession. An Essay* (Dublin, 1879). Rivington catalogued a series of blatant deceptions, some leading to prosecutions under the *Medical Act*. One rascally "Dr" John William Foster adopted the name of a real physician whose credentials had somehow remained on the Register after his death. But one now risked a £20 fine – a hefty sum – for illegally using the title "Doctor".

Bogus "medical botanists", homeopaths, and itinerant healers are all in Rivington's book, but one of the nastiest – who escaped prosecution, for he did not claim to be a doctor – was a folk quack who convinced a Spanish sailor that his hernia was cancer. The unfortunate sailor was treated with strong corrosives, which destroyed his abdominal wall and eventually killed him. By 1879 most hernias could in fact be easily cured surgically, using chloroform and Lister's carbolic spray, or non-surgically with a truss.[1]

HOMEOPATHS, WATER-CURERS, AND VICTORIAN ALTERNATIVE MEDICINE

Mid nineteenth-century professional medicine was exercised not so much by the matter of who was and who was not a proper doctor, as by what a formally trained physician should or should not do. Should he distance himself as much as possible from the homeopaths and the hydropathic water-curers? One sees a similar scenario developing in 1840 to that of the eighteenth century, when the term "quack" was often used by trained doctors of colleagues who innovated, used unorthodox techniques, or specialized, such as "Chevalier" John Taylor, the oculist.

The Liverpool Medical Institution, which was that burgeoning city's Medical Society, underwent a great deal of internal debate in the 1840s and 1850s about the scientific status of homeopathy in particular, especially as there were formally trained medical men within the Institution's ranks who already practised it. Dr Frederick Hervey Foster Quin, stated that he had achieved remarkable successes by administering minute doses of camphor and arsenic in cases of cholera, rather than the usual heavy doses of these and other metal-based drugs.

Homeopathy had been born out of academic medicine in the 1790s, for its founder Dr Samuel Hahnemann had received his training at Vienna, Leipzig, and Erlangen Universities, and was certainly no quack. But despairing at the often horrendous effects of the conventional heavy toxic drug usage of the day, he sought for a new way forward, especially when treating what were then mysterious s diseases such as fevers, diabetes, and mental illness. Hahnemann noticed that when, for instance, a healthy person was given quinine, it often generated symptoms that were similar to those of malaria, the disease it was used to cure. Should the physician, therefore, rather than blitzing a disease with toxic "counter-irritants" such as mercury-based calomel, arsenic, or antimony, administer a drug to the sick that, in a normal healthy person, would generate the symptoms of the illness, so that "like would cure like"? This was one of principles of homeopathy (from Greek *homoios*, "like", and *pathos*, "symptom" or "illness"). The other was that a drug increased in potency in direct relationship to its dilution in a given volume of water. Enough academically trained physicians were willing to try out, or practise, homeopathy by the early Victorian age as to cause ructions within the medical profession.

Another class of healer whose ideas provoked controversy were the water-curers or hydro-therapists: men who argued that large quantities of water, either passing through or enveloping the body, or both, were the key to health. This rationale lay at the heart of those therapeutic mineral springs resorts, such as those of Bath, England, and Baden-Baden, Germany. But better coaching roads by 1800 and the spreading railway network by 1845 added greatly to their number, popularizing places like Harrogate, Cheltenham, Leamington Spa, and Tunbridge Wells. There was also the new fashion of sea-bathing, and British seaside resorts, following in the wake of late eighteenth-century Brighton, simply burgeoned, creating the first beach culture. Whitley Bay, Yarmouth, Bognor Regis, Eastbourne, Torquay, Tenby, Southport, and Blackpool began to boom. Now a modest middle-class invalid, consumptive, or delicate person could go to the coast, walk, breathe bracing sea-air, swim, and *feel better* in the context of a holiday. By 1890, the industrial masses of Glasgow, Manchester, Leeds, and elsewhere were scraping together their shillings for third-class train tickets and cheap seaside digs.

Books were coming to be written on the medicinal virtues of specific resorts, depending, as in Hippocrates' *Airs, Waters, and Places*, on types of breeze, sea temperatures, or the mineral content of the waters. The impeccably qualified Dr Augustus Bozzi Granville, MD, FRS's *The Spas of England and Principal Sea-Bathing Places*,[2] dedicated to Her Majesty Queen Victoria, provided both a learned and a readable guide to English watering-places, complete with advice about railway connections!

Also in 1841, Sir James Clark, MD, FRS, "Physician in Ordinary to the Queen, and to the Prince Albert" published the third edition (1829) of his *The Sanative Influence of Climate*, or, as appeared on the spine of the 1841 volume, "Clark on Climate", describing the health-giving properties not only of English but also of Mediterranean, Atlantic, and even West Indies and Australasian resorts. With reputations like those of Clark and Granville behind it, who could question the curative power of airs and waters?

Why did homeopathy and watering-places work, medically speaking? Largely, I would suggest, because the patient's body was likely to respond favourably to a withdrawal of concentrated doses of heavy metal drugs, and the wider joys of a holiday resort. I think the status of homeopathy is more questionable today, when bacteriology, pharmacology, and psychotherapeutic drugs render many formerly "mysterious" diseases curable or controllable.

NURSING, THE NEW MEDICAL PROFESSION: SARAH PODGER, MARY SEACOLE, AND FLORENCE NIGHTINGALE

Florence Nightingale was, by any standards, a formidable lady. Born in Florence, Italy, in 1820, the daughter of secure, wealthy Hampshire and Derbyshire landowners, and with her own private allowance of £500 per annum, she was beautiful, brilliant, and unconventional. She had no wish to marry, her favourite reading was medical books and tables of statistics, and she enjoyed manipulating legions of facts and figures. Ladylike socializing bored her. As a girl, she wanted to be a nurse, but such a low-class ambition was beyond the pale to William and Frances, her parents, even though, as she always claimed, her vocation had come to her as a personal call from God.

We will return to Florence's career shortly, but let me make it clear that this formidable gentlewoman entertained no doubt of the fact that in reforming nursing, just as in the British Army, it was the duty of the genteel officer-class to lead, and the "other ranks" to obey. Florence had no time for disobedient, rough-and-ready lower-class nurses with ideas of their own, any more than she had for "blue-stockings" (i.e. feminists) or "advanced thinking" radical women of any class. Florence was a fascinating combination of the daring "one-off" original and the instinctive traditionalist; and heaven help anyone who got in her way – woman, man, or Satan himself!

One woman of whom Florence would certainly have disapproved, one suspects, was Sarah Podger, the "Receiving" (or senior accident and emergency) nurse in Aesculapius Scalpel's (Edward Berdoe's) autobiographical novel *St Bernard's*; or, to be more exact, she would have disapproved of the real-life 1850s' woman on whom Sarah was clearly based.

Aesculapius, whom we met when looking at body-snatching, tells us that Mrs Podger was over 50, small, tough as they come, and loved her tipple: "... lor' how are yer goin' to nuss [nurse] a capital hoperation on tea...?" She was jolly, kindly, and popular – yet thoroughly efficient in her way.[3]

44. A "Nightingale" Nursing Sister. (*The Family Physician* (*c.* 1905), Vol. IV, p. 292. A. Chapman collection.)

Sarah Podger ruled her large and important division of "St Bernard's" with an iron hand. She was instinctively deferential to the young gentlemen of the hospital, yet taught them more of how to do off-the-cuff diagnoses, manage trauma, bandage gushing wounds, and handle broken bones than most of the academic lecturers. "No member of the hospital staff was more interested than she in the pass list of the College of Surgeons; she felt it as a reproach against her teaching when any of them failed", so Aesculapius tells us.[4]

Florence Nightingale would have regarded Sarah Podger, with her jolly familiarity and partiality to a glass of something, as an undisciplined, uppity servant in need of putting in her place. But Sarah was one of a breed, and it was practical, salt-of-the-earth women like her who kept our hospitals functioning on a day-to-day basis, as their better-trained and educated great-granddaughters still do today.

One real-life woman of whom Florence did not really approve (in spite of her faintly praising remarks to the French chef Alexis Soyer) was Mary Seacole, who clearly had a good bit of Sarah Podger in her. Mary was born in Jamaica in 1805, the daughter of a free woman of African descent and James Grant, a Scottish officer. She was clearly an entrepreneur with a natural instinct for looking after people. Like her mother, Mary was a "doctress", or native herbal folk-healer. By the outbreak of the Crimean War in 1853, she was proprietress of a successful hotel and restaurant, New Blundell Hall, at Kingston, Jamaica, with cash of her own, including a handsome sum made from a South American gold-mining venture.

Now Mary Seacole – following her marriage – was a staunch British patriot, with a love of her now-deceased father's homeland and of the British Army, especially its brave Scottish regiments. It was her reading of the plight of British soldiers in the Crimea in 1854, combined with a need to secure her gold-mining assets in London, that prompted the entrepreneurial Mary to act. Seeking further challenges, she sailed for her beloved Britain – at her own expense – and tried to establish some useful contacts. This was, in fact, her second visit, for she had spent most of 1821 in London. Eventually, however, she made her own way to the Crimea and, digging further into her own pocket, established her patriotically named "British Hotel", primarily for wounded officers but also for ordinary soldiers.

This was to make her a legend and a popular heroine back in Britain. Here, "Dame Seacole" and her assistants did not simply nurse the British wounded back to health in a purely clinical fashion, but provided home comforts as well. She had no objection if a convalescent crippled hero wanted his 'baccy, a bottle of brandy, a game of cards, and a jolly song in the makeshift ward. We today might call it good therapeutic psychology; after all, she was a British officer's daughter and knew the type.

Mary in her own distinctive way became a darling of the British Army and the subject of a popular song published in *Punch*: "Dame Seacole was a kindly old soul…"[5] Her style of nursing, however, was more akin to that of Sarah Podger than that of "the Lady with the Lamp".

Florence Nightingale had to fight hard to pursue her perceived God-ordained vocation – most of all against her socialite mother and older sister Parthenope, though one senses that her scholarly, more solitary father understood her better. It had been her father who made sure that his conspicuously intelligent daughter received an academic training in classical and modern languages, history, and philosophy, which was unusual for a girl at that time. She had also been permitted to travel to Kaiserswerth, Germany, to learn nursing with a Lutheran sisterhood of charitable ladies, and had nursed genteel patients in London. But her high social status and fearlessness gave Florence social access to members of government and senior civil servants. As the Crimean chaos began to unfold, she began to use her contacts to effect, so that in the wake of the war's medical organizational scandal, in October 1854, she led a team of 38 nurses, including 15 Roman Catholic nuns, to the Crimea. Some of the nurses were ordinary nurses, while others were ladies. They lived and dined separately, as would have been the case in any household in England.

What awaited Florence in the vast, rambling makeshift barracks hospital at Scutari was appalling. Filth, sewage-contaminated drinking water, raging fevers, and gangrene abounded – and that was where men were already under formal medical, surgical, and army orderlies' care. Knowing empirically about dirt and contamination, though well over a decade before Pasteur, Lister, and Koch, she set to work, virtually terrorizing the medical establishment.

Her trained army of nurses, living under Florence's relentless discipline, set about reorganizing Scutari. Infection-related death rates fell from 40 per cent to 2 per cent. And while her wounded soldier patients enjoyed none of the free and easy joys of Mary Seacole's "British Hotel" (Florence, after all, had hundreds more to care for than Mary), she won their (literally) *undying* devotion. Clean water, clean bedding, clean bandages, systematic washing, the careful handling of wounded amputation stumps, and round-the-clock care for every wounded man, be he an officer or a private, did the trick. It was a lifesaving system of care no less stringent than Lister's antiseptic system would be after 1865. The slipshod or the irregular were now anathema.

When it came to quality of nursing care, Florence the high-born lady was a relentless egalitarian. This in itself may have derived from her own – for the time – unorthodox take on Anglican theology. She appears to have been convinced that *all* people would be reconciled to God after death, and could not believe that prostitutes or criminals would burn eternally, any more than would a bishop. Florence loathed the theological and sectarian censoriousness of her day, and could not believe that a loving God could be as small-minded as many religious people tried to make him. She even preferred non-religious hospitals (though Christian in ethos), feeling that religious ones were often too obsessed with saving souls to give adequate attention to proper care. Christ alone was the one who saved, not theologians. It may seem curious that this tough-minded English gentlewoman was also an instinctive mystic. With Roman Catholic nuns in Bermondsey (another charitable sisterhood in the East End), Florence would later study Christian mystical writers such as Mother Julian of Norwich and St Angela, seeing a mystical relationship with God as transcending dogmatic theology.

Following her return from the Crimea, Florence found herself famous. The Nightingale Fund, which had been established and subscribed to in her honour in November 1855, placed almost £45,000 at her disposal soon after her return to England. She used it in July 1856 to establish the Nightingale Training School, attached to St Thomas's Hospital. This would not only develop into the world's first major school of nursing, but would set out to create a new type of nurse: young women who would now be well educated and thoroughly trained in all aspects of their profession, with a big focus upon cleanliness and *compassion*. Her *Notes on Nursing* (1859) would become the bedrock of the new nursing profession.

Fond of Nurse Podger as he had been, Aesculapius Scalpel, as his career developed over the 25 years or so covered by his novel, became aware of the limitations of such "nusses". Writing reflectively of a time more consonant with the 1880s than the 1850s, Aesculapius confessed: "In nothing have hospitals improved more of late years than in their nursing arrangements",[6] and here he had Florence Nightingale and women of her calibre in mind. One of the problems with the great London hospitals, he reflected in the 1880s, was their growing focus upon academic teaching and scientific research rather than healing the sick poor.

The "Nightingale Hospital" in London's impoverished East End, and others run by sisterhoods of non-sectarian Christian ladies, in London's Commercial Road and elsewhere, focused upon good practical care for the poor and not upon writing scientific research papers. This was no longer some Lady Bountiful making her token visit to an approved, scrubbed, respectable pauper hovel, but thoroughly committed ladies, usually trained in Nightingale methods: true sisters of mercy, living among and tending the poor, in a truly Christian fashion, theology notwithstanding. Nurse Podger was effectively dismissed from "St Bernard's", to be replaced by nurses of the new Nightingale type, bringing in a higher level of dedication and turning a job into a profession. It is hard to overestimate the impact that Florence Nightingale had, and continues to have, upon the care for the sick and vulnerable.

But just as nursing was being transformed after 1860, so was the profession of physician. Aesculapius tells us that the great surgeon Sir Astley Cooper was reported to have said to a student audience at Guy's Hospital, sometime before his death in 1840, that he regarded purely *medical* (not surgical) treatment as "infamous and disgraceful", for "The art of medicine is founded on conjecture and improved by murder."[7] Yet this bleak medical landscape had itself been transformed by 1910.

SIR WILLIAM OSLER ON THE NEW PHYSICIANS

A new breed of physician was beginning to emerge in the late nineteenth century, as the germ theory, cellular pathology, new drugs, biochemistry, and much more were transforming the often "infamous and disgraceful", conjectural art of medicine into something more of a humane science. One shining example of this new breed was the Canadian Sir William Osler. Born in 1849, of Cornish maritime stock, Osler trained as a doctor at McGill University, Montreal, before moving on to chairs at Johns Hopkins, Baltimore, and finally becoming Regius Professor of Medicine at Oxford.

Osler was a great teacher and a great clinician. Fully aware of the advances going on around him, he still taught students how to be acutely observant at the bedside; in an age when instrumental diagnostic tests were still limited, the doctor should learn how to *read* his or her patient and pick up clues.[8] But the work done at the bedside must be backed up by the resources of the hospital laboratory and research library. The patient was not a *specimen*, but a distressed human being, and Osler's combination of humanity, astute observation, and hard science – spiced with humour – was to form a new generation of doctors in the twentieth century.

In some ways, what epitomized the speed of medical advance in Osler's lifetime was his own death during the terrible Spanish 'flu epidemic of 1919. Osler died in Oxford; less than half a mile away from his house, in the Dunn School of Pathology, a discovery would be made in 1940–41 which would fundamentally transform the treatment of infectious diseases and give new life to billions of people across the globe.

CHAPTER 26

The Wonderful Century

Over the past century, medicine in all of its branches has progressed beyond anything that anyone in 1900 could reasonably have imagined. Yet in so many ways, the key foundation stones for that rapid progress were already securely in place by the end of the Victorian age. What happened, therefore, in the twentieth century (and continues into the twenty-first) was really a fusion process, as techniques, technologies, and lines of research coalesced to yield an ever-expanding harvest of treatments and cures that, between them, have transformed the human condition wherever Western science had spread.

THE DRUGS THAT HIT THE SPOT

Central to the growth of post-1880 medicine has been the ability to discover the biochemistry of a given disease, and then to design and engineer a drug that would hopefully cure it. This approach goes back to Edward Jenner's purely empirical vaccination against smallpox, and came to find a firmer scientific footing in Pasteur's and Koch's work on developing successful (for rabies) and still unsuccessful (for cholera and tuberculosis) vaccines. But the first engineered drug to truly "hit the spot" was Paul Ehrlich's arsphenamine "Salvarsan 606" for the treatment of syphilis. Ehrlich had long been impressed by the way that a living body could generate antigens that would destroy an invading bacterium without harming the host. So could such "magic bullets" be engineered in the laboratory?

Soon after 1900, Ehrlich and his team at Frankfurt-am-Main were working on the *Spirochaeta pallida* (now called *Treponema pallidum*) bacterium which causes syphilis. They then heard that colleagues at the Liverpool School of Tropical Medicine had discovered that an organic arsenic compound, atoxyl, was effective in destroying certain related infections in laboratory animals, but that after a while the pathogen, the parasitic *trypanosome*, a corkscrew-like protozoon (Greek *trupän*, "to bore"), developed a resistance to atoxyl and proliferated. Yet, thought Ehrlich, if a compound could be developed that would destroy *all* the body's trypanosomes in one overwhelming dose, or series of doses, then they would never live long enough to develop immunity, and the disease would be obliterated with no harm done to the patient. This was the logic behind a whole new way of thinking about immunity and resistance, and would become the earliest form of chemotherapy.

By 1909, Ehrlich and his colleagues Alfred Bertheim and Sahachiro Hata had meticulously prepared hundreds of biochemically modified versions of the organic arsenical compound, and by 1911, their magic bullet finally hit its target: the long, threadlike spiral of *Spirochaeta pallida* at the heart of syphilis. Their successful preparation was number 606, and the resulting drug would become "Salvarsan 606".

Here, in Salvarsan 606, one finds many of those features that would become standard practice in medical science's war against invading bacteria, and later, viruses. This was a "cat and mouse" process, involving meticulous laboratory biochemistry, immunology, antigen studies, microbiology, and, once the magic bullet had been isolated in the test tube, the wholesale up-scaling of precision industrial manufacture essential to mass availability, and all working smoothly together. It was a practical art form no less beautiful, harmonious, and integrated as an intellectual structure than Michelangelo's Sistine Chapel, a Beethoven symphony, or Milton's *Paradise Lost*. And the same procedures pioneered in the treatment of syphilis would be used to wage war on all manner of subsequent infections, including tuberculosis, AIDS, and, most recently, Ebola.

Another great breakthrough would be penicillin.

438

PENICILLIN AND ANTIBIOTICS

While one can see great creative science, medical and otherwise, as similar in its intellectual elegance to great art, so it also has parallels to a good detective story. Just like a good "whodunit", medical research invariably involves making sense of a tangle of hidden clues, false leads, blind alleys, and sudden revelations. Who, in 1925, would have dreamt that 20 years down the line, mouldy bread would help to trigger a fundamental transformation in the cure of bacterial infections?

In his biochemical "engineering" of atoxyl arsenical compounds to attack the *Spirochaeta* of syphilis, Paul Ehrlich had shown how a "magic bullet" could be made to hit a specific protozoon, thereby inventing chemotherapy. But Salvarsan 606 was powerless against bacterial groups such as the *cocci*, which lay at the heart of a multitude of infections, including the often deadly pneumonia which can be caused by *Streptococcus*. So could a chemotherapeutic agent be engineered that would hit this family of bacilli?

Sir Alexander Fleming, professor of bacteriology at St Mary's Hospital, London, went on his summer holiday in 1928. On returning, he found that several glass dishes in which he was cultivating *cocci* bacteria for experimental purposes had been blighted by a mould that had somehow gained admittance. He identified the mould as *Penicillium notatum*, which was related to common bread mould. Fleming discovered that this mould produced a fluid that was capable of destroying the bacterial strains which caused some of the most common infections. This *anti-biotic* substance he named "penicillin". In the late 1920s, however, Fleming was unable to produce the penicillin in sufficient concentrations to be of any value in practical therapeutics. That would happen in Oxford a dozen years into the future.

Then in the mid 1930s, Gerhard Domagk in the IG Farben Laboratories, Germany, discovered that laboratory mice injected with the synthetic dye "protosil red" acquired a resistance to streptococci bacteria. It seems that the protosil red, once in an animal body (but less so *in vitro*, or in the test tube) turned into sulphanilamide, one of the sulphonamide group of chemicals. As things stood at the time, sulphonamides were found to have only a limited therapeutic effect, although in 1939 Domagk received the Nobel Prize for Medicine for his ground-breaking work. Medical chemistry was entering an almost "Alice in Wonderland" world where organic compounds were found to have strange new powers when ingested by a living creature, such as altering the course of dangerous diseases.

Recognizing the curative potential of chemotherapeutic agents following the discovery of the sulphonamides, Sir Howard Florey, Ernst Boris Chain, and the biochemist and ingenious apparatus-deviser, Dr Norman Heatley (PhD Cambridge), in Oxford's Dunn School of Pathology, began to pay new attention to penicillin. Penicillin's antibiotic power and low toxicity made it worth pursuing, especially as World War II had just broken out. The great drawback, however, lay in concentrating the drug from its natural mould in sufficient quantities to make it viable, and this is where skilled laboratory biochemistry was crucial.

On display in the Museum of the History of Science, Oxford, is a motley collection of old hospital bedpans, modified biscuit tins, and, my favourite, a large tin canister whose surviving label proclaims its original contents: "The Noted Sheep & Lamb KYLEMAL Drench". These vessels had all been adapted by the ingenious Heatley to collect and concentrate the first penicillin supply.

On 25 May 1940, sufficient penicillin had been concentrated for an *experimentum crucis*. Eight laboratory mice were injected with the deadly streptococcus bacterium, and only four of them with penicillin. When Florey and Chain returned to the lab after dinner in Heatley's Lincoln College, later that evening, the four penicillin-injected mice were well, whereas their uninjected brethren were dead. It was a decisive experiment. Recognizing its importance, the kindly Heatley took the dead mice home, wrapped them in little shrouds, and gave them burial in his garden: four little heroes of scientific progress! Then on 12 February 1941, penicillin was tried on a seriously ill human being in Oxford's Radcliffe Infirmary.

440

This was a reserve policeman, Albert Alexander, whose seemingly minor cut from a rose thorn had turned very nasty. The drug had a dramatic effect upon him, though the penicillin could not be concentrated in a sufficient quantity to save his life. (Around 1995, I had a chance conversation with Robert Smith, a long-retired Oxford policeman, who back in 1941 had known PC Alexander – an interesting point of contact with penicillin's first experimental subject.)

Since the war was underway, however, the industrialized manufacture of penicillin moved to the safety of the USA, and by the time of the D-Day landings in June 1944 enough penicillin was available to treat every wounded Allied soldier. Then penicillin, and its subsequent developments, began to transform the prospects of patients suffering from diseases such as double pneumonia. (Soon afterwards, the new antibiotic streptomycin would do the same for tuberculosis sufferers.)

The world's debt, from patients in great teaching hospitals to the sick child in the African village, to the discoverers of penicillin and subsequent antibiotics is incalculable. I had the honour to know, and to dine with, an eminent Oxford cellular pathologist and oncologist who had known several of the early penicillin researchers in the Dunn School. This was Sir Henry Harris FRS, who died in 2014, aged 89. In 1999, Sir Henry summed up penicillin history before the Royal Society, stating: "Without Fleming, no Florey or Chain; without Chain, no Florey; without Florey, no Heatley; without Heatley, no penicillin."[1]

Unfortunately, Norman Heatley was not included in Florey's and Chain's Nobel Prize in 1945. But in 1990, Oxford University helped to straighten the record by awarding the 79-year-old biochemist – who was not a physician – its prestigious DM (*Doctoris in Medicina*) degree. I was present at the June Encaenia ceremony in Oxford's Sheldonian Theatre, and when the chancellor conferred Heatley's DM, the audience erupted into spontaneous, sustained applause!

CANCER: RADIOLOGY, CHEMOTHERAPY, AND
BODY SCANS

There would be a long wait between Virchow's, Hodgkin's, Hughes Bennett's, and others' recognition of the cellular basis of cancer and the medical profession's ability to produce significant remissions or cures. Numerous new collateral discoveries would have to be made, in specialist sciences unimagined in 1860 – such as radiology, molecular immunology, chemical engineering, and even digital computer technology – before serious therapeutic inroads into cancer could be made. None could have predicted in 1860 the existence of, and biologically fundamental role played by, *deoxyribonucleic acid,* or DNA. Four Cambridge graduates, Francis Crick, James Watson, Maurice Wilkins, and Rosalind Franklin, made this momentous discovery, which led to a fundamentally new understanding of the chemical and molecular structure of living organisms in the 1950s; and pretty well all branches of medicine continue to develop fresh therapeutic insights on the strength of their discoveries.

One question continues to haunt us today, although DNA science has enabled us to make enormous progress: What actually causes cancer cells to proliferate in the body? Well into the twentieth century, many physiologists believed that bacterial infections were to blame for initiating cancer cell metastasis. Then later, they blamed viruses, which in a curious way pushed research in the right direction for the wrong reason. That direction led to the discovery and exploration of the then proto-science of radiology.

Iodized radium salts had been tried on throat cancers in France as early as the 1870s, but it was Wilhelm Conrad Röntgen's discovery of X-ray radiation – "Dr Röntgen's Rays" – that really began radiology as a medical science. We have seen how X-rays had been used as a new weapon in the surgeon's armoury, and the potential of radiological agents in anti-cancer treatments was soon seized upon. One reason for this was the efficacy of X-ray bursts at curing or reducing facial lupus lesions: a tubercular-related and hence a bacterial disease. Then, still in France, X-rays were found by 1905 to induce limited remissions in cases of stomach cancer.

So if X-ray radiation destroyed bacteria and also changed or even destroyed cell structures, could it not be used to curb cancer?

One great breakthrough came in 1898, when Marie Curie and her physicist husband Pierre succeeded in isolating the new elements polonium (named in honour of her native Poland) and, most importantly, radium, from pitchblende and other substances. For these discoveries she, her husband Pierre, and their colleague Henri Becquerel received the Nobel Prize for Physics in 1903. Radium's effects upon the human body had become evident in 1901, when Becquerel discovered mysterious burns on his stomach: he had been going around Paris with a vial of radium salts in his waistcoat pocket!

So what was this baffling stuff "radioactivity": a powerful atomic particle stream emitted by Röntgen's X-ray apparatus, along with radium and uranium salts, detectable on sensitive electrical instruments in the physics laboratory, and which did strange things to living tissue? An explanatory model became available following Joseph John ("J. J.") Thomson's discovery of the electron in 1903. Atomic physics was now beginning to impact upon medicine.

Much was tried, and much was claimed, especially for radium, in the first decades of the twentieth century. Vials of radium salts were attached to cancerous lesions, and no one at the time had any idea how much collateral damage was going to be caused by splashing radiation of various kinds around as though it were water. But it was Henri Coutard in Paris after 1922 who was the real father of "radiotherapy". Coutard came to realize that while a blitz dose of X-rays was likely to be counter-productive, short, sharp bursts of radiation, extending over a period of time, had beneficial effects upon cases of laryngeal cancer. He was claiming a 23 per cent success rate by 1935 – resounding success for the time – and his "fractionated" method of delivering radiation doses was to become standard.

From this approach, proton beam therapy would develop. A powerful electrically generated magnetic field would be used to fire a focused proton beam at a tumour: *protons* being electrically charged subatomic particles. This would bring about molecular changes within the cancer cells, yet could minimize the collateral damage to normal healthy cells caused by normal X-ray bursts. By destabilizing the cancer cells, one could give the healthy cells a chance of overcoming the aggressive invaders and hopefully destroying them. When we routinely speak of "the battle against cancer", we often do not realize how close to the mark we are: for modern cancer research and therapy are just as rooted in logistics, strategies, supply systems, and knock-out tactics as is any general's campaign when confronting an invading military force.

After radiology furnished the first prong in the strategic attack on cancer, chemotherapy in the late 1940s supplied the other, in an example of how one humanitarian tragedy provided the key to saving innumerable lives thereafter. On the night of 2 December 1943, the Luftwaffe sprung a surprise raid on Bari harbour in southern Italy, where American and Allied ships were unloading supplies for the liberation of Italy from Mussolini's *Fascisti*. A bomb exploded in a cargo ship, the *John Henry*, carrying toxic nitrogen mustard gas, killing and injuring many Italian civilians who were downwind of the gas cloud. Yet when Lt Colonel Dr Stewart F. Alexander of the US Army examined blood from bodies *post-mortem*, he saw something interesting under his microscope: a lymphoid and myeloid suppression, or a diminution of white blood cells.

Could nitrogen mustard gas, therefore, inhibit cell division; especially of fast-dividing malignant cells, such as caused lymphoma? Louis S. Goodman and Alfred Gilman subsequently found that nitrogen mustard compounds could shrink induced lymphomas in laboratory mice. It now seemed that the highly toxic gas which had killed and maimed countless thousands of men in the trenches of World War I could become a global lifesaver, if its complex molecular structure could be re-engineered to minimize its toxicity to patients. Before long, wonderful remissions were being reported for lymphoma and leukaemia patients upon whom the re-engineered nitrogen mustard molecule was tried. Yet sadly – evolution being what it is – the fast-dividing cancer cells evolved ways of reasserting themselves, terminating the remission, and killing the patient. But cancer research was clearly on the right track. The hunt was now on

444

for other anti-cancer compounds. Two women scientists, working separately, Lucy Wills and then Jane C. Wright, found respectively that folic acid compounds and methotrexate were effective, and in 1951, Jane C. Wright used methotrexate to bring about remissions in solid breast tumours.

By the mid 1960s, additional chemical discoveries had made it possible to devise new anti-cancer strategies: the use of cocktails of drugs, working together and in sequence, that could hit the biochemistry of cancer cells from different directions. By 2000, a sophisticated strategy of "adjuvant" therapy was being used in the war against cancer, as surgery, chemotherapy, and radiotherapy would be used together to first take out the cancer surgically, then clean up the affected site, using chemicals and radiation bursts, in a precise shock strategy reminiscent of an SAS strike upon a terrorist stronghold.

Fighting on the side of the cancer patient was a growing body of new technologies. Perhaps the first of these was epidemiological: the use of statistics as a part of preventative medicine. Sir Richard Doll, of London then of Oxford, pioneered the use of mass statistical surveys of smoking habits in 1950, and soon came to the stark conclusion that cigarette smoking and a high risk of lung cancer went together. Doll also found similar statistical connections between cancer and asbestos.

Then there was the digital revolution from the 1960s onwards. Computer technology changed our understanding not just of cancer, but of virtually every other branch of medicine as well, beginning with mass number crunching, then, from the 1970s, increasingly sophisticated diagnostic imaging. Magnetic Resonance Imaging (MRI) and Computer Tomography (CT) scans now made it possible to image the inside of the human body with minimal risk to the patient. Tomography (from Greek *tomos*, "slice") made it possible for X-ray "slices" to be made through a living human body, which the computer could then assemble into a moving image on a screen. Between them, MRI and CT scanning made it possible to see all sorts of things, from babies in the womb to metastasizing cancer cells in parts of the body far from the original tumour site.

These technologies are invaluable not just for primary diagnoses, but as guides to the physician or surgeon in the very process of administering treatment.

Sophisticated diagnostics apart, all of these treatment techniques, radical surgery, chemotherapy, and radiotherapy, along with bone marrow and platelets transplantation, are all massively invasive, and while they do work, with increasing efficiency, their impact upon the bodies and minds of both patients and loved ones is often traumatic. In many ways, the new movement in cancer treatment is towards genetic, immunological, and epidemiological modes of therapy, rather than the long-haul healthy tissue-damaging toxic treatments of the past century. New therapies are aimed at stimulating the body's own defences to overcome aberrant cells: "magic bullets" rather than the "shotgun-blast" followed by a long convalescence of many present-day radiological and chemotherapeutic techniques.

ADJUSTING THE BODY'S OWN CHEMISTRY, PHYSICS, AND ENGINEERING

In addition to paving the way towards antibiotics and cancer treatments, the nineteenth century's priceless gifts to twentieth- and twenty-first-century medical science included an ability to use laboratory science first to understand, and then to treat, a variety of problems in bodily chemistry and physics.

Diabetes had been recognized, symptomologically, since antiquity, and Thomas Willis had described – and named – *diabetes mellitus* (Greek *mellitus*; "honey-sweet"), a type of diabetes, in *Pharmaceutice Rationalis*, 1674. Yet it was Frederick Banting and his colleague Charles Best, in Toronto in 1921, who elucidated the mechanism of the disease and then went on to develop a cure. As always in scientific achievement, however, they stood on the shoulders of earlier researchers – in this case, Paul Langerhans, whose 1869 discovery of puzzling groups of cells in the pancreas led to their being subsequently named the "Islets of Langerhans". It was known that when a laboratory dog had its pancreas removed surgically, the animal could no longer digest fats and proteins,

446

developed diabetes, and died. By 1921, however, it was already known that a key organic chemical was present in the Islets of Langerhans, and Edward Sharpey-Shafer had called it "Insuline" from the islet clumps (Latin *insula*, "island") from which he had extracted it.

Banting, Best, and James Collip next prepared a refined extract from the Insuline, and found that when it was injected into a diabetic dog, the dog immediately got better. The new "insulin" preparation enabled the dog's body to metabolize carbohydrate and remain well so long as the injections continued. Then, on 11 January 1922, insulin was injected into young Leonard Thompson, a diabetic boy close to death in the Toronto General Hospital. He made a seemingly miraculous recovery and was soon sent home in good health, though in need of lifelong insulin injections. Just like Edward Jenner's James Phipps and Joseph Lister's James Greenlees, Leonard Thompson joined that band of children whose amazing recovery would proclaim to the world that a major life- and world-changing medical discovery had been made. Yet when Banting, Best, and John Macleod, Toronto's professor of physiology, shared the 1923 Nobel Prize, James Collip, the technically adept biochemist (as with Heatley with the 1945 Prize for penicillin), was not included.

Nutrition chemistry, in many ways, went back to William Prout's and Justus von Liebig's early nineteenth-century work on what would be styled proteins, fats, and carbohydrates. But what was wrong when a diet failed to nourish adequately in spite of its apparent substance? This was not a new question, for as Dr James Lind discovered in 1747 following the world's first proper clinical trial conducted aboard HMS *Salisbury*, sick sailors quickly recovered from scurvy when given lemon juice, although their diet was otherwise rich in dried pork, beef, and ship's "biscuit" bread. So what were these hidden substances in food that gave vitality and life?

It would be the London-based Polish-born Casimir Funk, standing on the prior research of Axel Holst and Frederick Gowland Hopkins, who would identify a biochemical component in rice husks possessing the ability to prevent beriberi. Funk would style it a "vitamine", from the Latin word *vita*, "life". Once it was realized that healthy foodstuffs contained key, *vital*, chemicals, the hunt was on to identify more.

It was Sir Edward Mellanby who made what was perhaps the breakthrough discovery in therapeutic vitamin chemistry. Mellanby found that when he fed growing puppies on a bland diet of porridge, they developed rickets: a bone-growth abnormality first described clinically by Francis Glisson and John Mayow back in the seventeenth century. Mellanby found, however, that when he included cod liver oil in the puppies' diet, they grew up with normal limb bones. This led to the identification of a key component in the oil, which after 1922 came to be known as Vitamin D. It would also be discovered that Vitamin D was absorbed not just through foods, but also from sunlight through the skin, suggesting an explanation of why rickets, or bandy legs, were much more common among the poor, where children often lived in dark, overcrowded slums, and were sent to work in factories on leaving school at 12. (My late grandfather, Albert Platt, went to work in the Manchester cotton mills aged 12 in 1903, and had bandy legs, though otherwise fit and strong, dying in 1984.)

Vitamins introduced a new concept into medicine, however: the realization that a disease did not have to be caused by a pathogen or cell-malfunction, but could also derive from a trace-substance deficiency in a seemingly adequate diet. But once the significance of the first vitamins came to be recognized, and Adolf Windhaus of Göttingen began to elucidate their complex structures, others were rapidly identified by the sophisticated biochemistry of the twentieth century. These included Vitamin C, which was the biologically active component in those citrus fruit juices which Dr Lind had discovered cured scurvy in sick sailors. By the 1950s, free school milk, orange juice, balanced school dinners, and cod liver oil for children, dispensed through Great Britain's new National Health Service, had effectively eradicated rickets, scurvy, and undernourishment from the post-World War II baby boom generation.

The late twentieth century also saw the dramatic impact of both bio-technologies and bio-engineering as part of the armoury of practical therapeutics. Who, in 1950, could have imagined that DNA, genetics, tissue-typing, new anti-rejection drugs, and a variety of ancillary surgical techniques, such as computer-combined scans, would make it possible to routinely replace worn-out or diseased body parts with transplanted or synthetic organs? The world was first alerted to these potential wonders when, in December 1967, Dr Christian Barnard in Cape Town, South Africa, transplanted the heart of a young road-accident victim, Denise Darvall, into the chest of a 53-year-old man with terminal heart disease.

The University of Minnesota Surgery Department, with which I have as a medical historian had the honour of a long lecturing association, became a pioneer in the new technology of organ transplantation, including the development of "antilymphocyte globulin" (ALG) and other anti-rejection drugs, at the hands of people like Dr John Najarian. Where Barnard first showed the way with the heart in 1967, so other surgeons now routinely transplant lungs, livers, and other organs. The tragedy involved here is the invariably sudden accidental death of a young person – somewhere in the world – who has the right bio-compatible tissue-type organ for a seriously sick person in, say, Minnesota, Stockholm, or Tokyo. A patient might be told at 10 hours' notice that a new liver, or pancreas, packed in ice, is currently being flown over from Venezuela or New Zealand. The transplant surgeons are ready to spring into action the moment it arrives in the hospital – even if that is 3.00 am. (I was once at a surgeons' dinner in Minneapolis St Paul, where one woman said she would have to leave immediately, as her bleeper told her that a liver had just arrived from Latin America, and she was off to fit it into her patient.)

In many ways, however, the future may lie with wholly artificial organs, made of bio-compatible polymers and titanium, or even bio-engineered animal parts, such as the aortic and pulmonary valves of pigs. All manner of such devices are already routinely in use, such as sliding-ball heart valves, stents to open up a blocked vein or artery, and polymer-based artificial lenses for people whose eyes have become clouded by cataracts.

There is, too, a growing array of prosthetic joints for people who have become crippled by arthritis. Complex joints, such as that of the knee, can require a long convalescence, but ingenuity being what it is, this in turn has led to innovation. Oxford became a major centre for such research after 1976, as the Irish engineering scientist John O'Connor and the surgeon John Goodfellow began to work together at the Nuffield Orthopaedic Centre, Oxford, to develop a minimally invasive procedure for knee replacements. Detailed X-rays would indicate beforehand exactly where the knee-bones were worn or damaged. Then, instead of replacing the whole bone structure and installing especially engineered titanium joint, relatively small titanium plates would be attached to the worn knee-bones through a keyhole incision, taking the pressure off the damaged bone, to make metal move painlessly upon metal. The recipients of the "Oxford Knee" could now be up and about within days, rather than weeks, thus minimizing stress and trauma, saving time and money – and making everyone happy.[2]

It is hardly surprising that some surgeons now joke that they are really up-market plumbers and motor mechanics. But what about mental illness?

WHO AM I? SCIENTIFIC MEDICINE AND THE SOUL

One of the most value-laden branches of medicine has always been that of the relationship between the body, mind, and soul. We saw how Thomas Willis, the devoutly Christian founder of neurology in the seventeenth century, like René Descartes and, into the early twentieth century, Sir Charles Sherrington (the discoverer of the synapses which lie at the heart of the brain's information processing system), worked within a broadly "dualist" model, in which the conscious, volitional "mind" interacted with the anatomical and physiological hardware of the body, to explain who we are and how we act. However, others, such as Julien Offray de La Mettrie, Auguste Comte, and, in the present day, Baroness Susan Greenfield,

450

took a different line, seeing the soul as an "epiphenomenon" or side-effect of a purely physical neurophysiology.

Brain scans and digital technologies over the last half-century have revealed things about the living brain which no one could have imagined in 1900, as electrical engineering, computer imaging, neurochemistry, and psychology have come together to tell us what goes on inside our heads when we think, feel, sleep, suffer a stroke, or develop dementia.

Yet the question remains: "Who am I?" I do not *feel* that my mind (or "soul") is just an illusion, generated by a busily working cortex. Nor do I think that brain-evolution-based models of morality are especially convincing. Why, instead of saving our own Western skins, and gloating over the destruction of the less fortunate, do we feel compelled to risk our own physical and economic security to help the distant stranger whom we are never likely to meet, such as Ebola victims in East Africa or starving children whose communities have been blighted by famine or barbaric sectarian wars?

This is a *vast* subject, where deeply held personal, religious, and ideological beliefs, for and against the soul, meet the facts of modern scientific research. But it is also an area where the *interpretation* of those facts depends so heavily on the worldview of the interpreter. Neuroscience is not black and white in the way that dealing with bacteria and cancer cells and transplanting a heart valve are. Suffice to say, as far as my own personal beliefs go, I can find no better summing up than to quote the Victorian Prime Minister Benjamin Disraeli, who, on another subject in a debate in 1864, declared: "I, my lord, I am on the side of the Angels."[3]

We saw in previous chapters how most "mad-doctors", up to the nineteenth century, attempted to treat the mentally ill within the context of returning a temporarily disturbed mind to normality. In the late nineteenth and twentieth centuries, however, this model was frequently sidelined, or even dismissed, especially by the profoundly influential Sigmund Freud and the Vienna-based "alienist" school of psychoanalysis. In this often aggressively secular interpretation of personhood, people are rather like broken mirrors: shattered pieces, some capable of hanging together to form functioning individuals and others fragmented into dysfunctional neuroses and psychoses. Bad parental relationships, sexual repression, and awful inner tensions ruin each one of us to one degree or another, and our *subconscious* mind (Freud's truly great realization) becomes a cauldron of dark forces which bubble up in our dreams, and lead to odd behaviour or full-blown mental illness. So to Freud, we are all damaged goods: but some more damaged than others.

Freud's younger Swiss colleague Carl Jung went on to develop his own system of psychoanalytic philosophy around 1912, based on universal symbols, motifs, and a collective subconscious, which was less bleak than Freud's. Both Freud's and Jung's psychological systems claimed to have empirical foundations, based upon what people on the psychiatric couch had said; yet with no precise yardsticks of measurement, it was by no means impossible that patient's utterances could be interpreted in the context of the analyst's own personal beliefs, fears, and traumas. But Freud's analytic ideas in particular exerted an enormous influence on social and psychological policy on both sides of the Atlantic over many decades of the twentieth century.

What really changed the Western world's treatment and management of mental illness, and for the better, were the new psychoactive drugs that had the capacity to affect a person's brain chemistry. These drugs could bring about mood and other psychological changes, and make it possible for many people who would otherwise have lived out their days confined in lunatic asylums to live relatively normal lives in the community. Drugs such as chlorpromazine (1951–53), an anti-psychotic used to treat psychoses and schizophrenia; imipramine (Tofranil) (1958) and Prozac (1987), both antidepressants; and lithium (1949–54) for bipolar disorder all changed the whole landscape of psychiatric treatment. These and other drugs could have nasty side effects, while others, such as amphetamines, could also lead to erratic behaviour when not used under medical supervision. Yet the development of psychoactive drug chemistry after 1950 would be transformative to the lives of many mentally ill people, and

I know several personally who have derived great benefit from these drugs.

(A consultant anaesthetist friend of mine once remarked when we were discussing psychoactive drugs: "Ah, lithium: the oldest drug in the universe. The element formed in the first few minutes after the Big Bang.")

Yet irrespective of all the psychological and neurological developments of the twentieth century, from the neuroses of the Viennese School to the post-1950 brain scans and psychotherapeutic drugs, through all the mists of psychoses, dementia, and drug-induced states of mind, there is something ticking away inside each human being. And only when that elusive thing which we call a "soul" is set free from its fleshly bonds does a person become a cadaver.

CONCLUSION: OUR MODERN DUTY OF CARE

Good medical care has always been expensive, be it in a medieval hospital with its buildings, nursing staff, and cooks; a Victorian scientific teaching hospital; or an integrated National Health Service research, treatment, and aftercare facility. Whether such facilities come from private charitable benefactions, or from the taxpayer, all need substantial endowments to pay for buildings, upkeep, equipment, salaries for the staff, and scholarships for training the rising generation. As therapeutic capacity has expanded beyond all recognition since 1850, so the costs have soared: sometimes at a faster rate than a nation's economic growth. Yet, as most people simply cannot afford to access modern medicine as a commercial product, how *do* we pay the doctors?

I would suggest that, in addition to the historical, predominantly Christian charity institutions mentioned in previous chapters, Great Britain has played a major innovative role here. This really began with the "self-help" movement of the early Victorian industrial era, when groups of working men set up their own "sick clubs" and savings banks, maybe organized through their trade union or chapel, to build up a modest capital sum. From this, the club might "buy" a certain amount of a local doctor's time, so that sick people could get "free" medical help, while at the same time a patient unable to work might draw a small sum from the fund with which to pay the rent and help feed his family.

But as costs rose, this was clearly inadequate, leading the charismatic Welsh wizard Chancellor of the Exchequer, David Lloyd George, (who had already introduced Old Age Pensions in 1908), to see the passage of the first National Insurance Act through Parliament in 1911. This Act required that all workers earning less than £3 per week should pay four pence into a government-backed fund. Their employers were also obliged by law to pay a further three pence into the same fund, while the government paid in another two pence. So each worker now had nine pence per week going into an official insurance fund.

Lloyd George's Act had a transformative effect across the board. It took poorer doctors "off bicycles and put them into Ford cars", thus providing a more secure career structure for young, slum-working professionals. Now the working man or woman could see a local panel doctor when ill, while at the same time receiving 10 shillings per week to help out with family necessaries, plus any separate sick club money for which they might be eligible. These were tiny sums, even for 1911, but hopefully they kept the workhouse at bay.

Over the decades, the scheme was improved upon, but it took an inevitable battering after the blitzing of London and many other British cities in World War II. After 1945, a new approach to health care was clearly needed, resulting, in 1948, in the creation of Aneurin Bevan's National Health Service, aimed at providing a "cradle-to-grave" health-care package for every Briton. In the early days, things seemed relatively straightforward: free spectacles, false teeth, trusses, cod liver oil and concentrated orange juice, nutritious powdered baby milk, splinted broken bones, the removal of infected teeth, tonsils or adenoids, and operations for hernia, appendicitis, or a prolapsed uterus would make the nation fit and well. But if you had advanced cancer, heart disease, chronic mental illness, or liver or kidney failure, then it was just palliative care.

Since 1948, all of these diseases have proved curable or manageable, so that a person saved from a heart attack in 1970 might be stabilized with kidney failure in 1990, and then need hospice care through a protracted terminal cancer in 2005, all at an escalating cost. Fortunately, growing national wealth has, in many cases, kept pace with improvements to the nation's health; but the basic fact remains that expanding medical capacity can never really be cost-effective in pure accountancy terms, as today's cures become tomorrow's support problems.

Politics and finance apart, this, quite simply, is the price we have to pay for being a civilized and humane society, where we not only recognize a need to look after our own sick and vulnerable, but where the charities that burgeoned after World War II (such as the Quaker-initiated Oxfam in 1942, Christian Aid after 1953, and secular Médecins sans Frontières, 1971), combined with Western governments' aid programmes, acknowledge a duty of care to the *very* distant stranger. This was seen most recently in the provision of combined resources to quarantine and help the victims of the 2014 West African Ebola epidemic, and, by July 2015, to develop a successful vaccine against that disease. We recognize an obligation to help and improve the lives of people so desperately poor, that there is no conceivable way in which they could ever reimburse, in purely economic terms, the countries working on their behalf.

Where does this urge in the West to help the vulnerable, both at home and abroad, come from, be they in St Bartholomew's or another European charity hospital in 1200, or in an isolation tent in Sierra Leone in 2014? Could it have something to do with the fact that all these nations, from Poland to California, are joint heirs of the Judeo-Christian system of ethics, values, and beliefs? Even though latter-day secular philosophies may deride the transcendent dimension of this tradition, the values originally expressed in ancient Jewish teachings about the duties of the rich towards the poor, and then in Christ's actions and parables, such as that of the Good Samaritan, have become part of the very fabric of Western thinking about human responsibility. Then when we add to the ethical the Greco-Roman pagan tradition of *scientia*, or organized

455

natural knowledge, as filtered through medieval European and Arabic scholarship, we get a sense of how our medical traditions came into being. These ethics of care, combined with the traditions of observation, systematic investigation, research, and an advancing science-based technology which itself stands upon a secure foundation of economic enterprise and resourcefulness, give us an understanding of how physicians and medical scientists went from plagues to progress, and on to create the modern world.

Cataract Operation Performed by a Traditional Shaman Surgeon in a Village to the East of Agra, Northern India, c. 2010

*T*he following account was kindly given to me by Prianka Bose, currently (2016) a pre-medical undergraduate student at Oglethorpe University in Atlanta, GA. Prianka studied with me on a course at St John's College, Oxford, over the summer of 2014, when still an undergraduate student at Rhodes College, Memphis, Tennessee.

At the time of the cataract operation, some years previously, she was visiting some very poor villages in northern India with her father, physician Vinay Bose, and helping to bring about some simple improvements to the quality of life of the villagers. "This village was devoid of anything past the late 17th century, but the people were happy, content with their way of life", Prianka writes.

The cataract operation was performed by an elderly shaman, who apparently read Sanskrit, for he consulted beforehand a leather-bound copy of the *Sushruta Samhita*, by the great sixth-century BC Indian surgeon Sushruta. His patient was a man of around 50 years old, and the operation took place in the shaman's small house. Prianka was brought in as a volunteer assistant to lay out the shaman's instruments on his tray, steady the unanaesthetized patient's head during surgery, and use her fingers to keep his eye wide open during the 25-minute procedure.

Beforehand, the patient's arms and legs were immobilized "with ropes and chains". "Then, the Shaman began to gesticulate weirdly. His strange movements made his fingers look [as though] they were appearing and disappearing in tandem with his chanting. The rhythms of his voice had a lulling, trancelike effect. The patient became still and I could feel the tautness of his muscles relax... I suspect that the Shaman used hypnotism as a sedative, and the procedure was quick and simple."

Then it was time for Prianka to hold the patient's eyes open, and surgery began. She writes:

1.   *The needle was inserted into the corneal limbus region of the eye at an external angle near the pupil, making a "popping" noise. Watery fluid left the eye.*

2.   *The needle was then used to break the cataract into little pieces, working the cataract into the posterior chamber of the eye.*

3.   *The tip of the needle incised the mature lens capsule, and then the patient blew out of one nostril while I held a metal syringe, that the Shaman used to suck the suffusion.*

4.   *The opaque cataract substance displaced the lens to the side, revealing an opening through the pupil that allowed more light through. Then the patient's eye was bound with a linen cloth soaked in a ghee, egg-white, and milk mixture.*

The patient was very poor, had no money, and clearly was not capable of paying for surgery at a hospital. Prianka assumed that he had an occupation which required clear vision, hence his willingness to submit to surgery. The shaman-surgeon appeared to have had an education, probably in traditional Indian medicine, and seemed able to read Sanskrit, and the walls of his tiny house cum operating theatre were "lined with trinkets and old, leather-bound books".

The villagers, however, were poor, materially speaking, distrusted modern technology (though Prianka's father was helping to improve the water supply to homes in the village by the installation of an aquafer and plumbing), and "lived lives built on mysticism, whimsy, and necessity".

45. Cataract couching needle, sucking needle, linen, milk, and egg white and ghee, used by a local folk-surgeon in a remote village in northern India, *c.* 2010. Drawn by Prianka Bose.

APPENDIX 2

Stents and Tents

*I*n modern-day surgery the term "stent" is applied to a generally flexible appliance inserted to widen a stenosed, or narrowed, natural passageway. Although "stents" are used in many departments of surgery, they are perhaps most commonly inserted into blocked or narrowed blood vessels with the intention of restoring a normal blood flow.

But where does the word "stent" come from? The received wisdom is that it derives from Charles Thomas Stent (1807–85), a London dentist who pioneered the use of a rubber-based "gutta-percha" compound to make plates for false teeth. Yet I have found no explicit connection between Mr Stent and the modern appliances which supposedly bear his name.

The term "tent" possesses a precise surgical connotation going back to at least 1403, when John Bradmore used a set of graded wooden, honey-dipped "tents" to reopen an arrow wound in the face of young Prince Hal (the future Henry V), so that a surgical instrument could be inserted to remove an embedded arrowhead (as described in Chapter 10 of this book). The *Oxford English Dictionary* cites several references to the surgical usage of the word "tent", such as from Lanfranc's *Cirurgie* (*c.* 1400) and John of Arderne's *Treatise on Fistula in Ano* (*c.* 1425).

All of these references are to an appliance being used to widen a narrowed aperture in the human body: in these medieval cases, an opening on the outside of the patient, as the internal cardiovascular system would not really become safely operable until the mid twentieth century.

Though not mentioning the word "stent" in an explicitly medical context, the *OED* cites usages of the word dating back to the fourteenth century, where a "stent" was used to expand or widen something. There was the use of poles upon which to stretch out fishing nets, or to lay out a sheet of something. "Stent" could also be used in taxation, as when describing the *extent* of a property to be assessed. Yet whether in medical, legal, or general usage, the words "tent" and "stent" imply some device employed to expand or open up something.

So could the name for an expanding device in surgery extend back long before Charles Thomas Stent was born?

Notes

Full bibliographical details for books and articles cited in the notes may be found in the "Further Reading" section under the relevant chapter heading.

CHAPTER 1
1.   Proverbs 30:19.
2.   See Breasted, *Edwin Smith Surgical Papyrus*.
3.   Ezekiel 30:2.
4.   2 Kings 9:20.
5.   Singer and Underwood, *Short History of Medicine* (see General Reading), p. 12.
6.   1 Samuel 16:14–16; Daniel 4:32–34.
7.   2 Kings 20:7, Isaiah 38:21.
8.   2 Chronicles 21.
9.   Numbers 21:5.
10.   *Hippocratic Writings* (London: Penguin), p. 262.
11.   Hippocrates III, *On Wounds, Surgery, and Fractures*, Section XXXV, p. 265.
12.   Aristotle, *De Generatione*, Bk. II, Section IV, 739, 22–7 (transl. Peck).

CHAPTER 2
1.   Wilmott, "Gladiators in ancient Rome…"; Bomgardner, *Story of the Roman Amphitheatre*.
2.   Mattern, *Prince of Medicine*, pp. 83–84.
3.   Jackson, *Doctors and Diseases*, p. 66.
4.   Jackson, *Doctors*, p. 93.
5.   Jackson, *Doctors*, p. 126.
6.   Matthew 9:20.
7.   Celsus, *De Medicina*, Bk. III, pp. 272 (Latin), 273 (English).
8.   Pliny, *Natural History* (London: Penguin), pp. 224–67.

CHAPTER 3
1.   *The Pardoner's Tale*, ed. F. N. Robinson (London, 1957), lines 889–90.
2.   Albucasis, *Surgery*, p. 154.
3.   Albucasis, *Surgery*, pp. 558–60.
4.   Albucasis, *Surgery*, p. 552.
5.   Hippocrates III, *On Wounds, Surgery, and Fractures*, Section XXXIII, pp. 259–63.
6.   Albucasis, *Surgery*, p. 784.

CHAPTER 4
1.   *The Works of Aristotle*, vol. VII, Bk. XV, Section 6, 911b–912b.
2.   Bacon, *Opus Majus*, ed. Jebb (1733), transl. R. Smith, *Compleat System of Opticks*, p. 352.
3.   See Appendix 1.

4. Albucasis, *Surgery*, pp. 252–56.
5. Albucasis, *Surgery*, p. 256.
6. Albucasis, *Surgery*, p. 254.
7. Barker, *The Brontës*, pp. 506–07.

CHAPTER 5

1. Chapman, *Gods in the Sky, Slaying the Dragons, Stargazers* (General Reading).
2. Full title in Further Reading.
3. De Vrend, *The Old English Herbarium*, pp. 1–233 and 234–72.
4. *Boke of Secretes*, Fol. 8 v., no. 13.
5. *Boke of Secretes*, Fol. 14 r.
6. J. H. White, *A Book of Beasts*.
7. Prologue, line 677.
8. Edition by Luisa Cogliati Arano, 1976.
9. Welsford, *The Fool*, p. 41.
10. Morley, *Bartholomew Fair*, pp. 1–15.
11. *The Stripping of the Altars*.
12. Vol. II, pp. 67–82.

CHAPTER 6

1. Acts 26:24.
2. Daniel 4:32–34.
3. 2 Kings 5.
4. *Hippocratic Writings* (London: Penguin), "The Sacred Disease", p. 237.
5. "Sacred Disease", p. 250.
6. "Sacred Disease", p. 246.
7. "Sacred Disease", p. 247.
8. "Sacred Disease", p. 244.
9. "Sacred Disease", p. 249.
10. Coghill translation (London: Penguin), p. 235.
11. Porter, *The Greatest Benefit*, p. 112
12. Act III, Sc. 3.
13. Act I, Sc. 2.
14. Clarke and Dewhurst, *History of Brain Function*, p. 23.
15. Clarke and Dewhurst, *History of Brain Function*, p. 16.

CHAPTER 7

1. 1 Samuel 5–6.
2. Duplaix, "Fleas, the Lethal Leapers", *National Geographic* 173, no. 5 (May 1988), pp. 672–94.
3. Ecclesiastes 8:15.
4. Coghill translation, p. 268.
5. *Magna Chirurgia* ed. McVaugh and Ogden, vol. 1, p. 119; Benedictow, *The Black Death*, p. 236.
6. 2 Kings 15:5; 2 Chronicles 26:21.
7. *The Oxford Book of English Verse*, no. 177, pp. 207–08.

CHAPTER 8

1.   Porter, *The Greatest Benefit*, pp. 117–18; Wangensteen and Wangensteen, *Rise of Surgery*, pp. 11–23, 312, etc.; Walker, *History of Neurological Surgery*, p. 224.

CHAPTER 9

1.   Thevenet, "Guy de Chauliac", *Annals of Vascular Surgery* 7, no. 2 (1993), p. 208.
2.   Porter, *The Greatest Benefit*, p. 117.
3.   Ziegler, *The Black Death*, pp. 23–24.
4.   Hippocrates III, *On Wounds, Surgery, and Fractures*, "Wounds to the Head", p. 263.

CHAPTER 10

1.   Blandy and Lumley, *The Royal College of Surgeons*, p. 3.
2.   D'Arcy Power (ed.), *Fistula in Ano*.
3.   Ms Harley, 1736, pp. 41–52.
4.   Blandy and Lumley, *Royal College of Surgeons*, p. 3.
5.   Blandy and Lumley, *Royal College of Surgeons*, p. 3; Appendix 2.
6.   Gardiner and Allen, *Before the Mast*, pp. 185–229.
7.   Druitt, *The Surgeon's Vade Mecum*, p. 185.
8.   Clowes *A Profitable and Necessarie Booke*, ch. 13; Chapman, "A History of Surgical Complications", in *Surgical Complications* (General Reading), pp. 18, 37.
9.   Blandy and Lumley, *Royal College of Surgeons*, pp. 3–6.

CHAPTER 11

1.   Saunders and O'Malley, *Anatomical Drawings of Andreas Vesalius*.
2.   Linker and Womack (transl.), *Ten Books of Surgery*.
3.   *Les Oeuvres D'Ambroise Paré, Conseiller et Premier Chirurgien du Roy*.
4.   Paget, *Ambrose Paré*, p. 31.

CHAPTER 12

1.   Aubrey, *Brief Lives* (ed. Dick), "William Harvey", pp. 118–33; p. 129 for George Ent.
2.   Aubrey, *Brief Lives*, "Harvey".
3.   Keele, *William Harvey*, pp. 37–38, 90.
4.   *De Motu Cordis* (transl. Franklin), p. 31.
5.   Kilgour, "William Harvey", p. 295.
6.   *De Motu Cordis*, ch. 2, p. 27.
7.   *De Motu Cordis*, p. 62.
8.   Aubrey, *Brief Lives*, "Harvey", p. 131.
9.   Slack, "Mirrors of Health", in Webster, *Health, Medicine, and Mortality*, pp. 237–73 (General Reading); Chapman, *Medicine of the People*, pp. 31–34.
10.   *The Surgeon's Mate* (1639), p. 172.
11.   Clowes, "A prooved Practise…", p. 33.

464

CHAPTER 13

1. Aubrey, *Brief Lives*, "Harvey", pp. 128–33.
2. Watkins, *Newes from the Dead*, p. 1, and Plot, *Natural History of Oxfordshire* (2nd edn, 1705), p. 203.
3. Plot, *Oxfordshire*, p. 202.
4. *Newes from the Dead*, p. 4.
5. *Newes from the Dead*, p. 7.
6. *Newes from the Dead*, p. 7.
7. Plot, *Oxfordshire*, pp. 202–03.
8. *Newes from the Dead*, p. 6.
9. *Newes from the Dead*, p. 8.
10. Cox, *Recollections of Oxford*, p. 23.
11. Wood, *Athenae Oxonienses*, vol. 2, cols. 549–51.
12. Aubrey, *Brief Lives* (ed. Clark), "Thomas Willis", pp. 302–04.
13. Aubrey, *Brief Lives* (ed. Clark), "Thomas Willis", pp. 302–04.
14. Aubrey, *Brief Lives* (ed. Clark), "Thomas Willis", pp. 302–04; Dewhurst (ed.) *Willis's Oxford Casebook*, p. 126.
15. Dewhurst (ed.), *Willis's Oxford Casebook* and *Thomas Willis's Oxford Lectures*.
16. Christopher Wren, *Parentalia*, p. 230.
17. Genesis 9:4.
18. Pepys, *Diary*, 14 November 1666.
19. I am indebted for this information to Dr Christopher Jewell, senior lecturer in epidemiology at Lancaster University Medical School.

CHAPTER 14

1. Aubrey, *Brief Lives* (ed. Clark), "Robert Burton", p. 165.
2. Wood, *Athenae Oxonienses*, vol. 2, cols. 627–28.
3. Burton, *The Anatomy of Melancholy*, Part 1, Section 2, Member 3, Subsection 4 (Everyman edn, vol. 1, p. 259).
4. Wood, *Athenae Oxonienses*, vol. 2, cols. 627–28.
5. Burton, *The Anatomy of Melancholy* (1621 edn), p. 4, "Democritus Junior to the Reader".
6. Burton, *The Anatomy of Melancholy*, vol. 2, p. 214.
7. Willis, *Cerebri Anatome* (Latin text), pp. 95–96.
8. Willis, *Cerebri Anatome* (Latin text), pp. 95–96; Hughes, *Thomas Willis*, p. 69.
9. Willis, *Cerebri Anatome* transl. Pordage, p. 119.
10. Wallis, "A Relation of an Accident of Thunder and Lightning in Oxford"; Plot, *Oxfordshire*, pp. 5–6.
11. Chapman, "The Scholar, the Thunderbolt, and the Anatomist", pp. 59–62.
12. Wallis, "A Relation of an Accident...", pp. 222–26.
13. Chapman, *Slaying the Dragons* (General Reading), pp. 33–37.
14. Dewhurst, *John Locke*, pp. 36–37.
15. Locke, *An Essay Concerning Human Understanding*, "Epistle to the Reader".
16. *An Essay Concerning Human Understanding*, Bk. 1, ch. 1, section 8.
17. *Pathologiae Cerebri*, "Caput 1", p. 3.

CHAPTER 15

1. Observation 16, pp, 101–06.
2. Mayow, *Physico-Medico Works*, pp. 71–73.
3. Mayow, *Physico-Medico Works*, p. 103.
4. *Philosophical Transactions* 63, 5 (12 September 1670), pp. 2035–56; Chapman, *England's Leonardo*, p. 28 and n. 38.
5. Mayow, *Physico-Medico Works*, p. 103.
6. Hooke, *Philosophical Transactions* 2, 28 (24 October 1667), pp. 539–40.
7. Sprat, *History of the Royal Society*, p. 232.
8. Chapman, *England's Leonardo*, p. 101.
9. Chapman, "From Alchemy to Airpumps", in Williams, Chapman, and Rowlinson, *Chemistry at Oxford*, pp. 25–30.
10. *The Life and Times of Anthony Wood*, vol. 1, p. 427; Chapman, "From Alchemy to Airpumps", Williams et al., *Chemistry at Oxford*, p. 48, n. 20.
11. Osborne, "The Medical Interests of the Oxford Chemists of the Late Seventeenth Century", pp. 41–46.

CHAPTER 16

1. John Ward, *Diary* (ed. Severn), pp. 8–9.
2. Ward, *Diary*, p. 12.
3. In Harriet Joseph, *Shakespeare's Son-in-Law*.
4. Ward, *Diary*, pp. 244–46.
5. Ward, *Diary*, p. 246.
6. Ward, *Diary*, p. 268.
7. Ward, *Diary*, p. 261.
8. Ward, *Diary*, p. 131.
9. Ward, *Diary*, p. 92.
10. Ward, *Diary*, p. 102.
11. Ward, *Diary*, p. 242.
12 Ward, *Diary*, p. 104.
13. Ward, *Diary*, p. 235.
14. Ward, *Diary*, p. 244.
15. *Primitive Physick*, "Preface", Section 7.
16. Edward Stone, "An Account of the Success of the Bark of the Willow", *Philosophical Transactions*, 53 (1763), p. 195.
17. Stone, "An Account", pp. 195–200.
18. Lind, *A Treatise of the Scurvy*; Harvie, *Limeys*, pp. 92–94.
19. Hales, *Statical Essays, containing Haemastaticks*, pp. 1–13.
20. Hales, *Statical Essays*, pp. 1–3, 9, 13, 18–21.
21. Virgin, *Sydney Smith*, p. 179.

CHAPTER 17

1. Porter, *The Greatest Benefit*, p. 299.
2. Krikler, "The Foxglove, 'The Old Woman of Shropshire', and William Withering", *Journal of the American College of Cardiology* 5, Suppl. A (May 1985), pp. 255–86; Sheldon, *The Life and Times of William Withering*.

3.   Amyand, "Of an Inguinal Rupture", *Philosophical Transactions* 443 (October 1736), pp. 329–42.
4.   Quist, *John Hunter*, pp. 77–80, 146–55.

Chapter 18

1.   Ned Ward, *The London Spy* (ed. Kenwick), p. 47.
2.   Ward, *London Spy*, p. 49.
3.   Ward, *London Spy*, p. 48.
4.   Act IV, Sc. 1.
5.   Salgardo, *The Elizabethan Underworld*, ch. 6, pp. 117ff.; ch. 10, pp. 183ff.
6.   Allan Ingram, *Voices of Madness: Four Pamphlets, 1683–1796* (Stroud: Sutton, 1997).
7.   Roberts, *Memoirs of the Life and Correspondence of Mrs. Hannah More*, vol. 2, p. 144; Porter, "Francis Willis", *Oxford DNB*.
8.   Macalpine and Hunter, *George III and the Mad-Business*, p. 53.
9.   Porter, *The Greatest Benefit*, p. 497.
10.   Courtney, "Samuel Warneford", *Oxford DNB*.
11.   E.g. *L'Assommoir* (ch. 13).

Chapter 19

1.   Chapman, *Medicine of the People*, p. 111.
2.   *The Diary of Robert Hooke*, 1672–1680, ed. Robinson and Adams.
3.   Chapman, *England's Leonardo*, pp. 111–17; Beier, "Experience and Experiment", in Hunter and Schaffer, *Robert Hooke*, New Studies, pp. 235–52.
4.   Uglow, *Hogarth*, p. 289: plate, "Company of Undertakers".
5.   Porter, *Quacks, Fakers, and Charlatans*, pp. 110–11.

Chapter 20

1.   Himes, *A Medical History of Contraception*.
2.   Thurston and Williams, "An Examination of John Fewster's Role in the Discovery of Smallpox Vaccination", *Journal of the Royal College of Physicians of Edinburgh* 45 (2015), 173–79.
3.   Shapter, *History of the Cholera in Exeter*, pp. 228–29.
4.   Braithwaite, *The Retrospect of Medicine* XX (1849), pp. 392–93.
5.   Braithwaite, *The Retrospect of Medicine* XX, p. 392.
6.   Dickens (ed.), *Household Words* 10 (1 June 1850), "Greenwich Weather Wisdom", pp. 222–25. Frederick Knight Hunt is the unnamed journalist (p. 225).
7.   Booth, *Darkest England and the Way Out*.

Chapter 21

1.   Southey, *Poetical Works*, pp. 457–58.
2.   Hood, *Poetical Works*, pp. 123–25.
3.   H. and M. Evans, *Andrew Ure galvanising a murderer's corpse*, plate, in *Sources in Illustration, 1500–1900*, p. 139, plate 62c.
4.   Aesculapius Scalpel, *St. Bernard's*, pp. 14–15.

5.   *The Memoirs of Berlioz*, pp. 46–49; O'Shea, *Music and Medicine*, pp. 233–35.
6.   Aesculapius Scalpel, *St. Bernard's*, p. 12.
7.   Cole, *Things for the Surgeon*, p. 114.

CHAPTER 22
1.   *Philosophical Transactions* 120 (1830), pp. 187–200.
2.   *Encyclopaedia Britannica*, 9th edn, vol. XVI (1883), p. 264.
3.   "Notice of some Microscopic Observations of the Blood and Animal Tissue", *Philosophical Magazine* N.S. 2 (1827), pp. 130–38.
4.   "Some Observations of the Structure and Functions of Tubular and Cellular Polypi", *Philosophical Transactions* 124 (1834), pp. 365–88.
5.   *The Edinburgh Medical and Surgical Journal* 64 (1 October 1845), pp. 413–23.
6.   Fisher, *Joseph Lister*, pp. 147–49.
7.   *The Lancet* 2 (1902), pp. 1397, 1476.
8.   "On the Alleged Increase of Cancer", *Proceedings of the Royal Society* 53 (1893), pp. 405–07.

CHAPTER 23
1.   *The Guinness Book of Records*, ed. Norris and Ross McWhirter (London, 1967), p. 23.
2.   Cole, *Things for the Surgeon*, p. 13.
3.   Thorwald, *Century of the Surgeon*, p. 109.
4.   *American Notes and Pictures from Italy* (Chapman and Hall, London, 1842, 1907), p. 22.
5.   Thorwald, *Century of the Surgeon*, p. 121.
6.   Full details in Further Reading.
7.   Porter, *The Greatest Benefit*, p. 368.

CHAPTER 24
1.   pp. 387–89.
2.   Ballance, "The Surgery of the Heart" (3 parts), *The Lancet*, 1920, 1: 3 January, pp. 1–6; 10 January, pp. 73–77; 17 January, pp. 134–39.
3.   Bland-Sutton, "Missiles and Emboli", *The Lancet*, 10 May 1919, pp. 773–5.
4.   Souttar, "Surgical Treatment of Mitral Stenosis", *British Medical Journal* 2 (1925), pp. 603–06.

CHAPTER 25
1.   Rivington, *The Medical Profession*, pp. 94–95.
2.   3 vols, 1841.
3.   Aesculapius Scalpel, *St. Bernard's*, p. 49.
4.   Aesculapius Scalpel, *St. Bernard's*, p. 51.
5.   *Punch*, "A Stir for Seacole", 6 December 1856; Robinson, *Mary Seacole*, pp. 163–65.
6.   Aesculapius Scalpel, *St. Bernard's*, p. 201.
7.   Aesculapius Scalpel, *St. Bernard's*, p. 200.
8.   *The Principles and Practice of Medicine* (1892).

CHAPTER 26

1. Harris, "Howard Florey and the discovery of penicillin", *Notes and Records of the Royal Society* 53, 2 (1999), p. 249.
2. Keating, *Great Medical Discoveries*, pp. 61–3.
3. Bradford, *Disraeli*, pp. 254–55.

Further Reading

GENERAL READING

Hedley Atkins, *The Surgeon's Craft* (Manchester: Manchester University Press, 1965).

Iain Bamforth (ed.), *The Body in the Library: A Literary Anthology of Modern Medicine* (London and New York: Verso, 2003).

Elizabeth Bennion, *Antique Medical Instruments* (London: Philip Wilson, 1979).

Robert Bud and Deborah Jean Warner (eds), *Instruments of Science: An Historical Encyclopaedia* (New York and London: Garland Publishing, 1998).

John Camp, *Magic, Myth, and Medicine* (London: Priory Press, 1973).

Allan Chapman, *Gods in the Sky: Astronomy, Religion, and Culture from the Ancients to the Renaissance* (London: Channel 4 TV and Macmillan, 2001).

——, "A History of Surgical Complications", in Nadey S. Hakim and Vassilios E. Papalois, *Surgical Complications, Diagnosis and Treatment* (London: Imperial College Press, 2007), pp. 1–40.

——, *Slaying the Dragons. Destroying Myths in the History of Science and Religion* (Oxford: Lion Hudson, 2013).

——, *Stargazers: Copernicus, Galileo, the Telescope, and the Church. The Astronomical Renaissance, 1500–1700* (Oxford: Lion Hudson, 2014).

Logan Clendening, *A Source Book of Medical History* (New York: Dover Press, 1960).

John Cule, *A Doctor for the People: 2000 Years of General Practice in Britain* (London: Update Books, 1980).

Maurice Daumas, *Scientific Instruments of the 17th and 18th Centuries and their Makers*, ed. and transl. by Dr Mary Holbrook (London: Batsford, 1972).

Sir Harold Ellis, *Surgical Case Histories from the Past* (London: Royal Society of Medicine, 1994). Reprints several key historic surgical papers.

Richard L. Gregory and O. L. Zangwill (eds), *The Oxford Companion to the Mind* (Oxford: Oxford University Press, 1997).

Howard W. Haggard, *Devils, Drugs, and Doctors: The Story of the Science of Healing, from Medicine Man to Doctor* (London: Harper, 1929).

G. Melvyn Howe, *Man, Environment, and Disease in Britain: A Medical Geography through the Ages* (London: Pelican, 1976).

R. W. Johnson, *Disease and Medicine* (London: Batsford, 1967).

Conrad Keating, *Great Medical Discoveries: An Oxford Story* (Oxford: Bodleian Library, 2013).

Roderick E. M. McGrew, *Encyclopaedia of Medical History* (London: Macmillan, 1985).

C. M. MacLaurin, *Post Mortems of Mere Mortals, Essays Historical and Medical* (New York: Sun Dial Press, 1930).

Jonathan Miller, *The Body in Question* (London: Jonathan Cape, 1978).

David S. Muckle, *A Doctor's Look at Life and History* (Newcastle upon Tyne: Oriel Press, 1970).

John O'Shea, *Music and Medicine: Medical Profiles of Great Composers* (London: Dent, 1990, 1993).

Kurt Pollak with E. Ashworth Underwood, *The Healers: The Doctor, Then and Now* (London: Thomas Nelson and Sons, 1968).

Roy Porter, *Bodies Politic: Disease, Death, and Doctors in Britain, 1650–1900* (New York: Cornell University Press, 2001).

——, *The Greatest Benefit to Mankind: A Medical History of Humanity from Antiquity to the Present* (London: Harper Collins, 1997).

—— (ed.), *The Cambridge Illustrated History of Medicine* (Cambridge: Cambridge University Press, 1996).

Roger Rolls, *Diseased, Douched, and Doctored. Thermal Springs, Spa Doctors and Rheumatic Diseases* (London Publishing Partnership, 2012).

Anton Sebastian, *A Dictionary of the History of Medicine* (New York: Parthenon, 1999).

R. Scott Stevenson, *Famous Illnesses in History* (London: Eyre and Spottiswoode, 1962).

J. Simmons, *Doctors and Discoveries: Lives that created today's medicine from Hippocrates to the present* (Boston: Houghton Mifflin, 2002).

H. M. Sinclair and A. H. T. Robb-Smith, *A Short History of Anatomical Teaching in Oxford* (Oxford: Oxford University Press, 1950).

Charles Singer and E. Ashworth Underwood, *A Short History of Medicine* (Oxford: Oxford University Press, 1962).

George Edward Trease, *Pharmacy in History* (London: Bailliere, Tindall, and Cox, 1964).

Sir John Nicholas Walton, Jeremiah A. Barondess, and Stephen Lock (eds), *The Oxford Medical Companion* (Oxford: Oxford University Press, 1994).

Owen H. Wangensteen and Sarah D. Wangensteen, *The Rise of Surgery from Empiric Craft to Scientific Discipline* (Minneapolis: University of Minnesota Press, 1978).

Charles Webster (ed.), *Health, Medicine, and Mortality in the Sixteenth Century* (Cambridge: Cambridge University Press, 1979).

Robert Wilkins, *The Fireside Book of Deadly Diseases* (London: Robert Hale, 1994).

Leo M. Zimmermann and Ilza Veith, *Great Ideas in the History of Surgery* (Baltimore: Williams and Wilkins, 1961).

CHAPTER 1

Aristotle, *The Basic Works of Aristotle*, ed. and introd. by Richard McKeon (New York: Random House, 1941).

——, *De Generatione Animalium* Book II, Sect. IV, 739:22–27, in *Aristotle's Generation of Animals*, transl. Arthur Leslie Peck (London: Heinemann, and Cambridge, MA: Harvard University Press, 1943).

——, *De Generatione Animalium*, in *The Works of Aristotle* ed. and transl. J. A. Smith and W. D. Ross (Oxford: Clarendon Press, 1912).

J. H. Breasted, *The Edwin Smith Surgical Papyrus*, 2 vols (Chicago: University of Chicago Press, 1930).

Don R. Brothwell, *Digging up Bones* (London: British Museum, 1972).

Don R. Brothwell and A. T. Sandison (eds), *Diseases in Antiquity: A Survey of the Diseases, Injuries, and Surgery of Early Populations* (Springfield, IL: C. C. Thomas, 1967).

Aidan Cockburn and Eva Cockburn, *Mummies, Diseases, and Ancient Cultures* (Cambridge: Cambridge University Press, 1980).

Morris Raphael Cohen and Israel Edward Drabkin, *A Source Book in Greek Science* (New York: McGraw Hill, 1948).

Rosalie David (ed.), *Mysteries of the Mummies* (London: Macmillan, 1978).

Gerhard Henrik Armauer Hansen and Carl Looft, *Leprosy in Its Clinical and Pathological Aspects*, English edn, transl. Norman Walker (London, 1895).

Hippocratic Writings (see under Chapter 2).

Hippocrates III, *Wounds, Surgery, and Fractures*, trans. E. T. Withington, Loeb Classical Library (London: Heinemann and Cambridge, MA: Harvard University Press, 1928).

Jamieson B. Hurry, *Imhotep the Vizier and Physician to King Zoser* (Oxford: Oxford University Press, 1928).

David Kohn, "Smoking Has Always Been Bad for You" [heart disease in Egyptian mummies], *The Guardian*, 31 August 2015.

C. D. Leake, *The Old Egyptian Medical Papyri* (Lawrence, KA: University of Kansas Press, 1952).

Ange-Pierre Leca, *The Cult of the Immortal. Mummies and the Ancient Egyptian Way of Death* (London: Souvenir Press, 1980; London: Anchor Books, 1981).

James Longrigg, *Greek Medicine: From the Heroic to the Hellenistic Age. A Source Book* (London: Duckworth, 1998).

E. D. Phillips, *Greek Medicine* (London: Thames and Hudson, 1973).

Henry E. Sigerist, *A History of Medicine*, vol. 1 *Primitive and Archaic* (Oxford: Oxford University Press, 1951).

Henry E. Sigerist, *A History of Medicine*, vol. 2 *Early Greek, Hindu, and Persian Medicine* (Oxford: Oxford University Press, 1961).

Heinrich von Staden, *Herophilus: The Art of Medicine in Early Alexandria* (New York: Cambridge University Press, 1989).

Barbara Watterson, *Women in Ancient Egypt* (New York: St Martin's Press and Stroud, UK: Alan Sutton Publishing, 1991).

Calvin Wells, *Bones, Bodies, and Diseases* (London: Thames and Hudson, 1964).

CHAPTER 2

Lawrence Bliquez, *The Tools of Asclepius: Surgical Instruments in Greek and Roman Times* (Leiden: Brill, 2014).

David Lee Bomgardner, *The Story of the Roman Amphitheatre* (London: Routledge, 2000).

Peter Brain, *Galen on Bloodletting* (Cambridge: Cambridge University Press, 1986).

Aulus Cornelius Celsus, *De Medicina*, transl. W. G. Spencer, Loeb Classical Library (London: Heinemann and Cambridge, MA: Harvard University Press, 1935, reprinted 1940, 1960).

Audrey Cruse, *Roman Medicine* (Stroud: Tempus, 2004).

Galen, *On the Natural Faculties*, transl. Arthur John Brock, MD, (Chicago, London, etc.: Encyclopaedia Britannica Inc., 1952).

Galen, *On Anatomical Procedures*, transl. Charles Singer (Oxford: Wellcome Museum and Oxford University Press, 1956). Contains Books I–IX.

Galen, *On Anatomical Procedures: The Later Books*, transl. W. L. H. Duckworth, M. C. Lyons, B. Towers (Cambridge: Cambridge University Press, 1962). Books X–XV.

Ralph Jackson, *Doctors and Diseases in the Roman Empire* (London: British Museum, 1988).

Hippocratic Writings, ed. and introd. by G. E. R. Lloyd, transl. by J. Chadwick, W. N. Mann, I. M. Lonie, and E. T. Withington (London: Penguin Classics, 1950, 1983).

Hippocratic Writings, transl. Francis Adams (Chicago, London, etc.: Encyclopaedia Britannica Inc., 1952).

Jan Johnston, *Galen on Diseases and Symptoms* (Cambridge: Cambridge University Press, 2006).

Fridolf Kudlien, "Galen", *Dictionary of Scientific Biography* ed. C. C. Gillespie, vol. V (New York: Scribner, 1972), pp. 227–32.

Donald G. Kyle, *Sport and Spectacle in the Ancient World* (Malden MA: Blackwell, 2007; 2nd edn Wiley Blackwell, 2014).

Susan P. Mattern, *The Prince of Medicine: Galen and the Roman Empire* (New York: Oxford University Press, 2013).

Gaius Plinius Secundus [Pliny the Elder], *Natural History*, a selection, transl. and introd. by John F. Healy (London: Penguin Classics, 1991).

John Scarborough, *Roman Medicine* (London, 1969).

Owsei Temkin, *Galenism: The Rise and Decline of a Medical Doctrine* (New York: Cornell University Press, 1974).

I. Veith, "Galen, the First Medical Autobiographer", *Modern Medicine* 27 (1959), pp. 232–45.

Tony Wilmott, "Gladiators in Ancient Rome: How Did They Live and Die?", BBC Magazine, May 2013.

Leonard G. Wilson, "Galen: Anatomy and Physiology", in *Dictionary of Scientific Biography*, ed. C. C. Gillespie, vol. V (New York: Scribner, 1972), pp. 233–37.

CHAPTER 3

Soheil M. Afnan, *Avicenna: His Life and Works* (London: Allen & Unwin, 1958).

Albucasis, *On Surgery and Instruments*, ed. and transl. from Arabic by M. S. Spink and G. L. Lewis (London: Wellcome Institute, 1973).

Alhazen (Ibn al-Haytham), *Opticae Thesaurus Alhazen Arabis* (Basilae, 1572). Books V, VI, and VII deal with glass cylinder refraction.

Edward G. Browne, *Arabian Medicine* (Cambridge, 1922).

J. Channing (ed.), *Rhazes de Variolis et Morbillis* (London, 1766).

L. I. Conrad, "Arab-Islamic Medicine", in W. F. Bynum and Roy Porter, *Companion Encyclopedia of the History of Medicine* (London: Routledge, 1993), pp. 676–727.

C. Elgood, *A Medical History of Persia and the Eastern Caliphate* (Cambridge: Cambridge University Press, 1951, 2010).

R. J. Forbes, *A Short History of the Art of Distillation* (Leiden: Brill, 1970).

O. C. Gruner, *A Treatise on the Canon of Medicine, Incorporating a Translation of the First Book* (London: Luzac & Co., 1930).

Hippocrates, *On Wounds, Surgery, and Fractures*, transl. E. T. Withington, Loeb Classical Library (London: Heinemann and Cambridge, MA: Harvard University Press, 1928), pp. 259–63.

Eric J. Holmyard, *Alchemy* (London: Penguin, 1957, 1968).

Eric J. Holmyard, *The Great Chemists* (London: Methuen, 1928).

Albert Z. Iskander, "Ibn al-Nafis", in *Dictionary of Scientific Biography* (New York: Scribner, 1969).

Amin Khairallah, *Outline of Arabic Contributions to Medicine and Allied Sciences* (Beirut, Lebanon: American Press, 1946).

M. S. Khan, *Islamic Medicine* (London: Routledge, 1986).

Martin Levey, *Early Arabic Pharmacology: An Introduction based on Ancient and Medieval Sources* (Leiden: Brill, 1973).

Max Meyerhof, "Ibn al-Nafis (XIIIth cent.) and His Theory of the Lesser Circulation", *Isis* 23 (1935), pp. 100–20.

Seyyed Hossein Nasr, *Islamic Science. An Illustrated Study* (London: World of Islam Festival Publishing Co. Ltd., 1976).

J. R. Partington, *A Short History of Chemistry* (London: Macmillan, 1937, 1965).

M. Ullman, *Islamic Medicine* (Edinburgh: Edinburgh University Press, 1978).

John B. West, "Ibn al-Nafis, the Pulmonary Circulation, and the Islamic Golden Age", *Journal of Applied Physiology* 105 (2008), pp. 1877–1900.

G. M. Wickens, *Avicenna: Scientist and Philosopher. A Millennary Symposium* (London: Luzac & Co., 1952).

CHAPTER 4

Albucasis, *On Surgery and Instruments* (see under Chapter 3).

Alhazen, *Opticae Thesaurus* (see under Chapter 3).

Aristotle, *Meteorologica*, transl. E. W. Webster, in *The Works of Aristotle*, vol. III (Oxford: Clarendon Press, 1931).

Aristotle, *Problemata*, ed. and transl. by E. S. Forster, in *The Works of Aristotle*, vol. VII (Oxford: Clarendon Press, 1927).

Roger Bacon, *Opus Majus*, ed. John Bridges, 2 vols (Oxford: Clarendon Press, 1897).

Roger Bacon's Opus Majus, ed. Samuel Jebb (London, 1733).

Juliet Barker, *The Brontës* (London: Weidenfeld and Nicolson, 1994).

B. Chance, *Ophthalmology* (New York: Hafner Publishing, 1962).

A. Chapman, "Divine Light", in *Mathematicians and Their Gods: Interactions between Mathematics and Religious Beliefs*, ed. Snezana Lawrence and Mark McCartney (Oxford: Oxford University Press, 2015), pp. 51–76.

Alistair Crombie, *Robert Grosseteste and the Origins of Experimental Science* (Oxford: Clarendon Press, 1953).

G. Gorin, *History of Ophthalmology* (Philadelphia: Lippincott, 1982).

S. H. Holbrook, *The Golden Age of Quackery* (New York: Macmillan, 1982).

R. Rutson James (ed.), *Studies in the History of Ophthalmology in England prior to the Year 1800* (Cambridge: Cambridge University Press, 1933).

David C. Lindberg, *Theories of Vision from Al-Kindi to Kepler* (Chicago: University of Chicago Press, 1976).

"Light", *Encyclopaedia Britannica*, 9th edn, vol. 14 (1882).

Seyyed Hossein Nasr, *Islamic Science* (see under Chapter 3), pp. 140–41.

Anton Sebastian, "Ophthalmology", in *A Dictionary of the History of Medicine* (see under "General Reading"), pp. 550–51.

A. Mark Smith, "Ptolemy's Theory of Visual Perception: An English Translation of the Optics", *Transactions of the American Philosophical Society* 86, Pt. 2 (1996).

A. Mark Smith (ed.), "Alhazen's Theory of Visual Perception", *Transactions of the American Philosophical Society* 91, Pts. 4–5 (2001).

Robert Smith, *A Compleat System of Opticks* (Cambridge, 1738).

CHAPTER 5

Luisa Cogliati Arano, *The Medieval Health Handbook Tacuinum Sanitatis* (London: Barrie and Jenkins, 1976).

Agnes Robertson Arber, *Herbals, the Origin and Evolution, a Chapter in the History of Botany 1470–1670* (Cambridge: Cambridge University Press, 1912).

Andrew Boorde, *Hereafter foloweth a compendyous Regyment of Helthe made in Mountpyllier* [Montpellier], *c.* 1545 (publication date unclear): Bodleian Library, Oxford, shelfmark "Crynes, 873".

H. P. Cholmley, *John of Gaddesden and the Rosa Medicinae* (Oxford: Clarendon Press, 1912), pp. 22–74.

Thomas Oswald Cockayne, *Leechdoms, Wortcunning, and Starcraft of Early England: Being a Collection of Documents, for the Most Part Never Before Printed Illustrating the History of Science in this Country Before the Norman Conquest* (London, 1864–65; reprinted 1965).

Dainton Courtney, "Bart's Hospital", *History Today*, vol. 28 (12 Dec. 1978), pp. 810–16.

Peter Dendle and Alain Touwaide (eds), *Health and Healing from the Medieval Garden* (Woodbridge: Boydell Press, 2015).

Eamon Duffy, *The Stripping of the Altars. Traditional Religion in England 1400–1580* (New Haven and London: Yale University Press, 1992).

Charles Graves, *The Story of St. Thomas's 1106–1947* (London: St. Thomas's Hospital, Faber and Faber, 1947).

Henry Morley, *Memoirs of Bartholomew Fair* (London, 1859; reprinted 1973).

J. F. Payne, *English Medicine in Anglo-Saxon Times* (Oxford: Clarendon Press, 1904).

Thomas Paynel, *Regimen Sanitatis Salerni* (London, 1541).

Regimen Sanitatis Salerni. The Schoole of Salernes most Learned and Iuditious Directorie (London, 1617).

James O. Robinson, "The Royal and Ancient Hospital of St. Bartholomew (Founded 1123)", *Journal of Medical Biography* 1 (February 1993), pp. 23–30.

Eleanor Sinclair Rohde, *The Old English Herbals* (London: Longmans, 1922).

Stanley Rubin, *Medieval English Medicine, AD 500–1300* (London: David and Charles; New York: Barnes and Noble, 1974).

"Salerno", *Encyclopaedia Britannica*, 14th edn, vol. 19 (1929).

Don C. Skemer, *Binding Words: Textual Amulets in the Middle Ages* (University Park, PA: Pennsylvania State University Press, 2006).

Hubert Jan de Vrend (ed.), *The Old English Herbarium and Medicina De Quadrupedibus*, Early English Texts Society (Oxford: Oxford University Press, 1984).

Enid Welsford, *The Fool: His Social and Literary History* (London: Faber and Faber, 1935).

Andrew Dickson White, *A History of the Warfare of Science with Theology in Christendom*, 2 vols (1896; reprinted New York: Dover, 1960).

T. H. White, *A Book of Beasts* (London: Jonathan Cape, 1969).

CHAPTER 6

Derek Baker (ed.), *Medieval Women* (Oxford: Blackwell, 1978).

Santha Bhattacharji, *God is an Earthquake: The Spirituality of Margery Kempe* (London: Darton, Longman and Todd, 1997).

Edwin Clarke and Kenneth Dewhurst, *An Illustrated History of Brain Function* (Oxford: Sandford, 1972).

Trudy Drucker, "The Malaise of Margery Kempe", *New York State Journal of Medicine* 72 (1972), pp. 2911–17.

Hippocrates, "The Sacred Disease", in *Hippocratic Writings* (see under Chapter 2), pp. 237–51.

Margery Kempe, *The Book of Margery Kempe*, ed. S. B. Meech and H. E. Allen, Early English Texts Society (Oxford: Oxford University Press, 1940, 1961).

Roy Porter, *A Social History of Madness: Stories of the Insane* (London: Weidenfeld & Nicolson, 1987). Chapter 6, "Mad Women", pp. 103–12.

Sr. Benedicta Ward, *Miracles and the Medieval Mind: Theory, Record, and Events, 1000–1215* (Aldershot: Scholar, 1982).

CHAPTER 7

Richard Barber (ed.), *The Pastons* (London: Penguin Lives and Letters, 1984).

W. G. Bell, *The Great Plague of London in 1665* (London: John Lane, 1924).

Ole Jørgen Benedictow, *The Black Death 1346–1353. The Complete History* (Woodbridge: Boydell, 2004).

Giovanni Boccaccio, *The Decameron*, transl. John Payne (New York: Random House, 1946). Pp. 88–92 for Florence plague.

Anna M. Campbell, *The Black Death and Men of Learning* (New York: Columbia University Press, 1931).

Ann G. Carmichael, "Universal and Particular: The Language of Plague, 1348–1500", *Medical History*, Suppl. 27 (2008), pp. 17–52.

Geoffrey Chaucer, *The Canterbury Tales*, modern English transl. by Sir Nevil Coghill (London: Penguin, 1960).

Guy de Chauliac, *Inventarium sive Chirurgia Magna*, ed. Michael R. McVaugh and Margaret S. Ogden (Leiden: Brill, 1997).

Charles Creighton, *A History of Epidemics in Britain*, 2 vols (Cambridge: Cambridge University Press, 1891).

Michael W. Dols, *The Black Death in the Middle East* (Princeton, NJ: Princeton University Press, 1977).

Nicole Duplaix, "Fleas, the Lethal Leapers", *National Geographic* vol. 173, no. 5 (May 1988), pp. 673–94.

"Epidemiological Notes and Reports, Plague Pneumonia – California", *Morbidity and Mortality Weekly Report* (Massachusetts Medical Society), *New England Journal of Medicine* vol. 30, no. 34 (31 August 1984), pp. 481–3.

Francis Aidan Gasquet, *The Great Pestilence, 1348–1349* (London, 1893).

Ian and Jennifer Glynn, *The Life and Death of Smallpox* (London: Profile, 2005).

L. Fabian Hirst, *The Conquest of Plague. A New Study in the Evolution of Epidemiology* (Oxford: Clarendon Press, 1953), p. 13 for John Clyn account.

"The Land of the Flea and the Home of the Plague", unspecified newspaper article from New Mexico, USA, 1984, sent to the author.

W. H. McNeil, *Plagues and Peoples* (London: Penguin, 1994).

Thomas Nashe, "In Time of Pestilence", in Sir Arthur Quiller-Couch (ed.), *The Oxford Book of English Verse, 1250–1918* (Oxford: Clarendon Press, 1939). Poem no. 177, pp. 207–08.

Paul Slack, "Mortality Crises and Epidemic Disease in England, 1485–1610", in Webster (ed,), *Health, Medicine, and Mortality* (see under General Reading).

J. F. D. Shrewsbury, *A History of Bubonic Plague in the British Isles* (Cambridge: Cambridge University Press, 1970).

J. F. D. Shrewsbury and C. Morris, "The Plague in Britain", *The Historical Journal* XIV, no. 1 (1971), pp. 205–15.

Philip Ziegler, *The Black Death* (New York: Harper, 1969).

Hans Zinsser, *Rats, Lice, and History* (London: Routledge, 1935, 1937).

CHAPTER 8

Mark Infusino, Dorothy Win, and Y. V. O'Neill, "Mondino's book of the human body", *Vesalius* 1, no. 2 (1995), pp. 71–6.

Lanfranc, *Science of Chirurgie*, Early English Texts Society (London: Kegan Paul, 1894).

Mondino [Mundinus] de Liuzzi, *Anathomia* (c. 1326). Like many medieval works that circulated in copied manuscript versions before printing, Mondino's *Anatomy* (printed in Venice, 1493, then elsewhere, 1513, 1541, etc.) is bibliographically complicated, owing to variations between editions. For a good modern facsimile of the 1493 Venetian printing see *Fasciculus Medicinae*, ed. Charles Singer (Florence, 1925).

Walton O. Shalick III, "Neurological Conditions in the European Middle Ages", in Stanley Finger, Francois Boller, and Kenneth L. Tyler, *History of Neurology* (Edinburgh and New York: Elsevier, 2010), pp. 79–90.

Charles Singer, *A Short History of Anatomy from the Greeks to Harvey* (New York: Dover, 1957).

Theodoric, *The Surgery of Theodoric*, transl. E. Campbell and J. Colton (New York: Appleton-Century-Crofts, 1955).

A. Earl Walker, *A History of Neurological Surgery* (Baltimore: Walker and Watkins, 1951).

Wangensteen and Wangensteen, *The Rise of Surgery* (see under General Reading). Praise for Mondeville, Lanfranc, etc.: pp. 11–23, 312.

CHAPTER 9

Guy de Chauliac, *The Cyrurgie of Guy de Chauliac*, ed. Margaret S. Ogden (Early English Texts Society, Oxford University Press, 1971).

——, *Inventarium sive Chirurgia Magna* (see under Chapter 7).

——, *Guydos Questions, Newly Corrected* (London, 1679; reprinted facsimile, Amsterdam: Theatrum Orbis Terrarum, 1968).

Thomas F. Glick, *Medieval Science, Technology, and Medicine: An Encyclopedia* (London: Routledge, 2005).

Edward Grant, *A Source Book in Medieval Science* (Cambridge, MA: Harvard University Press, 1974).

L. R. Lind, *Studies in Pre-Vesalian Anatomy: Biography, Translations, Documents* (Philadelphia: American Philosophical Society, 1975).

David C. Lindberg, *Science in the Middle Ages* (Chicago: University of Chicago Press, 1978).

André Thevenet, "Guy de Chauliac (1300–1370): Father of Modern Surgery", *Annals of Vascular Surgery* 7, no. 2 (1993), pp. 208–12.

Philip Ziegler, *The Black Death* (see under Chapter 7).

CHAPTER 10

John of Arderne, *Treatise on Fistula in Ano*, ed. D'Arcy Power, Early English Texts Society (Oxford: Oxford University Press, 1910).

R. T. Beck, *The Cutting Edge: Early History of the Surgeons of London* (London: Lund Humphries, 1974).

J. P. Blandy and John S. P. Lumley (eds), *The Royal College of Surgeons of England. 200 Years of History at the Millennium* (London: Royal College of Surgeons and Blackwell Science, 2000).

William Clowes, *A Profitable and Necessarie Booke for all those that are burned with the flame of Gunpowder* (London, 1596; 3rd edn 1637). Chapter 13 for arrow experiment.

William Clowes, *A Prooved Practise for all Young Chirurgians Concerning Burning with Gunpowder* (London, 1588). (Copy consulted in The Queen's College Library, Oxford.)

H. Cole and T. Lang, "The Treating of Prince Henry's Arrow Wound, 1403", *Journal of the Society of Archer-Antiquaries* 49 (2003), pp. 95–101.

A. M. Cooke, "Dr John Caius 1510–1573", *Journal of the Royal College of Physicians of London* 7, no. 4 (July 1973), pp. 365–71.

Robert Druitt, *The Surgeon's Vade Mecum* (London, 1873).

Julie Gardiner and Michael J. Allen, *Before the Mast: Life and Death aboard the Mary Rose*, *Archaeology of the Mary Rose* vol. 4 (Portsmouth: Mary Rose Trust, 2005).

John Gerard, *The Herball, or Generall Historie of Plants Gathered by John Gerarde of London Master in Chirurgerie* (London, 1597, 1633).

John Noble Johnson, *Life of Thomas Linacre* (London, 1835).

J. J. Keevill, C. Lloyd, and J. L. S. Coulter, *Medicine in the Navy 1200–1900*, 4 vols (Edinburgh: E. and S. Livingstone, 1957–63).

Sheila J. Lang, "John Bradmore and His Book 'Philomena'", in *A Social History of Medicine* 5 (1992), pp. 121–30.

Sir T. Longmore, *Richard Wiseman, Surgeon and Sergeant-Surgeon to Charles II: A Biographical Study* (London: Longmans, 1891).

Francis Maddison, Margaret Pelling, and Charles Webster, *Linacre Studies: Essays on the Life and Work of Thomas Linacre, c. 1460–1524* (Oxford: Clarendon Press, 1977).

Graham Martin, "Prince Rupert and the Surgeons", *History Today*, December 1990, pp. 38–43.

Jack Edward McCallum, *Military Medicine* (London: ABC-CLIO, 2008).

G. Parker, *The Early History of Surgery in Great Britain* (London: A. and C. Black, 1920).

Margaret Pelling and Charles Webster, "Medical Practitioners", in *Health, Disease, and Mortality in the Sixteenth Century*, ed. Charles Webster (Cambridge: Cambridge University Press, 1979), pp. 164–235.

J. F. South, *Memorials of the Craft of Surgery in London* (London: Cassell, 1886).

C. H. Talbot, *Medicine in Medieval England* (London: Oldbourne, 1967).

John Woodall, *The Surgeon's Mate, or A Treatise... of the Surgeon's Chest* (London, 1617, 1638; reprint 1978).

S. Young, *The Annals of the Barber Surgeons of London* (London, 1890).

CHAPTER 11

R. Brearley, "Padua and the dawn of scientific medicine", *Proceedings of the Royal Society of Medicine* 55 (1962), p. 217.

M. T. Gnudi and J. P. Webster, *The Life and Times of Gaspare Tagliacozzi, Surgeon of Bologna* (New York: Herbert Reichner, 1950).

Wallace B. Hamby, "Ambrose Paré", *Dictionary of Scientific Biography*, vol. X (New York: Scribner, 1974), pp. 315–17. Cites "Je le pansai, Dieu le guérit", p. 316.

Kenneth D. Keele, *Leonardo da Vinci on the Movement of the Heart and Blood* (London: Harvey & Blythe, 1952).

Edward MacCurdy, *The Notebooks of Leonardo da Vinci*, 2 vols (London: Reprint Society, 1938, 1954). Anatomy and physiology in vol. 1.

Charles Donald O'Malley, *Andreas Vesalius of Brussels, 1514–1564* (Berkeley, CA: University of California Press, 1964).

F. P. Packard, *The Life and Times of Ambrose Paré* (New York, 1926).

Stephen Paget, *Ambrose Paré and his Times, 1510–1590* (New York and London: Putnam, 1897). P. 31 for "wounds" quotation.

Ambrose Paré, *Les Oeuvres D'Ambroise Paré Conseiller et Premier Chirurgien du Roy* (Lyon, 1641).

Ambrose Paré, *Ten Books of Surgery*, transl. R. W. Linker and W. Womack (Athens, GA: Georgia University Press, 1969).

Jean-Pierre Poirer, *Ambrose Paré* (Paris: Pygmalion, 2006), p. 42 for wounds.

John B. de C. M. Saunders and Charles Donald O'Malley (eds), *Andreas Vesalius Bruxellensis. The Bloodletting Letter of 1539. An Annotated Translation* (London: Heinemann Medical, 1947).

John B. de C. M. Saunders and Charles Donald O'Malley, *The Anatomical Drawings of Andreas Vesalius: With Annotations on Translations...* (New York: Random House, 1982).

Charles Singer, *The Evolution of Anatomy* (London: Kegan Paul & Co., 1925).

The "Epitome" of Andreas Vesalius, transl. by L. R. Lind (New York: Macmillan, 1949).

Willy Wiegand and Samuel W. Lambert, *Three Vesalian Essays to Accompany the "Icones Anatomicae" of 1934...* (New York: Macmillan, 1952).

CHAPTER 12

John Aubrey, "William Harvey", in *Brief Lives*, ed. Oliver Lawson Dick (London, 1949, 1975).

Allan Chapman, "Astrological Medicine"; in Webster, *Health, Medicine and Mortality* (see under General Reading), pp. 275–300.

Allan Chapman, "William Harvey and The Circulation of the Blood", Seminars of the University of Minnesota, *The Journal of Laboratory and Clinical Medicine* 126, no. 5 (November 1995), pp. 423–27.

Mohamed Elmaghawry, "The Discovery of the Pulmonary Circulation: From Imhotep to Harvey", *Global Cardiology Science and Practice* (June 2014).

Kenneth J. Franklin, *William Harvey, Englishman* (London: MacGibbon & Kee, 1961).

William Harvey, *The Works of William Harvey M.D.*, transl. Robert Willis (London: Sydenham Society, 1847).

William Harvey, *De Motu Cordis* (1628), translated by Kenneth J. Franklin as *The Movement of the Heart and Blood in Animals. An Anatomical Essay by William Harvey* (Oxford: Blackwell, 1957).

Kenneth D. Keele, *Leonardo da Vinci on the Movement of the Heart and Blood* (see under Chapter 11).

Kenneth D. Keele, *William Harvey* (London: Nelson, 1965).

Frederick G. Kilgour, "William Harvey and His Contributions", *Circulation*, XXXIII (February 1961), pp. 286–96. Cites 1616 lecture for water bellows and clacks, p. 292.

Sir T. Longmore, *Richard Wiseman* (see under Chapter 10).

Charles Donald O'Malley, F. N. Poynter, and K. F. Russell, *William Harvey, Lectures on the Whole of Anatomy* (Berkeley and Los Angeles: University of California Press, 1961).

D. Ribatti, "William Harvey and the Discovery of the Circulation of the Blood", *Journal of Angiogenesis Research*, Sept. 2009.

Paul Slack, "Mirrors of Health and Treasures for Poor Men: The Uses of the Vernacular Medical Literature of Tudor England", in Webster, *Health, Medicine and Mortality* (see under General Reading), pp. 237–73.

Charles Webster, *The Great Instauration: Science, Medicine, and Reform 1626–1660* (London: Duckworth, 1975).

John B. West, "Marcello Malpighi and the Discovery of the Pulmonary Capillaries and Alveoli", *American Journal of Physiology: Lung, Cellular, and Molecular Physiology* 304, no. 6 (15 March 2013), pp. 383–90.

Gwyneth Whitteridge, *The Anatomical Lectures of William Harvey* (Edinburgh: E. & S. Livingstone, 1964).

Robert Willis, *An Anatomical Examination of the Body of Thomas Parr, 1699* (London, 1847).

Thomas Wright, *Circulation and William Harvey's Revolutionary Idea* (London: Chatto and Windus, 2012).

CHAPTER 13

John Aubrey, *Brief Lives*, chiefly of contemporaries, set down by John Aubrey between 1669 and 1696, vol. II, ed. Andrew Clark (Oxford: Clarendon Press, 1898): pp. 302–04 for Thomas Willis.

Robert Hooke (?), "An Account of an Experiment of *Transfusion*, practised upon a *Man* in *London*", *Philosophical Transactions* 2, 30 (9 December 1667), pp. 557–59.

Robert Hooke (?), "The Method observed in Transfusing the Bloud [*sic*] out of one Animal into another", *Philosophical Transactions* 1, 20 (Monday 17 December 1667), pp. 353–57.

George Valentine Cox, *Recollections of Oxford* (London, 1870). See 'Anatomy School Stories', p. 23, for Anne Greene rhyme.

Kenneth Dewhurst (ed.), *Thomas Willis's Oxford Lectures* (Oxford: Sandford, 1980).

Kenneth Dewhurst (ed.), *Willis's Oxford Casebook* (Oxford: Sandford, 1981).

Robert G. Frank, Jnr, *Harvey and the Oxford Physiologists: A Study of Scientific Ideas* (Berkeley, CA: California University Press, 1980).

J. T. Hughes, *Thomas Willis 1621–1675. His Life and Work* (Oxford: Rimes House, 2009).

Hansruedi Isler, *Thomas Willis (1621–1675), Doctor and Scientist* (New York and London: Hafner, 1968).

G. Keynes, *Blood Transfusion* (Bristol, 1949).

Samuel Pepys, *The Diary of Samuel Pepys*, a *new and complete transcription*, vols. I-XI, ed. Robert Latham and William Matthews (London: Bell and Hyman, 1970–83).

Samuel Pepys, *The Shorter Pepys*, selected and edited by Robert Latham (London: Book Club Associates with Bell and Hyman, 1986).

Robert Plot, *The Natural History of Oxfordshire*, 2nd edn (Oxford, 1705). Anne Greene.

Richard Watkins, *Newes from the Dead. Or a True and Exact Narration of the Miraculous Deliverance of ANNE GREENE... by a Scholar of Oxford* (Oxford, 1651). Bodleian Library, Oxford, copy in collection "Wood. S16".

Thomas Willis, *Cerebri Anatome* (Oxford, 1664), transl. Samuel Pordage, *The Anatomy of the Brain* (London, 1681).

———, *Diatribae... De Fermentatione* (London, 1659).

———, *De Anima Brutorum* (Oxford, 1672), transl. Samuel Pordage, *Two Discourses concerning the Soul of Brutes* (London, 1683).

———, *Pathologiae Cerebri* (Oxford, 1667).

Anthony Wood, "Thomas Willis", *Athenae Oxonienses*, vol. II, 2nd edn (London, 1723), pp. 549–50.

Christopher Wren [son of Sir Christopher], *Parentalia; or, Memoirs of the Family of Wrens* (London, 1750), p. 230.

Lee Zimmermann and K. M. Howell, "History of Blood Transfusion", *Annals of Medical History* N.S. 4 (1932), p. 415.

CHAPTER 14

John Aubrey, "Robert Burton", *Brief Lives*, ed. A. Clark (see under Chapter 13).

Robert Burton, *The Anatomy of Melancholy. What it is, With All Kindes, Causes, Symptomes, Prognostications, and severall cures of it* (Oxford, 1621). Christ Church, Oxford, Library, C. 45.c.30.

Robert Burton, *The Anatomy of Melancholy* (Oxford, 1628; 3 vols, London: Everyman, Dent, 1932, 1948).

Allan Chapman, "The Scholar, the Thunderbolt and the Anatomist", *The Wadham College Gazette*, January 1993, pp. 59–62.

Kenneth Dewhurst, *John Locke (1632–1704) Physician and Philosopher. A Medical Biography, With an Edition of the Medical Notes in* his *Journals* (Wellcome Historical Medical Library, London, 1963).

——, *Willis's Oxford Casebook* (see under Chapter 13).

——, *Willis's Oxford Lectures* (see under Chapter 13).

J. T. Hughes, *Thomas Willis* (see under Chapter 13).

John Locke, *An Essay Concerning Human Understanding* (1690), ed. A. D. Woozley (Fontana Library, Collins, 1964).

Sir William Osler, "John Locke as a physician", *The Lancet* II (1900), p. 10. Reprinted in *An Alabama Student and other Autobiographical Essays* (Oxford, 1926), pp. 69–74.

Robert Plot, *The Natural History of Oxfordshire* (see under Chapter 13), pp. 5–6 for account of death by lightning, May 1666.

Zbigniew Szydlo, *Water which Does not Wet the Hands: The Alchemy of Michael Sendivogius* (Warsaw: Polish Academy of Sciences, 1994).

John Wallis, "A Relation of an Accident of Thunder and Lightning in Oxford", *Philosophical Transactions of the Royal Society* 1 (1666), pp. 222–6.

Anthony Wood, *Athenae Oxonienses*, vol. II (see under Chapter 13).

Anthony Wood, *The Life and Times of Anthony Wood, Antiquary of Oxford,1632–1695 described by himself*, ed. Andrew Clark, vol. 2 (Oxford: Clarendon Press, 1892), p. 77.

CHAPTER 15

Allan Chapman, *England's Leonardo: Robert Hooke and the Seventeenth-Century Scientific Revolution* (Bristol and Philadelphia: Institute of Physics, 2005).

Allan Chapman, "From Alchemy to Airpumps: The Foundation of Oxford Chemistry to 1700", in *Chemistry at Oxford: A History from 1600 to 2005*, ed. Robert J. P. Williams, Allan Chapman, and John S. Rowlinson (Cambridge: Royal Society of Chemistry, 2009), Ch. 2, pp. 17–51.

Carole Brookes, "Experimental Chemistry at Oxford, 1449-c.1700. Its Techniques, Theories, and Personnel" (unpublished Oxford University M.Chem. thesis, 1985; copy in Bodleian Library, Oxford).

Kenneth Dewhurst, *John Locke: Physician and Philosopher* (see under Chapter 14).

Kenneth Dewhurst, *Dr Thomas Sydenham (1624–1689): His Life and Original Writings* (London: Wellcome Historical Medical Library, 1966).

W. N. Hargreaves-Mawdsley, *Oxford in the Age of John Locke* (Norman, OK: Oklahoma University Press, 1973).

Robert Hooke, "An Account of the Rise and Attempts to conveigh Liquors immediately into the Mass of Blood", *Philosophical Transactions*. 1, no. 7 (4 Dec. 1665), pp. 128–30.

Robert Hooke, "An Account of a Dog dissected by Mr Hook [*sic*]", in Thomas Sprat, *History of the Royal Society* (London, 1667), p. 232.

——, "A new Contrivance of Wheel-Barometer, much more easy to be prepared, than that, which is described in Micrography; imparted by the Author of that Book", *Philosophical Transactions* 1, no. 13 (4 June 1666), pp. 218–19.

——, "Of an Experiment [on a dog] made by Mr Hook [*sic*], of Preserving Animals alive by Blowing through their Lungs with Bellows", *Philosophical Transactions* 2, no. 28 (24 Oct. 1667), pp. 539–40.

Geoffrey Keynes, "The History of Blood Transfusion 1628–1914", *The British Journal of Surgery* XXXI, no. 38 (1943), pp. 38–50.

John Mayow, *Physico-Medical Works*, being a translation of "Tractatus Quinque Medico-Physici" (Edinburgh and London: Alembic Club, 1957).

Michael Osborne, "The Medical Interests of the Oxford Chemists of the Late Seventeenth Century" (unpublished Oxford University M.Chem. thesis, 2002; copy in Bodleian Library, Oxford).

Sir William Osler, "John Locke as a Physician" (see under Chapter 14).

Philosophical Transactions no. 63 (12 Sept. 1670), pp. 2035–56 (fresh sheep's blood put into an airpump).

Patrick Romanell, *John Locke and Medicine: A New Key to Locke* (Buffalo, NY: Prometheus, 1984).

Anthony Wood, *The Life and Times of Anthony Wood, Antiquary of Oxford, 1632–1695 described by himself*, ed. Andrew Clark, vol. 1 (Oxford: Clarendon Press, 1891), p. 472.

Zbigniew Szydlo, *Water which Does not Wet the Hands* (see under Chapter 14).

John B. West, "Robert Hooke: Early Respiratory Physiologist, Polymath, and Mechanical Genius", *Physiology* 29, no. 4 (2014), 222–333.

CHAPTER 16

D. G. C. Allan and R. E. Schofield, *Stephen Hales: Scientist and Philanthropist* (London: Scolar Press, 1980).

Allan Chapman, "Wadham's Medical Rector: The Reverend Edward Stone, Aspirin, and Eighteenth-Century Medical Practice", *Wadham College Gazette*, Pt. 1, 2008, pp. 113–26; Pt. 2, 2009, pp. 96–100.

Stephen Hales, *Statical Essays, containing Haemastaticks, or, An Account of some Hydraulick and Hydrostatical Experiments made on the Blood and Blood-Vessels of Animals* (London, 1733).

Stephen Hales, *Vegetable Staticks…* (London, 1727).

David I. Harvie, *Limeys, The Conquest of Scurvy* (Stroud: Sutton, 2005).

Harriet Joseph, *Shakespeare's Son-in-Law: John Hall, Man and Physician*, containing facsimile of Dr Hall's *Select Observations on English Bodies*, 2nd edn, 1679 (Stratford-upon-Avon, 1976).

Joan Lane with medical commentary by Melvin Earles, *John Hall and his Patients. The Medical Practice of Shakespeare's Son-in-Law* (Shakespeare Birthplace Trust, Sutton, 1994).

James Lind, *A Treatise of the Scurvy* (Edinburgh, 1753).

H. D. Rolleston, "James Lind, Pioneer of Naval Hygiene", *Journal of the Royal Naval Medical Service* 1 (1915), p. 181.

W. J. Sheils (ed.), *The Church and Healing* (Oxford: Blackwell, 1982).

Edward Stone, "An Account of the Success of the Bark of the Willow in the Cure of Agues…", *Philosophical Transactions* 53 (1763), 195–200.

Peter Virgin, *Sydney Smith* (London: Harper Collins, 1994).

Ralph Waller, *John Wesley. A Personal Portrait* (London: SPCK, 2003).

John Ward, *The Diary of the Revd. John Ward M.A. Vicar of Stratford-upon-Avon. Extending from 1648 to 1679. From the original MS. preserved in the Library of the Medical Society of London*, ed. Charles Severn (London, 1839). (Now in the Folger Shakespeare Library, Washington DC)

John Wesley, *Primitive Physick: or, An Easy and Natural Method of Curing most Diseases* (London 1747). Reprinted as *Primitive Remedies* (Santa Barbara, CA, 1975).

CHAPTER 17

Erwin H. Ackerknecht, *Medicine at the Paris Hospitals, 1749–1848* (Baltimore: Johns Hopkins Press, 1967).

Claudius Amyand, "Of an Inguinal Rupture with a Pin in the *Appendix Coeci*, incrusted with Stone: and some Observations on Wounds in the Guts", *Philosophical Transactions* 443 (Oct. 1736), pp. 329–42.

"Asylums" in *The Oxford Companion to the Mind* (see under General Reading).

William F. Bynum and Roy Porter (eds), *William Hunter and the Eighteenth-Century Medical World* (Cambridge: Cambridge University Press, 1985).

H. C. Cameron, *Mr Guy's Hospital, 1726–1948* (London: Longman, 1954).

John D. Comrie, *History of Scottish Medicine* (London: Wellcome Medical Museum, 1927).

Sir Zachary Cope, *William Cheselden 1688–1752* (Edinburgh: E. & S. Livingstone, 1953).

Sir Zachary Cope, *The Royal College of Surgeons of England: A History* (London: Anthony Blond, 1959).

John Druett, *Rough Medicine: Surgeons at Sea* (New York: Routledge, 2001).

John Fleetwood, *History of Medicine in Ireland* (Dublin: Browne & Nolan, 1951).

Andrew Fyfe, *A Compendium of the Anatomy of the Human Body: Intended principally for the use of students*, 3 vols (Edinburgh, 1807).

Terry Gould and David Uttley, *A Short History of St. George's Hospital, and the Origins of its Ward Names* (London: Continuum, 1997).

Fiona Haslam, *From Hogarth to Rowlandson: Medicine in Art in Eighteenth-Century Britain* (Liverpool: Liverpool University Press, 1966).

Ernest Heberden, "William Heberden (1710–1801)", *Oxford Dictionary of National Biography* (Oxford: Oxford University Press, 2004).

Campbell R. Hone, *The Life of Dr. John Radcliffe, 1652–1714, Benefactor of the University of Oxford* (London: Faber and Faber, 1950).

John Hunter, *A Treatise on the Blood, Inflammation, and Gun-shot Wounds* (London, 1794).

Derek Jarrett, *England in the Age of Hogarth* (London: Hart-Davis, MacGibbon, 1974).

L. S. King, *The Medical World of the Eighteenth Century* (Chicago: University of Chicago Press, 1958).

Dennis M. Krikler, "The Foxglove, 'the Old Woman of Shropshire', and William Withering", *Journal of the American College of Cardiology* 5 (Suppl. A) (May 1985), pp. 255–86.

M. Laignel-Levastine and R. Molinery, *French Medicine* (New York, 1934).

G. M. Lloyd, "The Life and Works Of Percivall Pott", *St. Bartholomew's Hospital Reports* 66 (1933), p. 291.

Irvine Loudon, *Medical Care and the General Practitioner 1750–1850* (Oxford: Clarendon Press, 1986).

W. Macmichael, *The Gold-Headed Cane* (London, 1854).

Drewry Ottley, *Life of John Hunter* (London, 1835).

Brenda Parry-Jones, *The Warneford Hospital, Oxford* (Oxford: Holywell Press, 1976).

Roy Porter, *English Society in the Eighteenth Century* (London: Penguin, 1982).

George Quist, *John Hunter, 1728–93* (London: Heinemann Medical, 1981).

Philip Rhodes, *Doctor John Leake's Hospital. A History of the General Lying-in Hospital, York Road, Lambeth, 1765–1971* (London: Davies-Poynter, 1977).

Roger Rolls, *The Hospital of the Nation: The Story of Spa Medicine and the Mineral Water Hospital, Bath* (Bath: Bird, 1988).

Lisa Rosner, *Medical Education in the Age of Improvement: Edinburgh Students and Apprentices 1760–1826* (Edinburgh: Edinburgh University Press, 1991).

Andrew Scull, *Museums of Madness* (London: Allen Lane, 1979).

Peter Sheldon, *The Life and Times of William Withering: His Work and Legacy* (Studley: Brewin Books, 2004).

D. A. Stansfield, *Thomas Beddoes M.D. 1760–1808, Chemist, Physician, Democrat* (Dordrecht: Reidel, 1984).

K. B. Thomas, "John Hunter and an Amputation under Analgesia in 1784", *Medical History* 2 (1958), p. 53.

A. Logan Turner, *The Story of a Great Hospital: The Royal Infirmary of Edinburgh* (Edinburgh: Oliver and Boyd, 1937).

Jenny Uglow, *Hogarth: A Life and a World* (London: Faber and Faber, 1997).

Cecil Wall, H. Charles Cameron, and E. Ashworth Underwood, *A History of the Worshipful Society of Apothecaries of London. Vol. 1, 1617–1815* (Oxford: Wellcome and Oxford University Press, 1963).

William Withering, *An Account of the Foxglove and some of its Medical Uses etc. With Practical Remarks on Dropsy and other Diseases* (London, 1785).

John Woodward, *To do the Sick no Harm: a Study of the British Voluntary Hospital System to 1875* (London: Routledge, 1974).

CHAPTER 18

Patricia Allderidge, "Management and Mismanagement at Bedlam, 1547–1633", in Webster, *Health, Medicine, and Mortality* (see under General Reading), pp. 141–61.

John B. Bamborough, "Robert Burton, 1577–1640", *Oxford Dictionary of National Biography* (Oxford: Oxford University Press, 2004).

Michael Collie, *Henry Maudsley: Victorian Psychiatrist* (Winchester: St Paul's Bibliographies, 1988).

W. P. Courtney, revised H. C. G. Matthew, "Samuel Warneford, 1763–1855", *Oxford Dictionary of National Biography* (Oxford: Oxford University Press, 2004).

Anne Digby, *Madness, Morality, and Medicine: A Study of the York Retreat 1796–1914* (Cambridge: Cambridge University Press, 1985).

Vivian Green, *The Madness of Kings: Personal Trauma and the Fate of Nations* (Stroud: Sutton, 1993).

Richard Hunter and Ida Macalpine, *Three-Hundred Years of Psychiatry, 1535–1860* (Oxford: Oxford University Press, 1963).

Allan Ingram (ed.), *Voices of Madness: Four Pamphlets, 1683–1796* (Stroud: Sutton, 1997).

Kathleen Jones, *Lunacy, Law, and Conscience 1744–1845: The Social History of the Care of the Insane* (London: Routledge, 1955).

Julia Keay, *Alexander the Corrector: The Tormented Genius who Unwrote the Bible* (London: Harper Perennial, 2004).

Ida Macalpine and Richard Hunter, *George III and the Mad-Business* (London: Pimlico, 1993, 1995).

Alexandra Medcalf, "The York Lunatic Asylum Scandal" (Borthwick Institute, University of York, 2014).

Roy Porter, "Francis Willis, 1718–1807", *Oxford Dictionary of National Biography* (Oxford: Oxford University Press, 2004).

——, *Mind Forg'd Manacles: A History of Madness in England from the Restoration to the Regency* (London: Penguin, 1987).

——, *A Social History of Madness: Stories of the Insane* (London: Weidenfeld and Nicolson, 1987).

William Roberts, *Memoirs of the Life and Correspondence of Mrs Hannah MORE*, 4 vols (London, 1834). Vol. 2, p. 144, for Willis.

Gämini Salgädo, *The Elizabethan Underworld* (Totowa, NJ: Rowman & Littlefield, 1977).

Edward Shorter, *A History of Psychiatry: from the Era of the Asylum to Prozac* (New York: Wiley, 1997).

R. S. Stephenson, "The insanity of George III", in *Famous Illnesses in History* (see under General Reading), pp. 154–71.

Vaughan Thomas, *Christian Philanthropy exemplified in a Memoir of the Rev. Samuel Wilson Warneford* (1855).

Samuel Tuke, *Description of the Retreat* (London, 1813; reprinted 1996).

Ned [Edward] Ward, *The London Spy* [1698–1703], ed. K. Fenwick (London: Folio Society, 1955).

Simon Winchester, *The Surgeon of Crowthorne* (London: Penguin, 1999).

Sarah Wise, *Inconvenient People: Lunacy, Liberty, and Mad-doctors in Victorian England* (London: Vintage Publishing, 2013).

Émile Zola, *L'Assommoir* (1877), transl. Margaret Mauldon, World's Classics edn (Oxford: Oxford University Press, 1995).

CHAPTER 19

J. M. Adair, *Essays on Fashionable Disorders* (London, 1790).

Phineas Taylor Barnum, *The Humbugs of the World* (London, 1866).

Lucinda McCray Beier, "Experience and Experiment: Robert Hooke, Illness, and Medicine", in *Robert Hooke: New Studies*, ed. Michael Hunter and Simon Schaffer (Woodbridge: Boydell Press, 1989).

V. Berridge and G. Edwards, *Opium and the People* (London: Allen Lane, 1981).

Marc Bloch, *The Royal Touch: Sacred Monarchy and Scrofula in England and France* (London: Routledge and Kegan Paul, 1973).

British Medical Association, *Secret Remedies: What they cost, and what they contain* (London, 1909).

P. S. Brown, "The Vendors of Medicines Advertised in Eighteenth-Century Bath Newspapers", *Medical History* XIX (1975), pp. 352–69.

William F. Bynum and Roy Porter (eds), *Medical Fringe and Medical Orthodoxy 1750–1850* (London: Croom Helm, 1987).

John Camp, *Magic, Myth, and Medicine* (see under General Reading).

W. A. Campbell, "Portrait of a Quack: Joshua Ward 1685–1761", *University of Newcastle Medical Gazette*, 1964.

Allan Chapman, *The Medicine of the People. A History of Popular Medicine before the National Health Service* (Chichester: Aeneas Press, 2001).

G. Coats, "The Chevalier Taylor", in R. Rutson James (ed.), *Studies in the History of Ophthalmology* (see under Chapter. 4).

Peter Dance, *Animal Fakes and Frauds* (Maidenhead: Sampson Low, 1976).

Robert Hooke, *The Diary of Robert Hooke 1672–1680*, ed. H. W. Robinson and W. Adams (London, 1935).

K. Dewhurst, *The Quicksilver Doctor: The Life and Times of Thomas Dover* (Bristol: John Wright & Sons, 1957).

D. M. Jackson, "Bach, Handel, and the Chevalier Taylor", *Medical History* XII (1968), pp. 385–93.

Eric Maple, *Magic, Medicine, and Quackery* (London: Robert Hale, 1968).

W. B. Ober, "Bach, Handel, and 'Chevalier' John Taylor M.D., Ophthalmiator", *New York State Journal of Medicine* LXIX (1969), pp. 1797–1806.

Philip Pinkus, *Grub Street Stripped Bare* (London: Constable, 1968).

Roy Porter (ed.), *Patients and Practitioners: Lay Perceptions of Medicine in Pre-industrial Society* (Cambridge: Cambridge University Press, 1985).

Roy Porter, *Quacks, Fakers, and Charlatans in English Medicine* (Stroud: The History Press, 2001).

Quacks and Quackery by a Medical Practitioner (London, 1844).

A. L. Rowse, *Simon Forman: Sex and Society in Shakespeare's Day* (London: Weidenfeld and Nicolson, 1974).

John Taylor, *An Exact Account of 243 different diseases to which the Eye and its Coverings are exposed* (Edinburgh, 1759).

John Taylor, *The History of the Travels and Adventures of the Chevalier John Taylor* (London, 1760).

Jenny Uglow, *Hogarth* (see under Chapter 17).

H. B. Wheatley, *Hogarth's London* (London, 1909).

A. D. Wright, "The quacks of John Hunter's time", *Transactions of the Hunterian Society* XI (1952–53).

CHAPTER 20

John Baron, *The Life of Edward Jenner*, 2 vols (London, 1838).

Derrick Baxby, "Edward Jenner (1749–1823)", *Oxford Dictionary of National Biography* (Oxford: Oxford University Press, 2004).

"Sir Joseph Bazalgette (1819–1891)", *Oxford Dictionary of National Biography* (Oxford: Oxford University Press, 2004).

William Booth, *Darkest England and the Way Out* (London and New York, 1890).

W. Braithwaite, *The Retrospect of Medicine, Being A Half-Yearly Journal* (London): vol. XVII (1846), vol. XX (1849), vol. LXI (1870).

Charles Dickens (ed.), *Household Words*, 10, (1 June, 1850), "Greenwich Weather Wisdom", pp. 222–25.

Thomas Dormandy, *The White Death: A History of Tuberculosis* (London: Hambledon, 1999).

Richard B. Fisher, *Edward Jenner* (London: André Deutsch, 1991).

Dorothy Fisk, *Dr Jenner of Berkeley* (London: Heinemann, 1959).

Ian and Jennifer Glynn, *The Life and Death of Smallpox* (see under Chapter 7).

Sandra Hempel, *The Medical Detective: John Snow, Cholera, and the Mystery of the Broad Street Pump* (London: Granta, 2006).

Norman E. Himes, *Medical History of Contraception* (London: Unwin, 1936).

Frederick Knight Hunt, "Greenwich Weather Wisdom"; see Charles Dickens above.

Edward Jenner, *An Inquiry into the Causes and Effects of the Variolae Vaccinae* (London, 1798).

Steven Johnson, *The Ghost Map: The Story of London's Most Terrifying Epidemic* (New York: Riverhead, 2007).

Thomas Malthus, *An Essay on the Principle of Population* (1798), ed. Geoffrey Gilbert (Oxford: Oxford University Press, 2008).

Richard Mead, *A Discourse on Small Pox and Measles* (London, 1748).

William Peterson, *Malthus* (London: Heinemann, 1979, 1999).

Thomas Shapter, *The History of the Cholera in Exeter in 1832* (1849; republished Wakefield and London: S. R. Publishers Ltd., 1971).

John Snow, *On the Mode of Communication of Cholera* (London, 1849).

L. Thurston and G. Williams, "An Examination of John Fewster's role in the Discovery of Smallpox Vaccination", *Journal of the Royal College of Physicians of Edinburgh*, 45 (2015), pp. 173–79.

R. Trench and E. Hillman, *London Under London: A Subterranean Guide* (London: Murray, 1984).

Ben Weinreb and Christopher Hibbert (eds), *The London Encyclopaedia* (London: Macmillan, 1983). See "Index of Persons" under "Bazalgette, Sir Joseph": 17 references to his metropolitan engineering works.

CHAPTER 21

Aesculapius Scalpel [Edward Berdoe], *St. Bernard's, The Romance of a Medical Student* (London, 1888).

James Blake Bailey, *The Diary of a Resurrectionist* (London, 1896).

James Moores Ball, *The Sack-'em-up Men. An Account of the Rise and Fall of the Modern Resurrectionists* (Edinburgh: Oliver & Boyd, 1928).

Hector Berlioz, *The Memoirs of Berlioz*, transl. D. Cairns, (London: Gollancz, 1969, 1977).

R. C. Brock, *The Life and Work of Sir Astley Cooper* (Edinburgh: E. & S. Livingstone, 1952).

Hubert Cole, *Things for the Surgeon: A History of the Resurrection Men* (London: Heinemann, 1964).

Bransby Blake Cooper, *Life of Sir Astley Cooper* (London, 1843).

Hilary and Mary Evans, *Andrew Ure Galvanising a Murderer's Corpse, 4th November 1818'*, plate, in *Sources in Illustration 1500–1900* (Bath, 1971), p. 139, 62c. Original in Louis Figuier's *Merveilles de la Science* (1867).

Judith Flanders, *The Invention of Murder: How the Victorians Revelled in Death and Detection and Created Modern Crime* (London: Harper Press, 2011).

History of the London Burkers (London, 1832).

Mrs Basil Holmes, *The London Burial Grounds* (London, 1896).

Thomas Hood, *The Poetical Works of Thomas Hood* (London, New York, 1890).

Henry Lonsdale, *Life of Robert Knox* (London, 1870).

George MacGregor, *History of Burke and Hare* (Glasgow, 1884).

Caroline McCracken-Fleshner, *The Doctor Dissected: A Cultural History of the Burke and Hare Murders* (Oxford: Oxford University Press, 2012).

John O'Shea, *Music and Medicine: Medical Profiles of Great Composers* (London: Dent, 1990).

Report from the Select Committee on Anatomy (1828).

Thomas Southwood Smith, *The Use of the Dead to the Living* (undated, c. 1835).

Robert Southey, *The Poetical Works of Robert Southey* (London, 1876).

S. S. Sprigge, *Life of Thomas Wakeley* (London, 1897).

John Struthers, *Historical Sketch of the Edinburgh Anatomical School* (Edinburgh, 1867).

J. L. Thornton, *John Abernethy, a Biography* (London, 1953).

Edward Warren, *Life of John Collins Warren* (Boston, 1860).

CHAPTER 22

Richard Barber, *Joseph Lister 1827–1912* (London, 1977).

John Hughes Bennett, "A Case of Hypertrophy of the Spleen and Liver in Which Death Took Place from Suppuration of the Blood", *Edinburgh Medical and Surgical Journal*, 1 October 1845.

T. D. Brock, *Robert Koch: A Life in Medicine and Bacteriology* (Washington DC: American Society of Microbiology Press, 1999).

Reginald S. Clay and Thomas H. Court, *The History of the Microscope* (London: C. Griffin & Co., 1932).

P. Debré and E. Forster, *Louis Pasteur* (Baltimore: Johns Hopkins University Press, 1998).

René J. Dubos, *Louis Pasteur: Free Lance of Science* (London: Victor Gollancz, 1951).

Richard B. Fisher, *Joseph Lister, 1827–1912* (New York: Stein and Day, 1977).

H. Gest, "The Discovery of Microorganisms by Robert Hooke and Antoni van Leeuwenhoek, Fellows of the Royal Society", *Notes and Records of the Royal Society* 58, 2 (2004), pp. 187–201.

Walter Gilbert Hartley, *The Light Microscope. Its Use and Development* (Oxford: Senecio, 1993).

T. Hodgkin and J. J. Lister, "Notice of some Microscopic Observations of the Blood and Animal Tissue", *Philosophical Magazine*, N.S. 2 (1827), pp. 130–38.

Robert Hooke, *Micrographia, Or some Physiological Descriptions of Minute Bodies made by Magnifying Glasses* (London, 1665).

A. M. Kass and E. H. Kass, *Perfecting the World: The Life and Times of Dr Thomas Hodgkin 1798–1866* (Boston: Harcourt Brace Jovanovich, 1988).

George King and Arthur Newsholme, communicated by Dr J. S. Bristow, FRS: "On the Alleged Increase of cancer", *Proceedings of the Royal Society* 53 (4 May1893), pp. 405–07.

William Arbuthnot Lane, Sir, "Resuscitation in Syncope Due to Anaesthetic and Other Conditions by Rhythmical Compression of the Heart", *The Lancet*, 2 (1902), pp. 1397, 1476.

J. J. Lister, "On some Properties of Achromatic Object-Glasses Applicable to the Improvement of the Microscope", *Philosophical Transactions* 120 (1830), pp. 187–200.

J. J. Lister, "Some Observations of the Structure and Functions of Tubular and Cellular Polypi", *Philosophical Transactions* 124 (1834), pp. 365–88.

Elie Metchnikoff, *The Founders of Modern Medicine: Pasteur, Koch, Lister* (New York: Walden, 1939; Classics of Medicine Library, 2006).

"Microscope", *Encyclopaedia Britannica* XVI, 9th edn (1883), pp 258–78.

498

Robert Reid, *Men and Microbes* (London: BBC, 1974).

Moira Davison Reynolds, *How Pasteur Changed History: The Story of Louis Pasteur and the Pasteur Institute* (Bradenton, FL: McGuinn & McGuire, 1994).

E. G. Ruestow, *The Microscope in the Dutch Republic* (Cambridge: Cambridge University Press, 1996).

Gerard L'Estrange Turner, *God Bless the Microscope: A History of the Royal Microscopical Society over 150 Years* (Oxford: Royal Microscopical Society, 1989).

——, *The Great Age of the Microscope: The Collection of the Royal Microscopical Society through 150 Years* (Bristol: Adam Hilger, 1989).

——, "Joseph Jackson Lister 1786–1869", *Oxford Dictionary of National Biography* (Oxford: Oxford University Press, 2004).

Rudolf Virchow, *Cellular Pathology*, transl. by Frank Chance (1863; reprinted New York: Dover, 1971).

CHAPTER 23

Jennifer Beinart, *A History of the Nuffield Department of Anaesthetics, Oxford, 1937–1987* (Chapter 7 by M. K. Sykes) (Oxford: Oxford University Press, 1987).

Frederick Cartwright, *The Development of Modern Surgery* (New York: T. Y. Crowell, 1968).

M. H. Armstrong Davison, *The Evolution of Anaesthesia* (Altrincham: John Sherratt, 1965).

Barbara M. Duncan, *The Development of Inhalation Anaesthesia* (Oxford: Wellcome Historical and Oxford University Press, 1947).

J. Duns, *Memoir of Sir James Y. Simpson* (Edinburgh, 1873).

H. Laing Gordon, *Sir James Young Simpson and Chloroform (1811–1870)* (London: Unwin, 1897).

Thomas E. Keys, *The History of Surgical Anaesthesia* (New York: Dover, 1963).

Robert Paterson, *Memorials of the Life of James Syme* (Edinburgh, 1874).

Victor Robinson, *Victory over Pain: A History of Anaesthesia* (New York: Sigma Books, 1946).

Rennie B. Schoepflin, "Myth 14. That the Church denounced anaesthesia in childbirth on biblical grounds", in *Galileo goes to Jail and other Myths about Science and Religion*, ed. Ronald L. Numbers (Cambridge, MA: Harvard University Press, 2009).

499

John Snow, *On Chloroform and Other Anaesthetics* (London, 1858).

K. Bryn Thomas, *The Development of Anaesthetic Apparatus* (Oxford: Blackwell Scientific, 1975).

Jürgen Thorwald, *The Century of the Surgeon* (New York: Pantheon, 1956, 1957).

 E. Ashworth Underwood, "Before and after Morton: A historical survey of anaesthesia", *British Medical Journal* II (1946), p. 525.

Edward Warren, *Life of John Collins Warren* (Boston, 1860).

John Collins Warren, "Inhalation of Ethereal Vapour for the Prevention of Pain in Surgical Cases", *Boston Medical and Surgical Journal*, 35 (1846), pp. 375–79.

CHAPTER 24

Sir Charles Ballance, "The Surgery of the Heart", *The Lancet*, 1, three parts: 3 January 1920, pp. 1–6; 10 January 1920, pp. 73–79; 17 January 1920, pp. 134–39.

Sir John Bland-Sutton, "Missiles and Emboli", *The Lancet*, 10 May 1919, pp. 773–75.

Robert Bud and Deborah Jean Warner, *Instruments of Science* (see under General Reading).

Frederick F. Cartwright, *Joseph Lister, the Man who made Surgery Safe* (London: Weidenfeld and Nicolson, 1963).

Richard B. Fisher, *Joseph Lister 1827–1912* (see under Chapter 22).

I. H. Flack, *Lawson Tait 1845–1899* (London: Heinemann, 1949).

Jerry L. Gaw, *"A Time to Heal": The Diffusion of Listerism in Victorian Britain* (Philadelphia: American Philosophical Society, 1999).

Sir Rickman Godlee, *Lord Lister*, 3rd edn (Oxford: Clarendon Press, 1924).

Henry Gray, *Anatomy Descriptive and Surgical*, with drawings by H. V. Carter MD (5th edn, London, 1869).

William Le Fanu, *A List of the Original Writings of Joseph Lord Lister O.M.* (Edinburgh: Livingstone, 1965).

Joseph Lister, "On a New Method of Treating Compound Fractures...", *The Lancet*, 1 (16 March 1867), pp. 326–29.

W. J. MacKay, *Lawson Tait, his Life and Work* (London: Wood, 1922).

A. Miles, *The Edinburgh School of Surgery before Lister* (London: A. and C. Black, 1918).

R. Paterson, *Memorials of the Life of James Syme* (Edinburgh, 1874).

Ruth Richardson, *The Making of Mr. Gray's Anatomy: Bodies, Books, Fortune, Fame* (Oxford: Oxford University Press, 2008).

Sir W. J. Sinclair, *Semmelweis, his Life and Doctrine* (Manchester: Manchester University Press, 1909).

Sir Henry Sessions Souttar, 'The Surgical Treatment of Mitral Stenosis', *British Medical Journal*, 2 (1925), pp. 603–06.

E. Ashworth Underwood, "Before and After Morton: A historical survey of anaesthesia", *British Medical Journal* 2 (1946), p. 525.

R. Vallery-Radot, *The Life of Louis Pasteur*, transl. R. L Devonshire (Garden City, NY: Doubleday, Page, and Co., 1923).

A. Logan Turner, *Joseph, Baron Lister. Centenary Volume, 1827–1927* (Edinburgh and London, 1927).

Abraham Wallace, *Reminiscences of Lister* (Manchester, 1928).

O. H. Wangensteen, S. D. Wangensteen, and C. F. Klinger, "Some Pre-Listerian and Post-Listerian Antiseptic Wound Practices and the Emergence of Asepsis", *Surgery, Gynaecology, and Obstetrics* 173, no. 4 (1973), pp. 677–702. [Since 1994, renamed *Journal of Surgery, Gynaecology, and Obstetrics*.]

CHAPTER 25

Aesculapius Scalpel (see under Chapter 21).

Mark Bostridge, *Florence Nightingale: The Woman and Her Legend* (London: Viking, 2008).

Allan Chapman, "Medical Art and Experimental Science in Early Victorian England" (the 150th Anniversary Lecture of the Liverpool Medical Institution, 28 May 1987), *Transactions and Report of the L.M.I., 1986–7* (Liverpool, 1987), pp. 26–39.

Trevor M. Cook, *Samuel Hahnemann, the Founder of Homeopathic Medicine* (Wellingborough: Thorsons, 1981).

John Cule, *A Doctor for the People* (see under General Reading).

Kenneth Dewhurst, *Oxford Medicine: Essays on the Evolution of the Oxford Clinical School 1770–1970* (Oxford: Sandford, 1970).

M. A. Elston, "Elizabeth Blackwell 1821–1910", *Oxford Dictionary of National Biography* (Oxford: Oxford University Press, 2004).

Anne Francis, *A Guinea a Box: A Biography of Thomas Beecham* (London: Hale, 1968). (Beecham of Beecham's Pills.)

Gillian Gill, *Nightingales: The Extraordinary Upbringing and Curious Life of Miss Florence Nightingale* (New York: Ballantine, 2004).

Augustus Bozzi Granville, *The Spas of England and Principal Sea-Bathing Places*, 3 vols (1841; Bath: Adams and Dart, 1971).

Sir Henry Harris, "Howard Florey and the Development of Penicillin", *Notes and Records of the Royal Society*, 53, no. 2 (1999), p. 249.

Irvine Loudon, *Medical Care and the General Practitioner 1750–1850* (see under Chapter 17).

Sir William Osler, *The Principles and Practice of Medicine* (New York and London: Appleton, 1892, 1915, etc.).

Walter Rivington, *The Medical Profession: An Essay* (Dublin: Fannin & Co., 1879).

Jane Robinson, *Mary Seacole: The Charismatic Black Nurse who became a Heroine of the Crimea* (London: Robinson, 2006).

Mary Seacole, *Wonderful Adventures of Mrs. Seacole in Many Lands* (London, 1857).

Jürgen Thorwald, *The Triumph of Surgery*, transl. Richard and Clara Winston (New York: Pantheon Books, 1960).

Ivan Waddington, *The Medical Profession in the Industrial Revolution* (Dublin: Humanities Press, 1984).

Cecil Woodham-Smith, *Florence Nightingale, 1820–1910* (London: Constable, 1950).

CHAPTER 26

Sarah Bradford, *Disraeli* (New York: Stein and Day, 1982).

Maurice Bruce, *The Coming of the Welfare State* (London: Batsford, 1961).

Allan Chapman, *The Medicine of the People* (see under Chapter 19).

Sir Zachary Cope, *Pioneers in Acute Abdominal Surgery* (London: Oxford University Press, 1939).

S. T. Crowe, *Halsted of Johns Hopkins, the Man and his Men* (Springfield, IL: Charles C. Thomas, 1957).

H. H. Dale, "Edward Mellanby 1884–1955", *Biographical Memoirs of Fellows of the Royal Society* 1 (1955), pp. 192–226.

George Frederick Drinka, *The Birth of Neurosis: Myth, Malady, and the Victorians* (New York: Simon and Schuster, 1984).

C. Eccles and W. C. Gibson, *Sherrington: His Life and Thought* (Berlin: Springer, 1979).

Michel Foucault, *The Birth of the Clinic: An Archaeology of Medical Perception*, transl. by Alan Sheridan (New York: Vintage Books, 1975).

John Grigg, *Lloyd George, the People's Champion, 1902–1911* (London, 1978; reissued Penguin, 2002).

Sir Henry Harris, "Howard Florey and the Development of Penicillin" (see under Chapter 25).

Seale Harris, *Banting's Miracle: The Story of the Discoverer of Insulin* (Philadelphia: J. B. Lippincott, 1946).

G. Melvyn Howe, *Man, Environment, and Disease in Britain* (see under General Reading).

Ruth G. Hodgkinson, *The Origins of the National Health Service: The Medical Services of the New Poor Law, 1834–1871* (London: Wellcome Historical Medical Library, 1967).

Robert Rhodes James, *Henry Wellcome* (London: Hodder and Stoughton, 1994).

Conrad Keating, *Great Medical Discoveries* (see under General Reading).

George King and Arthur Newsholme, "On the alleged increase in cancer" (see under Chapter 22).

"Edward Mellanby", *The Lancet* 268 (1955), pp. 309–10.

John S. Najarian, *The Miracle of Transplantation: The Unique Odyssey of a Pioneer Transplant Surgeon* (Beverley Hills: Medallion, 2009).

Eunice Schofield, *Medical Care of the Working Class in 1900* (Federation of Local History Societies in Lancashire, 1979).

Lloyd Stevenson, *Sir Frederick Banting* (Toronto: Ryerson Press, 1946).

G. Wolf, "The Discovery of Vitamin D: The Contribution of Adolf Windhaus", *Journal of Nutrition* 134, no. 6 (2004), pp. 1299–1302.

ADDENDUM TO CHAPTER 26

Some useful books, full of practical information about early twentieth-century medical and surgical techniques and practice.

John D. Comrie, *News Chronicle Home Doctor* (London, 1931). For popular domestic use.

A. H. Douthwaite, *Materia Medica: Pharmacy, Pharmacology, and Therapeutics* (London: J. & A. Churchill, 1949).

The Family Physician. A Manual of Domestic Medicine, 5 vols (London, Paris, and Melbourne: Cassell and Co., undated, *c.* 1910 from internal evidence). Probably intended for general practitioner doctors.

A. Bramble Green, *A Manual of Human Anatomy for Dental Students* (London: Benn Bros., 1923).

Russell Howard, *The House-Surgeon's Vade-Mecum* (London, 1911).

Edwin Lankester, MD, FRS, etc., *Haydn's Dictionary of Popular Medicine and Hygiene* (London: Ward Lock: undated, *c.* 1900).

J. D. Dixon Mann, *Forensic Medicine and Toxicology* (London, 1898).

The Practical Home Doctor, by a Harley Street specialist, illustr. by W. Thornton Shiells (London: Odhams, 1939).

Index